PROGRESS IN BOUNDARY ELEMENT METHODS

Progress in Boundary Element Methods

Volume 1

Edited by
C. A. Brebbia
School of Engineering,
University of Southampton

A HALSTED PRESS BOOK

JOHN WILEY & SONS
New York — Toronto

Published in the U.S.A. and
Canada by Halsted Press, a
Division of John Wiley & Sons,
New York

Library of Congress Cataloging in Publication Data
Main entry under title:

Progress in boundary element methods.

"A Halsted Press book."
Includes bibliographical references and index.
1. Boundary value problems. I. Brebbia, C. A.
TA347.B69P76 1981 620'.001'51535 81–6454
ISBN 0–470–27223–6 AACR2

Filmset Monophoto Times by Mid-County Press, London SW15
Printed in Great Britain by The Anchor Press Ltd and bound
by Wm Brendon & Son Ltd, both of Tiptree, Essex

Preface

It has now become clear to many engineers that the early claims of finite elements researchers were, in some cases, exaggerated. Finite elements, in many cases, constitutes an inaccurate and expensive technique. These problems have resulted in a renewal of interest in other numerical methods, amongst them the boundary solutions such as 'boundary integral techniques'. In their more general form these solutions consist of a series of surface elements, hence the name 'boundary elements' has now become widely accepted.

The boundary element method offers important advantages over 'domain' type solutions such as finite elements and finite differences. One of the most interesting features of the method is the much smaller system of equations and considerable reduction in the data required to run a problem. In addition, the numerical accuracy of the results is greater than that of finite elements. These advantages are more marked in three-dimensional problems. The method is also well suited to problems solving with infinite domains such as those frequently occurring in stress analysis for which the classical domain methods are unsuitable.

In spite of the current interest shown by analysts in boundary elements, there are presently very few books published on this topic. None of them deals in detail with non-linear, time dependent and complex problems for which the method is much more powerful than the currently used finite element programs.

This book presents the fundamentals of boundary element method in a simple fashion, interpreting the method as a particular case of weighted residual techniques. A chapter is dedicated to the important topic of fundamental solutions and how they can be obtained. Potential and elasticity problems are dealt with in detail in Chapters 3 and 4 with special applications to three-dimensional problems. Plasticity is an important topic in engineering and here again boundary elements have proved to be more economical and simple to use than finite elements. Chapter 5 reviews the fundamentals of plasticity theory and how it can be applied using boundary integral expressions. Chapter 6 is exclusively dedicated to time dependent problems for which the boundary element is a powerful alternative to finite elements. Engineers will also find interesting Chapter 7 which deals with the application of BEM in elastodynamics, especially for problems related to foundations.

The chapter on the combination of BE and FEM indicates how the technique could be applied in cases for which the problem cannot be solved using only boundary elements. This chapter also explains how symmetric matrices, similar to those used in FEM, can be obtained using BEM.

Finally Chapter 9 discusses some of the mathematical properties of BEM, including their accuracy and convergence.

Practising engineers, postgraduate students and researchers who want to be able to understand the new method will find this an invaluable text book.

Carlos A. Brebbia

Contributors

C. A. Brebbia	School of Engineering, University of Southampton, England	(Chapters 1, 5 and 6)
S. Walker	Structural Dynamics, Southampton, England	(Chapter 2)
E. Alarcón	Politechnic University, Madrid, Spain	(Chapters 3 and 7)
A. Martin	University of Seville, Spain	(Chapter 3)
F. Paris	Polytechnic University, Madrid, Spain	(Chapter 3)
F. Hartmann	Dortmund University Germany	(Chapter 4)
J. C. F. Telles	School of Engineering, University of Southampton, England	(Chapter 5)
L. C. Wrobel	School of Engineering, University of Southampton, England	(Chapter 6)
J. Dominguez	University of Seville, Spain	(Chapter 7)
M. Margulies	Schlumberger-Doll Research Connecticut, USA	(Chapter 8)
W. L. Wendland	Darmstadt University, Germany	(Chapter 9)

Contents

1

Basis of boundary elements

by C. A. Brebbia

1.1 INTRODUCTION

Engineers and physical scientists have in recent years become very conversant with numerical techniques of analysis. These numerical techniques are based on the approximate solution of an equation or set of equations describing the physical problem. The most widely known approximate method is finite differences which approximates the governing equations of the problem using local expansions for the variables, generally based on truncated Taylor series. The technique can be interpreted as a special case of the more general weighted residual methods (Brebbia[1]).

The finite element method has attracted the attention of the analysts largely due to its property of dividing the continuum into a series of elements, which can be associated with physical parts. The existing literature on finite elements is by now very extensive and encompasses structural[2] as well as fluid flow[3] and other types of problems. The method can sometimes be based on variational principles or more generally on weighted residual expressions. The great interest that the method attracted at the beginning of the 1960s had two important consequences; (1) it stimulated an impressive amount of work in computational techniques and efficient engineering software; (2) substantial research into basic physical principles such as variational techniques and weighted residuals was originated.

The first of the above points came as a natural consequence of the emergence of new and powerful computers, i.e. second generation machines which were able to solve engineering problems involving large amounts of numerical storage and manipulation. For a while the continuous progress in computer technology including the development of third generation machines, distracted scientists from the development of mathematical methods and their basis principles, i.e. point (2) above. These methods can be traced to pre-computer times[4,5] and involve different ways of solving the governing equations of a problem, i.e. Galerkin, collocation, least-squares, line techniques, matrix progression or transfer, the combination of different techniques, etc. Fortunately they were not forgotten and they reappeared in the finite element literature, sometimes with different names, such as Galerkin finite elements, finite element strip-method, some time integration schemes, etc. Another important development in approximate analysis was the investigation of mixed principles and the realization that physical problems can be expressed and solved in many different ways in accordance with the part or equations of the problem that we need to approximate. These approximations are of fundamental importance for the computer implementation of the different numerical techniques. Mixed methods can be traced to Reissner[6] and more specifically for finite elements to Pian[7]. An excellent exposition of mixed methods in structural mechanics can be found in the book by Washizu[8].

Integral equation techniques were until recently considered to be a different type of analytical method, somewhat unrelated to approximate methods. They became known in Europe through the work of a series of Russian authors, such as Muskhelishvili[9], Mikhlin[10], Kupradze[11], and Smirnov[12] but were not very popular with engineers. A predecessor of some of this work was Kellog[13] who applied integral equations for the solution of Laplace's type problems. Integral equation techniques were mainly used in fluid mechanics and general potential problems and known as the 'source' method, which is an 'indirect' method of analysis; i.e. the unknowns are not the physical variables of the problem. Work on this method continued throughout the 1960s and 1970s in the pioneering work of Jaswon[14] and Symm[15], Massonet[16], Hess[17] and many others.

It is difficult to point out precisely who was the first one to propose the 'direct' method of analysis. It is found in a different form, in Kupradze's book[11]. It seems fair however from the engineering point of view to consider that the method originated in the work of Cruse and Rizzo[18] in elastostatics.

Since the early 1960s a small research group started working at Southampton University, on the applications of integral equations to solve stress analysis problems. Hadid's thesis[19] partly based on Hadjin's[20] work was published in 1964 and dealt with the use of integral equations in shell analysis. Unfortunately the presentation of the problem, the difficulty of defining the appropriate Green's functions and the parallel emergence of the finite element method all contributed to minimize the importance of this work. It was not until 1973 that the group produced another significant thesis, Watson[21], dealing with three-dimensional elastostatics problems. By then recent developments in finite elements had started to find their way into the formulation of boundary integral equations. The idea of using general curved elements originated in another Southampton University thesis presented by Lachat[22] in 1975. Still the question of how to effectively relate the boundary integral equations to other approximate techniques was unresolved. This was done by Brebbia who published a series of papers on the relationship of different approximate methods, culminating with the first book in 1978 for which the title 'Boundary Elements' was used[1]. More recently this work has been expanded to encompass time dependent and non-linear problems[23]. Two important International Conferences were held at Southampton University in 1978 and 1980. The edited Proceedings of these conferences — the only ones so far on this topic — are now standard references[24,25].

One of the most important recent application of boundary elements has been for the solution of time dependent problems. This has now opened the way for solving more complex problems, such as wave propogation, Navier–

Stokes flow, elastodynamics, etc. The first boundary integral solution for the diffusion equation was due to Rizzo and Shippy[26], who used the Laplace transform to remove the time dependence. This approach was discussed by Liggett and Liu[27] where some of its disadvantages are pointed out. Chang et al.[28] in 1973 solved the problem in the time domain using space and time piecewise constant values for the variables. This approach has been extended by Wrobel and Brebbia[29] who applied it to solve complex temperature problems. Brebbia and Walker[23] proposed a step by step finite difference scheme for the time derivatives. The use of a time dependent fundamental solution implies that one does not need to perform any finite differences on the time derivative and produces an accurate and efficient solution specially when higher order time interpolation functions are employed[30,31] (see also Chapter 6).

The first applications of boundary elements in elastodynamics can be traced to Rizzo and Cruse[32,33] and the technique is now widely accepted for solving soil–structure interaction problems (see Chapter 7 of this book). Cruse's work has been of fundamental importance for the development of boundary elements as he has contributed to its application not only in elastostatics and elastodynamics but also in fracture mechanics[34] where boundary elements are much more accurate than finite elements.

The boundary element method generally offers a series of advantages over more classical 'domain' type techniques such as finite differences or finite elements, amongst them the possibility of working with smaller systems of equations and obtaining more accurate results. Because of these advantages the technique is of interest for solving a range of non-linear problems as well as linear ones. The first formulation of boundary elements for plasticity was presented in 1971 by Swedlow and Cruse[35] in a purely theoretical paper which laid the foundations for all subsequent work. However, they did not present the expressions for initial stresses or strains which are needed in stepwise plasticity analysis. These expressions were presented by Mendelson[36] in 1973 for two- and three-dimensional bodies although the plane strain formulation was correctly presented by Mukherjee[37] only in 1977. None of the above authors gave the proper expressions for strains and stresses at internal points. It was Bui[38] who in 1978 published the correct expressions for strains and stresses for three dimensional bodies and finally in 1979, Telles and Brebbia[39] presented the complete formulation for boundary elements applied to two- and three-dimensional problems. Applications relevant to engineering practice are presented in reference 40 and more can be found in Chapter 5 of this book.

It may be convenient in many applications to combine boundary elements with finite elements, as some subregions of the problem may be more amenable to solution using the latter. This may be achieved in either of two ways: (1) by considering the whole problem using an equivalent BEM approach or (2) converting the BEM subregions into equivalent FEM. The two approaches are described in detail in Brebbia and Georgiou[41]. As the FEM is very well established, with computer codes readily available for its implementation, the consideration of the BE subregion as an additional equivalent finite element matrix seems most attractive. The formation of this 'equivalent' matrix used to model the BE regions presents however certain problems; at sharp geometric discontinuities, there are also discontinuities of surface tractions which require special attention and the equivalent FE matrix formed is not inherently symmetric, unlike the classical FE approach. A technique which overcomes these problems and provides an acceptable FE type formulation using the BEM method has been suggested by Georgiou and Brebbia[41]. The technique is based on an idea advanced by Chaudonneret[42] in 1978. An early paper involving a crude way of symmetrizing the matrices has been published by Zienkiewicz et al.[43]. This work has been largely superseded by a recent paper by Mustoe et al.[44] who presented an interesting way of overcoming some of the difficulties due to the non-symmetric way in which the integrations are carried out. It involves weighting the Somigliana's identity with the actual shape functions used for the boundary variables. This in effect means that instead of applying a point source to form each equation, a distributed source on either side of the node is applied and a numerical integration process which sums the 'influence' for a series of 'sources' as opposed to a single one has to be applied. Unfortunately this significantly increases the number of integrations involved.

1.2 THE BOUNDARY ELEMENT METHOD AS A WEIGHTED RESIDUAL TECHNIQUE

The engineering analyst usually tries to find an approximate solution to a mathematical problem expressed in function of a differential equation or set of equations. The equation or system of equations is applicable over a certain domain and has a set of boundary conditions. Let us write the equation as,

$$L(u_0)=0, \qquad \text{in } \Omega \text{ domain} \tag{1.1}$$

with boundary conditions,

$$\begin{aligned} S(u_0)&=s, \text{ essential boundary condition on } \Gamma_1 \\ G(u_0)&=g, \text{ natural boundary condition on } \Gamma_2 \end{aligned} \tag{1.2}$$

$$\text{Total boundary } \Gamma=\Gamma_1+\Gamma_2$$

One can then propose a discretization of the continuum into elements or cells and assume that over each of these subregions the u_0 unknown varies in accordance with certain predetermined shapes or modes. Notice that for the case of boundary elements only the surface of the domain needs to be discretized. These shape functions belong to a set of linearly independent functions with unknown coefficients, whose values are determined once the resulting algebraic system of equations is solved.

Let us write the approximations for u as a set of linearly independent functions φ_i multiplied by a series of unknown α_i coefficients i.e.

$$u_0 \simeq u = \alpha_1 \varphi_1 + \alpha_2 \varphi_2 + \alpha_3 \varphi_3 + \dots \tag{1.3}$$

Introducing this approximation to the u_0 function into Equations 1.1 and 1.2 will produce certain errors in the Ω domain and on the Γ boundary. The aim of the different types of approximate techniques is to make this error as small as possible. The errors can be made zero at certain points, collocation, finite differences, etc. or they can be distributed over a subregion (collocation by subregions, Galerkin, least squares, cell-collocation, finite elements, etc.). If the dimensions of the problems are reduced by one, then one may only be required to distribute the errors on the boundary of the region under consideration. The way in which the error is distributed produces different types of weighted residual techniques and hence approximate methods.

To understand how the distribution can be carried out let us consider a function w which is to be our weighting function and is defined by a set of known linearly independent functions ψ_i multiplied by arbitrary β_i coefficients, i.e.

$$w = \beta_1 \psi_1 + \beta_2 \psi_2 + \beta_3 \psi_3 + \dots \tag{1.4}$$

Consider in what follows that $L$is a linear operator for simplicity and that all boundary conditions are identically satisfied by the approximate function u, Equation 1.3. In this case we only need to weight Equation 1.1 by the w function and we can now identify the following techniques.

(1) *Method of point collocation*

In this case we simply collocate the $L(u)$ equation at a series of points 'i' and make it equal to zero at these points. We can represent this operation by the following inner product,

$$\langle L(u), \quad \Delta_i \rangle = 0, \qquad \text{for all } i \tag{1.5}$$

where Δ_i is a Dirac delta function whose integral is equal to 1 at the point 'i' and zero everywhere else. Hence w can be written as,

$$w = \beta_1 \Delta_1 + \beta_2 \Delta_2 + \beta_3 \Delta_3 + \dots \tag{1.6}$$

(2) *Method of moments*

The weighting function w is now assumed to be such that moments of the $L(u)$ operator can be taken, i.e.

$$\langle L(u), \quad \psi_i \rangle = 0 \tag{1.7}$$

where ψ_i is defined for a two dimensional problem for instance, by

$$w = \beta_1 1 + \beta_2 x + \beta_3 y + \beta_4 xy + \dots \tag{1.8}$$

(3) *Method of collocation by subregions*

In this case the integral of the error over a subregion cell or subdomain is set equal to zero. The finite difference technique can be interpreted as a subregion or cell collocation technique, Lau and Brebbia[45], for which the value of u is defined over a cell, i.e. the part of the domain connecting the centre point to the neighbouring points, and then we collocate the $L(u)$ operator using Equation 1.5. The main difference with Equation 1.5 is that the u functions of the cell–collocation method are valid only over a part or cell of the Ω domain. In addition to the Dirac delta we can propose over our subregion a simple function such as 1 or other more complex ψ_i weighting functions.

(4) *Generalized weighting*

In this case the functions φ_i and ψ_i belong to two different sets. The φ_i satisfy the boundary conditions of the problem and the ψ_i sometimes but not necessarily satisfy the homogeneous version of those conditions. This gives the following weighted residual statement

$$\langle L(u), \quad \psi_i \rangle = 0 \tag{1.9}$$

(5) *Galerkin method*

If the weighting functions are chosen to be equal to those in the approximating set, i.e. $\psi_i = \varphi_i$, we have a special version of Equation 1.9 called the Galerkin method. Here

$$\langle L(u), \quad \varphi_i \rangle = 0 \tag{1.10}$$

(6) *Least-square method*

The least-square technique is somewhat different to the previous ones. It consists of defining the square of the error function, i.e.

$$\varepsilon^2 = (L(u))^2 \tag{1.11}$$

and minimizing it with respect to the α_i coefficients of u. This gives for a linear operator L,

$$\langle L(u), \quad L(\varphi_i) \rangle = 0 \tag{1.12}$$

Hence the weighting functions here are not the original φ_i functions as in Galerkin's but the expressions found by applyinf the $L(\ )$ operator to them

(7) *Weak formulations*

Let us assume that the $L(\ )$ operator is symmetric, i.e. it can be decomposed as follows:

$$\langle L(u), \quad w \rangle = -\langle D(u), \quad D(w) \rangle + \langle G(u), \quad S(w) \rangle_\Gamma \tag{1.13}$$

where G and S indicate the natural and essential boundary condition operators. Consider for simplicity that the w functions are such that they satisfy the homogeneous version of the essential b.c., i.e. $S(w) \equiv 0$. This gives,

$$\langle L(u), \quad w \rangle = -\langle D(u), \quad D(w) \rangle \tag{1.14}$$

Notice that the order of derivation for u has decreased and the derivatives of w have correspondingly increased. $D(\ )$ is an operator of lower order than $L(\ )$.

The classical Rayleigh–Ritz technique can be interpreted as a particular case of Equation 1.14 for which $\psi_i = \varphi_i$. Hence the stationarity condition for a Rayleigh–Ritz function can be written as,

$$\langle D(w), \quad D(\varphi_i) \rangle = 0, \qquad \text{for all } i \tag{1.15}$$

It is now apparent that the least-square solution, Equation 1.12, implies a higher order weighting than Rayleigh–Ritz Equation 1.15. As it is to be expected, least-square solutions tend to satisfy the governing equation operator more closely than Rayleigh–Ritz. Both methods have the property of giving symmetric matrices for the above type of symmetric operator.

(8) *Boundary solutions*

In these cases the φ_i or the ψ_i set of functions identically satisfy the operator L but not the boundary conditions which are then approximated using any of the above discussed methods.

Consider that the functions of the φ_i set satisfy the governing equations but that ψ_i may not do so. The problem consists now of approximating boundary conditions 1.2, i.e.

$$\begin{aligned} S(u) &= s, \text{ on } \Gamma_1 \\ G(u) &= g, \text{ on } \Gamma_2 \end{aligned} \tag{1.16}$$

This can be done by substituting the approximation for u into Equation 1.16 and weighting the boundary error functions as follows:

$$\begin{aligned} \langle S(u) - s, \quad G(w) \rangle_{\Gamma_1} &= 0 \\ \langle G(u) - g, \quad S(w) \rangle_{\Gamma_2} &= 0 \end{aligned} \tag{1.17}$$

Equations 1.17 allow us to find a set of algebraic equations from which the unknown α_i coefficients can be obtained. Notice that the solution implies only boundary integrations, i.e. the order of the problem has been reduced by one.

Another possibility is to propose weighting functions w which satisfy the governing equations, i.e. $L(w) \equiv 0$. We can start with,

$$\langle L(u), \quad w \rangle = 0 \tag{1.18}$$

and integrate by parts until the $L(\ )$ operator passes to the w function. This gives,

$$\langle L(u),\quad w\rangle = \langle u,\quad L(w)\rangle + \langle G(u),\quad S(w)\rangle_\Gamma - \langle S(u),\quad G(w)\rangle_\Gamma = 0 \tag{1.19}$$

Or if $L(w) \equiv 0$

$$\langle L(u),\ w\rangle = \langle G(u),\ S(w)\rangle_\Gamma - \langle S(u),\ G(w)\rangle_\Gamma = 0 \tag{1.20}$$

We can also write Equation 1.20 using the following conditions; $G(u) = g$ on Γ_2, $S(u) = s$ on Γ_1. This gives,

$$\langle G(u),\ S(w)\rangle_{\Gamma_1} + \langle g,\ S(w)\rangle_{\Gamma_2} = = \langle S(u),\ G(w)\rangle_{\Gamma_2} + \langle s,\ G(w)\rangle_{\Gamma_1} \tag{1.21}$$

Notice that Equation 1.21 is a generalization of Equation 1.17.

We can use the above ideas to generalize the weighted residual expressions. Let us propose a more general statement corresponding to approximation of the boundary conditions as well as the domain equation, i.e.

$$\langle L(u),\ w\rangle = \langle G(u) - g,\ S(w)\rangle_{\Gamma_2} - \langle S(u) - s,\ G(w)\rangle_{\Gamma_1} \tag{1.22}$$

This gives, after integrating by parts,

$$\langle D(u),\ D(w)\rangle = \langle S(u) - s,\ G(w)\rangle_{\Gamma_1} + + \langle g,\ S(w)\rangle_{\Gamma_2} + \langle G(u),\ S(w)\rangle_{\Gamma_1} \tag{1.23}$$

Integrating again we obtain the inverse relationship, i.e.

$$\langle u,\ L(w)\rangle = \langle s,\ G(w)\rangle_{\Gamma_1} + \langle S(u),\ G(w)\rangle_{\Gamma_2} - - \langle g,\ S(w)\rangle_{\Gamma_2} - \langle G(u),\ S(w)\rangle_{\Gamma_1} \tag{1.24}$$

(9) *Singular functions*

In certain boundary problems it is convenient to choose weighting functions which are singular and represent the solution of Equation 1.1 without any boundary conditions, i.e. the solution of,

$$L(w^*) = \Delta_i \tag{1.25}$$

where w^* is this special type of singular function. It is identified with an asterisk to denote this and differentiate it from w. Δ_i is a Delta Dirac function whose integral is equal to 1 at the point under consideration and zero everywhere else. In this case Equation 1.19 becomes,

$$u_i + \langle G(u),\ S(w^*)\rangle_\Gamma = \langle S(u),\ G(w^*)\rangle_\Gamma \tag{1.26}$$

Taking this expression to the boundary produces a c coefficient in the limit, i.e.

$$c_i u_i + \langle G(u),\ S(w^*)\rangle_\Gamma = \langle S(u),\ G(w^*)\rangle_\Gamma \tag{1.27}$$

This expression reduces to Somigliana's identity and other similar integral relationships depending on the definition of the inner products and the type of physical problem under consideration.

Another possibility when using these w^* type functions is to consider that they are distributed over part of the boundary rather than concentrated at node 'i'. This distribution of the singularities may involve weighting the Somigliana's type identity with the distribution shape function proposed. This in effect means that to form each equation a distributed source on either side of the 'i' source node is applied. Some authors[44] have proposed to carry out this distribution using a numerical integration scheme. Unfortunately, this significantly increases the number of integrations required and is too expensive for practical applications, although it can be a useful technique to symmetrize BE matrices. Notice that if we have the fundamental solution for a distributed source between two points the double integrals would not be needed.

Finally, notice that we can also propose least-square boundary solutions to symmetrize the problem, starting with,

$$\begin{aligned} S(u) - s &= 0, \qquad \text{on } \Gamma_1 \\ G(u) - g &= 0, \qquad \text{on } \Gamma_2 \end{aligned} \tag{1.28}$$

The conditions to satisfy are now

$$\begin{aligned} \langle S(u) - s,\ S(w)\rangle_{\Gamma_1} &= 0 \\ \langle G(u) - g,\ G(w)\rangle_{\Gamma_2} &= 0 \end{aligned} \tag{1.29}$$

for a given w function which satisfied $L(w) \equiv 0$. The above formulation does not seem to have been applied in practice.

1.3 SUGGESTIONS FOR FURTHER RESEARCH

Several important topics related to boundary solutions or more specifically the boundary element method require further investigation. Boundary element method implies an approximate technique by which the problem dimensions have been reduced by one or in the case of time dependent problems by more than one. The problem becomes a boundary problem and the surface of the domain can be divided into elements over which some approximate technique can be applied. What differentiates boundary element from classical integral equation is its emphasis on elements and compatibility as well as its reliance on general principles such as weighted residuals. This is important as some classical boundary integral solutions suffer from lack of accuracy and convergence due to poor representation of the geometrical shapes. These problems appear not to have been sufficiently investigated.

An area of confusion amongst boundary integral researchers is the one related to geometry and function representations. It is evident to finite element practitioners that the degree of complexity assumed for the geometry and the one for the functions representing the boundary variables are interrelated. We know from variational techniques research that trying to represent the geometry with higher order functions while assuring simple functions for the variables will introduce considerable errors in the solution. This is an important point which seems to have escaped the attention of many researchers applying integral equations.

Although current boundary element practice is heavily dependent on using fundamental solutions based on solving the governing equations under concentrated loads and no boundary conditions, it is of the utmost importance to point out that other solutions can be used as well. These other solutions are illustrated by those used in such methods as hybrid finite elements and the trigonometric type solutions proposed by Atluri[46]. Analytical functions which have been obtained over a close domain with certain boundary conditions could also be used as fundamental solutions provided that our domain of interest is within the domain of the analytical solutions. This of course can be easily arranged by a change of dimensions. Once these ideas are better understood we will be able to solve many complex plate and shell and other problems using the wealth of solutions already available as weighting functions[47,48]. The possibilities of using these and other approximate fundamental solutions are endless.

More work is also needed in the general field of non-linear boundary elements. A good deal of research has been done on the applications of boundary elements to plasticity and the problem appears now to have been solved to a large extent (see Chapter 5). Other non-linear material problems are not yet as well understood although it seems clear how to treat problems such as viscoelasticity and non-tension materials. Geometrically non-linear problems however have not been studied, although it is an important topic of research.

The general field of boundary element non-linear problem solving requires further investigation. Up to now BEM researchers have concentrated on using iterative solution techniques similar to the seudoforce vectors applied on the right hand side of most FEM solutions to take into consideration the non-linearities. It has been assumed impossible to implement the tangent matrix approach in BEM but it may be worthwhile to investigate an incremental form of this approach in which the tangent matrix could be computed numerically. It may allow us to apply BEM to some non-linear problems which up to now have been the exclusive domain of the FEM.

Time dependent problems may be one of the most fruitful research topics in BEM. The method seems well suited to solve problems in time and space. FEM solutions in time and space are on the contrary, very uneconomical and have been largely abandoned.

BEM are also well suited for the small computers or microprocessors, which are nowadays being used in many engineering offices. BEM solutions tend to require less matrices and data storage than FE solutions while having a large number of repetitive operations such as numerical integration which can be economically carried out on a microcomputer.

Finally it is to be expected that a better understanding of the boundary element method will increase our knowledge of approximate techniques and open new frontiers to our research in the computer solution of engineering problems.

REFERENCES

1 Brebbia, C. A., *The Boundary Element Method for Engineers*, Pentech Press, London (1978)

2 Brebbia, C. A. and Connor, J. J., *Fundamentals of Finite Element Techniques for Structural Engineers*, Butterworths, London (1973)

3 Connor, J. J. and Brebbia, C. A., *Finite Element Method for Fluid Flow*, Butterworths, London (1977)

4 Kantorovich, L. V. and Krylov, V. I., *Approximate Methods of Higher Analysis*, Interscience, New York (1958)

5 Courant, R. and Hilbert, D., *Methods of Mathematical Physics*, Interscience (1953)

6 Reissner, E., 'A Note on Variational Principles in Elasticity' *Int. J. Solids and Sgructures*, **1**, pp. 93–95 and 357 (1965)

7 Pian, T. H. H. and Tong, P., 'Basis of Finite Element Method for Solid Continua' *Int. J. Numerical Methods Engng.*, **1**, pp. 3–28 (1969)

8 Washizu, K., *Variational Methods in Elasticity and Plasticity*, 2nd edn, Pergamon Press, New York (1975)

9 Muskhelishvili, N. I., *Some Basic Problems of the Mathematical Theory of Elasticity*, P. Noordhoff Ltd., Groningen (1953)

10 Mikhlin, S. G., *Integral Equations*, Pergamon Press, New York (1957)

11 Kupradze, O. D., *Potential Methods in the Theory of Elasticity*, Daniel Davey & Co., New York (1965)

12 Smirnov, V. J., 'Integral Equations and Partial Differential Equations', in *A Course in Higher Mathematics*, Vol. IV, Addison-Wesley (1964)

13 Kellogg, P. D., *Foundations of Potential Theory*, Dover, New York (1953)

14 Jaswon, M. A. 'Integral Equation Methods in Potential Theory, I' *Proc. R. Soc.*, Ser. A, p. 273 (1963)

15 Symm, G. T., 'Integral Equation Methods in Potential Theory, II' *Proc. R. Soc.*, Ser. A, p. 275 (1963)

16 Massonnet, C. E., 'Numerical Use of Integral Procedures in Stress Analysis', *Stress Analysis*. Zienkiewicz, O. C. and Holister, G. S. (eds), Wiley (1966)

17 Hess, J. L. and Smith, A. M. O., 'Calculation of Potential Flow about Arbitrary Bodies', *Progress in Aeronautical Sciences* Vol. 8, Kuchemann, D. (Ed.). Pergamon Press (1967)

18 Cruse, T. A. and Rizzo, F. J., 'A Direct Formulation and Numerical Solution of the General Transient Elasto-Dynamic Problem, I', *J. Math. Analysis Applic.*, **22** (1968)

19 Hadid, H. A., *An Analytical and Experimental Investigation into the Bending Theory of Elastic Conoidal Shells*, Ph.D. Thesis, University of Southampton (1964)

20 Hajdin, N., 'A Method for Numerical Solutions of Boundary Value Problems and its Application to Certain Problems in the Theory of Elasticity', Belgrade University Publication (1958)

21 Watson, J., *Analysis of Thick Shells with Holes by Using Integral Equation Method*, Ph.D. Thesis, University of Southampton (1973)

22 Lachat, J. C., *A Further Development of the Boundary Integral Technique for Elastostatics*, Ph.D. Thesis, University of Southampton (1975)

23 Brebbia, C. A. and Walker, S., *The Boundary Element Technique in Engineering*, Newnes–Butterworths, London (1979)

24 Brebbia, C. A. (Ed.), *Recent Advances in Boundary Element Methods, Proc. 1st Int. Conference Boundary Element Methods*, Southampton University, 1978, Pentech Press, London (1978)

25 Brebbia, C. A. (Ed.), *New Developments in Boundary Element Methods, Proc. 2nd Int. Conference Boundary Element Methods*, Southampton University, 1980, CML Publications, Southampton (1980)

26 Rizzo, F. J. and Shippy, D. J., 'A Method of Solution for Certain Problems of Transient Heat Conduction', AIAA J., **8**, No. 11, pp. 2004–2009 (1970)

27 Liggett, J. A. and Liu, P. L., 'Unsteady Flow in Confined Aquifers: A Comparison of Two Boundary Integral Methods', *Water Resources Res.*, **15**, No. 4, pp. 861–866 (1979)

28 Chang, Y. P., Kang, C. S. and Chen, D. J., 'The Use of Fundamental Green's Function for the Solution of Problems of Heat Conduction in Anistropic Media', *Int. J. Heat Mass Transfer*, **16**, pp. 1905–1918 (1973)

29 Wrobel, L. C. and Brebbia, C., 'The Boundary Element Method for Steady-State and Transient Heat Conduction', *1st Int. Conf. Numerical Methods in Thermal Problems*, Swansea, Taylor, C. and Morgan, K. (Eds.), Pineridge Press (1979)

30 Wrobel, L. C. and Brebbia, C., 'A Formulation of the Boundary Element Method for Axisymmetric Transient Heat Conduction', *Int. J. Heat Mass Transfer*, **24** (1981)

31 Wrobel, L. C. and Brebbia, C. 'Boundary Elements in Thermal Problems', *Numerical Methods in Heat Transfer*, Lewis, R. W. (ed.), Wiley, Chichester, England (1981)

32 Cruse, T. A. and Rizzo, F. J., 'A Direct Formulation and Numerical Solution of the General Elastodynamic Problem, I', *J. Math. Analysis Applic.*, **22** (1968)

33 Cruse, T. A. and Rizzo, F. J., 'A Direct Formulation and Numerical Solution of the General Transient Elastodynamic Problem, II', *J. Math. Analysis Applic.*, **22** (1968)

34 Cruse, T. A. and Vanduren, W., 'Three Dimensional Elastic Stress Analysis of a Fracture Specimen with Edge Crack', *Int. J. Fract. Mech.*, **7**, pp. 1–15 (1971)

35 Swedlow, J. L. and Cruse, T. A., 'Formulation of Boundary Integral Equations for Three-Dimensional Elasto-Plastic Flow', *Int. J. Solids Struct.*, **7**, 1673–1683, 1971

36 Mendelson, A., *Boundary Integral Methods in Elasticity and Plasticity*', Report No. NASA TN D-7418 (1973)

37 Mukherjee, S., 'Corrected Boundary Integral Equation in Planar Thermoelastoplasticity', *Int. J. Solids Struct.*, **13**, pp. 331–335 (1977)

38 Bui, H. D., 'Some Remarks about the Formulation of Three-Dimensional Thermoelastoplastic Problems by Integral Equations', *Int. J. Solids Struct.*, No. 14, pp. 935–939 (1978)

39 Telles, J. C. F. and Brebbia, C. A., 'On the Application of the Boundary Element Method to Plasticity', *Appl. Meth. Modelling*, **3**, pp. 446–470 (1979)

40 Telles, J. C. F. and Brebbia, C. A., 'The Boundary Element Method in Plasticity', in *New Developments in Boundary Element Methods*, Brebbia, C. A. (Ed.), CML Publications, England (1980)

41 Brebbia, C. A. and Georgiou, P., 'Combination of Boundary and Finite Elements in Elastostatics', *Applied Math. Modelling*, **3**, pp. 213–220 (1978)

42 Chaudonneret, M., 'On the Discontinuity of the Stress Vector in the Boundary Integral Equation Method for Elastic Analysis', 'Recent Advances in Boundary Element Methods', Brebbia, C. A. (Ed.), Pentech Press, London (1978)

43 Zienkiewicz, O. C., Kelly, D. W. and Bettess, P., 'The Coupling of the Finite Element Method and Boundary Solution Procedures', *Int. J. Num. Meth. Engng.*, **11**, 355–375

44 Mustoe, G., 'A Symmetric Direct Boundary Integral Equation Methof for Two-Dimensional Elastostatics', Paper presented at the 2nd Int. Symposium on Boundary Element Methods, Southampton University, March (1980)

45 Lau, P. and Brebbia, C., 'The Cell Collocation Method in Continuum Mechanics', *Int. J. Mech. Sci.*, **20**, pp. 83–85 (1978)

46 Grannell, J. J., Atluri, S. N. and Fitzgerald, J. E., 'Boundary Discretization Using the Edge–Function Method in 3-Dimensional Elasticity', in *Recent Advances in Boundary Element Methods*, Brebbia, C. A. (Ed.), Pentech Press (1978)

47 Lukasiewicz, S., *Local Loads in Plates and Shells*, Noordhoff Int. Publ., Leyden (1979)

48 Teodorescu, P. &., *Dynamics of Linear Elastic Bodies*, Abacus Press, England (1975)

2

Fundamental solutions

by S. Walker

2.1 INTRODUCTION

This chapter is an attempt to bring together a review of the methods of construction and the use of fundamental solutions, particularly with reference to their use in the Boundary Element Method.

We will start by considering the physical significance of fundamental solutions and study their mathematical properties. Later in the chapter a number of approximate and numerical fundamental solutions will be considered together with a number of methods of avoidance of the singularities inherent in these solutions.

At the end of the chapter will be found a table of the fundamental solutions for a selection of differential equations.

2.2 EIGENFUNCTIONS AND THE DIRAC DELTA FUNCTION

Consider a medium in which we have a physical property represented by the function u which may be a function of space and time. If we know the governing equation for u and this equation is linear we can write this equation as

$$L(u)=0 \tag{2.1}$$

throughout the medium which we shall for the moment consider to be finite in extent. Here L is a linear differential operator, for which we have

$$L(\alpha\varphi_1+\beta\varphi_2)=\alpha L(\varphi_1)+\beta L(\varphi_2) \tag{2.2}$$

where φ_1 and φ_2 are any two functions with sufficient continuity properties to make $L(\varphi_1)$ and $L(\varphi_2)$ meaningful, and α and β are constants.

The eigenfunctions φ_n corresponding to L are defined by

$$L(\varphi_n)=\lambda_n\varphi_n \tag{2.3}$$

where λ_n is a constant called the eigenvalue corresponding to the eigenfunction φ_n. The form of the eigenfunctions will depend on the coordinate system used.

For rectangular Cartesian coordinates we can choose the eigenfunctions to be the harmonic functions $\cos(nx)$, $\sin(nx)$, $\cos(ny)$, etc. or using complex notation $e^{\pm inx}$. The eigenvalues λ_n are then just polynomials involving powers of 'n' and 'in' for an equation with constant coefficients[1].

It is usual to choose our eigenfunctions to be a complete set, in the sense that for any function u a number N can be found such that

$$\left\| u - \sum_{1}^{N} \alpha_n \varphi_n \right\| < \varepsilon \tag{2.4}$$

where ε is any small number and α_n are constants. Here we have used '$\| \, \|$' to denote the norm of a function defined by

$$\|u\| = \sqrt{\int_{\Omega} u \hat{u} \, d\Omega} \tag{2.5}$$

where the integration is over all the variables in u on the problem region, Ω and $\hat{}$ denotes complex conjugation. Equation 2.4 means that we can expand any sufficiently continuous function u in terms of a series of eigenfunctions to any desired degree of accuracy, i.e.

$$u = \lim_{N \to \infty} \left(\sum_{n=1}^{N} \alpha_n \varphi_n \right) \tag{2.6}$$

In this chapter we shall use the notation

$$\langle u, v \rangle_{\Omega} = \int_{\Omega} u \hat{v} \, d\Omega \tag{2.7}$$

$$\langle u, v \rangle_{\Gamma} = \int_{\Gamma} u \hat{v} \, d\Gamma \tag{2.8}$$

to represent integration of the two functions u and v over the region Ω and boundary Γ respectively. The expression $\langle u, v \rangle_{\Omega}$ which will be a number, is referred to as the 'scalar product' of u and v.

The norm of a function u may now be written

$$\|u\| = \sqrt{\langle u, u \rangle_{\Omega}} \tag{2.9}$$

Given a complete set of eigenfunctions we may form an ortho-normal set which has the property that

$$\langle \varphi_m, \varphi_n \rangle_{\Omega} = \tilde{\delta}_{mn} \tag{2.10}$$

where $\tilde{\delta}_{mn}$ is the Krönecker delta defined by

$$\tilde{\delta}_{mn}=\begin{cases}0, & \text{if } m\neq n\\ 1, & \text{if } m=n\end{cases} \tag{2.11}$$

We may now define the Dirac delta function on the region in terms of our complete orthonormal set for this region, by

$$\delta_\Omega(\vec{x}-\vec{\xi})=\sum_{n=0}^{\infty}\varphi_n(\vec{x})\hat{\varphi}_n(\vec{\xi}) \tag{2.12}$$

where $\vec{x}$ and $\vec{\xi}$ are two points in space. Hereafter we shall refer to $\vec{\xi}$ as the 'source' point and $\vec{x}$ as the observation point.

We can see that as a 'forcing function' on the right-hand side of an equation the delta function would have for the solution the field due to a series of sources separated by twice our normalisation range.

Consider now an expansion of u in our region Ω in terms of our eigenfunctions φ_n given by:

$$u(\bar{x})=\sum_{n=-\infty}^{\infty}\alpha_n\varphi_n \tag{2.13}$$

(Note that in two dimensions we shall interpret the summation

$$\sum_{n=-\infty}^{\infty}\alpha_n\varphi_n$$

to mean the double summation

$$\sum_{n_1=-\infty}^{\infty}\sum_{n_2=-\infty}^{\infty}\mathrm{a}_{n_1n_2}\varphi_{n_1n_2}$$

where $\vec{n}=(n_1, n_2)$. For simplicity we shall omit the vector sign from the suffices.) Multiplying by $\hat{\varphi}_m$ and integrating over Ω we obtain:

$$\begin{aligned}\langle u, \varphi_m\rangle_\Omega &= \sum_{n=-\infty}^{\infty}\langle\alpha_n\varphi_n,\varphi_m\rangle_\Omega\\ &=\sum_{n=-\infty}^{\infty}\alpha_n\langle\varphi_n, \varphi_m\rangle_\Omega\\ &=\sum_{n=-\infty}^{\infty}\alpha_n\tilde{\delta}_{nm}=\alpha_m\end{aligned} \tag{2.14}$$

$$\alpha_m=\langle u, \varphi_m\rangle_\Omega \tag{2.15}$$

an expression for our expansion coefficients α_m.

Note that for the case $\varphi_m=(2\pi)^{-1}e^{i\pi mx}$ (the eigenfunctions for the one dimensional Laplace equation, over the interval $-\pi$ to π) the left-hand side corresponds to a finite Fourier transform of u giving the Fourier coefficient α_m.

Now consider the expression

$$\langle u(\vec{x}), \quad \delta_\Omega(\vec{x}-\vec{\xi})\rangle_\Omega \tag{2.16}$$

We may write both terms as expansions in terms of our φ_n:

$$\left\langle \sum_{n=0}^{\infty} \alpha_n\varphi_n(\vec{x}), \sum_{m=0} \varphi_m(\vec{x})\hat{\varphi}_m(\vec{\xi}) \right\rangle_{\Omega(\vec{x})}$$

$$= \sum_{n=0}^{\infty} \alpha_n \sum_{m=0}^{\infty} \varphi_m(\vec{\xi})\langle\varphi_n(\vec{x}),\varphi_m(\vec{x})\rangle_{\Omega(\vec{x})}$$

$$= \sum_{n=0}^{\infty} \alpha_n \sum_{m=0}^{\infty} \varphi_m(\vec{\xi})\tilde{\delta}_{nm}$$

$$= \sum_{n=0}^{\infty} \alpha_n\varphi_n(\vec{\xi}) = u(\vec{\xi})$$

Hence

$$\langle u(\vec{x}), \ \delta_\Omega(\vec{x}-\vec{\xi})\rangle_{\Omega(\vec{x})} = u(\vec{\xi}) \tag{2.17}$$

We have now derived the selective property of a delta function which we shall use extensively in what follows. For real functions the hat ^ denoting complex conjugation becomes superfluous.

The properties of fundamental solutions defined on finite regions are discussed in detail in Chapter 4 of Brebbia and Walker[2] together with a number of examples of their use. We have seen that we may expand any function within a domain by putting

$$u(x_1, x_2) = \sum_{n=-\infty} \alpha_n\varphi_n \tag{2.18}$$

The eigenfunctions depended on the coordinate system chosen and the shape of the region.

If we now wish to represent a function over all space we cannot use an expansion similar to (2.18), in this case we have to construct a set of eigenfunctions normalised over all space. This set will now depend on a continuously varying parameter $\vec{\kappa}$, say, rather than just being denoted by a discrete set of subscripts such as n. We shall denote these eigenfunctions by $\varphi(\vec{v}, \vec{x})$. These functions will now depend only on the coordinate system used.

κ will be a vector with the same number of components as x, i.e. in two dimensions the two components κ_1, κ_2 will correspond to the subscripts, n_1 and n_2 in $\varphi_{n_1 n_2}$. We shall see that in infinite space the summations will be replaced by integrations and our discrete indices will be replaced by the now continuously varying parameter $\vec{\kappa}$.

For an infinite region our orthonormalisation condition becomes

$$\int_{\text{All space}} \varphi(\vec{\kappa}, \vec{x})\hat{\varphi}(\vec{\kappa}', \vec{x})\mathrm{d}\vec{x} = \langle \varphi(\vec{\kappa}, \vec{x}), \varphi(\vec{\kappa}', \vec{x}) \rangle_{\vec{x}} = \delta(\vec{\kappa} - \vec{\kappa}') \tag{2.19}$$

where the delta function is now the classical Dirac delta function defined for all space, which is zero when $\vec{\kappa} = \vec{\kappa}'$ but infinite at $\vec{\kappa} = \vec{\kappa}'$. To define δ uniquely we require that

$$\langle \varphi(\vec{\kappa}, \vec{x}), \delta(\vec{\kappa} - \vec{\kappa}') \rangle_{\vec{\kappa}} = \varphi(\vec{\kappa}', \vec{x}) \tag{2.20}$$

which is the replacement property of the delta function with respect to our eigenfunctions. One consequence of Equation 2.20 is that

$$\langle 1, \delta(\vec{\kappa} - \vec{\kappa}') \rangle_{\vec{\kappa}} = 1 \tag{2.21}$$

This function is now expressed in terms of our eigenfunctions by,

$$\delta(\vec{x} - \vec{\xi}) = \langle \varphi(\vec{\kappa}, \vec{x}), \varphi(\vec{\kappa}, \vec{\xi}) \rangle_{\vec{\kappa}} \tag{2.22}$$

The brackets $\langle \ \rangle_{\vec{\kappa}}$ now denote an integration over all $\vec{\kappa}$ space, i.e.

$$\langle \varphi(\vec{\kappa}, \vec{x}), \varphi(\vec{\kappa}, \vec{\xi}) \rangle_{\vec{\kappa}} = \int_{\text{All space}} \varphi(\vec{\kappa}, \vec{x})\hat{\varphi}(\vec{\kappa}, \vec{\xi})\mathrm{d}\vec{\kappa} \tag{2.23}$$

The expansion of any bounded function will now be given by

$$u(\vec{x}) = \langle \alpha(\vec{\kappa}), \varphi(\vec{\kappa}, \vec{x}) \rangle_{\vec{\kappa}} \tag{2.24}$$

where by an argument similar to Equations 2.13 to 2.15 we have

$$\alpha(\vec{\kappa}) = \langle \varphi(\vec{\kappa}, \vec{x}), u(\vec{x}) \rangle_{\vec{x}} \tag{2.25}$$

Equations 2.24 and 2.25 are a generalised transform pair.

If we now consider the expression corresponding to Equation 2.16 for the infinite domain

$$\langle u(\vec{x}), \delta(\vec{x} - \vec{\xi}) \rangle_{\vec{x}} \tag{2.26}$$

and expand both in terms of eigenfunctions we obtain

$$\langle\langle\alpha(\vec{\kappa}),\ \varphi(\vec{\kappa},\ \vec{x})\rangle_{\vec{K}},\langle\varphi(\vec{\kappa},\ \vec{x}),\ \varphi(\vec{\kappa}',\ \vec{\xi})\rangle_{\vec{K}'}\rangle_{\vec{X}} \tag{2.27}$$

assuming convergence of the integrals we may rearrange the above to give

$$\langle\langle\langle\varphi(\vec{\kappa},\ \vec{x}),\ \varphi(\vec{\kappa}',\ \vec{x})\rangle_{\vec{X}},\ \hat{\alpha}(\vec{\kappa})\varphi(\vec{\kappa}',\ \xi)\rangle_{\vec{K}'}\rangle_{\vec{K}}$$

$$=\langle\langle\delta(\vec{\kappa}-\vec{\kappa}'),\ \alpha(\vec{\kappa})\varphi(\vec{\kappa}',\ \vec{\xi})\rangle_{\vec{K}'}\rangle_{\vec{K}}$$

using Equation 2.22

$$=\langle\alpha(\vec{\kappa}),\ \langle\delta(\vec{\kappa}-\vec{\kappa}'),\ \varphi(\vec{\kappa}',\ \vec{\xi})\rangle_{\vec{K}'}\rangle_{\vec{K}}$$

$$=\langle\alpha(\vec{\kappa}),\ \varphi(\vec{\kappa},\ \vec{\xi})\rangle_{\vec{K}}$$

using Equation 2.20

$$=u(\vec{\xi})\ \text{(by Equation 2.24)} \tag{2.28}$$

Hence

$$\langle u(\vec{x}),\ \delta(\vec{x}-\vec{\xi})\rangle_{\vec{X}}=u(\vec{\xi}) \tag{2.29}$$

This relation is fundamental to the boundary element method.

2.3 CONSTRUCTION OF FUNDAMENTAL SOLUTIONS

The fundamental solution in infinite space corresponding to an isolated point source at the point $\vec{\xi}$ is defined by

$$L(u^*(\vec{x},\ \vec{\xi}))=\delta(\vec{x}-\vec{\xi}) \tag{2.30}$$

Expanding u^* and δ in terms of eigenfunctions, i.e. putting

$$u^*(\vec{x},\ \vec{\xi})=\langle\varphi(\vec{\kappa},\ \vec{x}),\ \hat{\alpha}(\vec{\kappa},\ \vec{\xi})\rangle_{\vec{K}} \tag{2.31}$$

and

$$\delta(\vec{x}-\vec{\xi})=\langle\varphi(\vec{\kappa},\ \vec{x}),\ \varphi(\vec{\kappa},\ \vec{\xi})\rangle_{\vec{K}} \tag{2.32}$$

we write Equation 2.30,

$$L(\langle\varphi(\vec{\kappa},\ \vec{x}),\ \hat{\alpha}(\vec{\kappa},\ \xi)\rangle_{\kappa})=\langle\varphi(\vec{\kappa},\ \vec{x}),\ \varphi(\vec{\kappa},\ \vec{\xi})\rangle_{\vec{K}} \tag{2.33}$$

$$\therefore\ \langle L(\varphi(\vec{\kappa},\ \vec{x})),\ \hat{\alpha}(\vec{\kappa},\ \vec{\xi})\rangle_{\vec{K}}=\langle\varphi(\vec{\kappa},\ \vec{x}),\ \varphi(\vec{\kappa},\ \vec{\xi})\rangle_{\vec{K}} \tag{2.34}$$

$$\therefore\ \langle\lambda(\vec{\kappa})\varphi(\vec{\kappa},\ \vec{x}),\ \hat{\alpha}(\vec{\kappa},\ \vec{\xi})\rangle_{\vec{K}}=\langle\varphi(\vec{\kappa},\ \vec{x}),\ \varphi(\vec{\kappa},\ \vec{\xi})\rangle_{\vec{K}} \tag{2.35}$$

which can be written

$$\langle \varphi(\vec{\kappa}, \vec{x}), \hat{\lambda}(\vec{\kappa})\hat{\alpha}(\vec{\kappa}, \vec{\xi})\rangle_{\vec{K}} = \langle \varphi(\vec{\kappa}, \vec{x}), \varphi(\vec{\kappa}, \vec{\xi})\rangle_{\vec{K}} \tag{2.36}$$

This is an identity for all $\vec{x}$ and $\vec{\xi}$, hence we may write

$$\alpha(\vec{\kappa}, \vec{\xi}) = \frac{\hat{\varphi}(\vec{\kappa}, \vec{\xi})}{\lambda(\vec{\kappa})} \tag{2.37}$$

and we can write our fundamental solution

$$u^*(\vec{x}, \vec{\xi}) = \langle \varphi(\vec{\kappa}, \vec{x}), \varphi(\vec{\kappa}, \vec{\xi})/\lambda(\vec{\kappa})\rangle_{\vec{K}} \tag{2.38}$$

This procedure indicates the general method of finding fundamental solutions by transform methods. We may also use the fundamental solution to solve problems like

$$L(u) = P \tag{2.39}$$

where P is some function of $\bar{x}$ in our region.

If we put

$$u(\vec{x}) = \langle u^*(\vec{x}, \vec{\xi}), P(\vec{\xi})\rangle_{\Omega(\vec{\xi})} \tag{2.40}$$

we obtain Equation 2.39 by substituting for u using

$$L(u^*(\vec{x}, \vec{\xi})) = \delta(\vec{x} - \vec{\xi})$$

Hence Equation 2.40 gives the solution of Equation 2.39. Furthermore if we have

$$L(u) = M(P) \tag{2.41}$$

where M is some linear differential operator, then

$$u(x) = M(\langle u^*(\vec{x}, \vec{\xi}), P(\vec{\xi})\rangle_{\Omega}) \tag{2.42}$$

is the solution of Equation 2.41 as can also be verified by substitution.

Consider a simple example which illustrates these points.

Example

Consider the one-dimensional reduced wave equation given by

$$\frac{d^2u}{dx^2} + \kappa^2 u = 0 \quad \text{(a)}$$

The orthonormal eigenfunctions we choose correspond to our one-dimensional Cartesian system and are given by

$$\varphi(\kappa', x) = \frac{1}{\sqrt{2\pi}} e^{-i\kappa' x} \tag{b}$$

and the eigenvalues

$$\lambda(\kappa') = \kappa^2 - \kappa'^2$$

We wish to find the fundamental solution defined by

$$\frac{d^2 u^*}{dx^2} + \kappa^2 u^* = \delta(x - \xi) \tag{c}$$

We may write

$$\begin{aligned} \delta(x-\xi) &= \langle \varphi(\kappa, x), \varphi(\kappa, \xi) \rangle_\kappa \\ &= \frac{1}{2\pi} \int_{-\infty}^{\infty} \exp[-i\kappa(x-\xi)] d\kappa \end{aligned} \tag{d}$$

and

$$\begin{aligned} u^*(x, \xi) &= \langle \tilde{u}^*(\kappa, \xi), \varphi(\kappa, x) \rangle_\kappa \\ &= \frac{1}{\sqrt{2\pi}} \int_{-\infty}^{\infty} \tilde{u}^*(\kappa, \xi) e^{-i\kappa x} d\kappa \end{aligned} \tag{e}$$

$\tilde{u}^*(\kappa, \xi)$ is the Fourier transform of the fundamental solution given by

$$\tilde{u}^*(\kappa, \xi) = \frac{1}{\sqrt{2\pi}} \int_{-\infty}^{\infty} u^*(x, \xi) e^{i\kappa x} dx \tag{f}$$

from Equation 2.26, which is the inverse relationship corresponding to Equation (e).

The fundamental solution is now given by

$$\frac{d^2}{dx^2}\left(\frac{1}{\sqrt{2\pi}} \int_{-\infty}^{\infty} \tilde{u}^*(\kappa', \xi) e^{-i\kappa' x} d\kappa'\right) + \kappa^2 \frac{1}{\sqrt{2\pi}} \times$$

$$\times \int_{-\infty}^{\infty} u^*(\kappa', \xi)^{-i\kappa' x} \mathrm{d}\kappa'$$

$$= \frac{1}{2\pi} \int_{-\infty}^{\infty} \exp\left[i\kappa'(-x+\xi)\right] \mathrm{d}\kappa' \tag{g}$$

$$\therefore \int_{-\infty}^{\infty} (\kappa^2 - \kappa'^2)\tilde{u}^*(\kappa', \xi)^{-i\kappa' x} \mathrm{d}\kappa'$$

$$= \frac{1}{\sqrt{2\pi}} \int_{-\infty}^{\infty} \exp\left[i\kappa'(-x+\xi)\right] \mathrm{d}\kappa' \tag{h}$$

or

$$\int_{-\infty}^{\infty} \left((\kappa^2 - \kappa'^2)\tilde{u}^* - \frac{1}{\sqrt{2\pi}} \mathrm{e}^{i\kappa' x} \right) \mathrm{e}^{-i\kappa' x} \mathrm{d}\kappa' = 0 \tag{i}$$

Note that the expression in large parentheses is the equation resulting from taking the Fourier transform of both sides of Equation (c).

Note that Equation (i) is an identity for all values of ξ and we may write simply

$$(\kappa^2 - \kappa'^2)\tilde{u}^* = \frac{1}{\sqrt{2\pi}} \mathrm{e}^{i\kappa' \xi} \tag{j}$$

which gives the Fourier transform of the solution directly.

Alternatively we may exploit the orthonormalisation of our eigenfunctions to retrieve the same result. We have therefore

$$\tilde{u}^*(\kappa', \xi) = \frac{1}{(\kappa^2 - \kappa'^2)} \frac{1}{\sqrt{2\pi}} \mathrm{e}^{i\kappa' \xi} \qquad \kappa \neq \kappa' \tag{k}$$

and hence using Equation (f), we can write

$$u^*(x, \xi) = \frac{1}{2} \int_{-\infty}^{\infty} \frac{1}{(\kappa^2 - \kappa'^2)} \exp[i\kappa'(\xi - x)] \mathrm{d}\kappa' \tag{l}$$

which by simple contour integration gives (taking the real part for our actual solution)

$$u^*(x, \xi) = \frac{1}{2\kappa} \sin(\kappa|x-\xi|) = \frac{1}{2\kappa} \sin(\kappa r) \qquad \text{(m)}$$

where r is the (positive) distance between x and ξ.

The above example demonstrates the procedure used to obtain the fundamental solution in Cartesian coordinates. The main steps of the analysis are given below:

(1) Transform the equation a sufficient number of times in order to obtain an algebraic equation in terms of the transformed function $\tilde{u}^*(\vec{\kappa}, \vec{\xi})$.

(2) Solve this equation in terms of $\tilde{u}^*$.

(3) Invert the transformations to obtain the fundamental solution $u^*(\vec{x}, \vec{\xi})$.

In practice it may not be necessary to perform the full number of transformations as the resulting transformed equation at some point may have a known fundamental solution. The inverse transforms may then be performed on this function to obtain the fundamental solution for the original equation.

As an example of the non-trivial use of this technique, the fundamental solution for a concentrated load P applied at the point $\vec{\xi}$ in an anisotropic elastic space will now be deduced starting from Navier's equations in the form[3]

$$C_{pqrs}u_{r,sq} = 0 \qquad (2.43)$$

where C_{pqrs} is the tensor relating stresses to strains and involves the elastic constants of the material, u_r is the displacement in direction x_r and $_{,sq}$ denotes the differentiation $\partial^2/\partial x_s \partial x_q$. The summation convention has been used here, a repeated suffix is assumed to be summed.

It is convenient to define a fundamental tensor U^*_{ji} by

$$u_i^* = P_j U^*_{ji} \qquad (2.44)$$

We then require the fundamental solution corresponding to Equation 2.43, i.e.

$$C_{pqrs}U^*_{tr,sq} = \delta_t(\vec{x}, \vec{\xi}), \qquad (p=1, 2, 3) \qquad (2.45)$$

where the subscript t indicates that the delta function δ_t is the tth component of some vector delta function; t then represents the direction of the applied force at $\vec{\xi}$. We may take $\vec{\xi}$ to be the origin by a change of axes. Taking the triple Fourier transform of Equation 2.45 we obtain

$$F_{pr}(\vec{\kappa})U^*_{rt}(\vec{\kappa}) = \delta_{pt} \qquad (2.46)$$

where δ_{pt} is the Kronecker delta as before

$$\tilde{U}^*_{rt}(\vec{\kappa}) = \frac{1}{(2\pi)^{3/2}} \int_{-\infty}^{\infty} U_{rt}(\vec{x}) e^{i\vec{\kappa}\cdot\vec{x}} d\vec{x} \qquad (2.47)$$

and

$$F_{pr}(\vec{\kappa}) = C_{pqrs}\kappa_q\kappa_s \tag{2.48}$$

Inverting the matrix equation (2'46) to give us the triple Fourier transform of the fundamental tensor we have

$$U^*{}_{rt}(\vec{\kappa}) = F_{tr}^{-1}(\vec{\kappa}) \tag{2.49}$$

Inverting the Fourier transforms we obtain the usual form for the fundamental solution in component form as

$$U^*_{rt}(\bar{x}) = \frac{1}{(2\pi)^{3/2}} \int_{-\infty}^{\infty} (C_{pqrs}\kappa_q\kappa_s)^{-1} e^{-i\vec{\kappa}.\vec{x}} d\vec{\kappa} \tag{2.50}$$

more usually written

$$U^*_{rt}(\bar{x}) = \frac{1}{(2\pi)^{3/2}} \int_{-\infty}^{\infty} F_{tr}^{-1} e^{-i\vec{\kappa}.\vec{x}} d\vec{\kappa} \tag{2.51}$$

This can be transformed to a simpler form by taking spherical polar coordinates and integrating twice to give

$$U^*_{ij}(\vec{x}) = \frac{1}{8\pi^2 r} \int_{|\eta|=1} F_{ij}^{-1}(\eta) ds \tag{2.52}$$

The integral is now taken round a unit circle which has centre at the source point (the origin here) and lying in the plane perpendicular to the line joining $\bar{x}$ to the origin at a distance r. For a general source point ξ, Equation 2.52 becomes

$$U^*_{ij}(\vec{x}, \vec{\xi}) = \frac{1}{8\pi^2|\vec{x} - \vec{\xi}|} \int_{|\eta|=1} F_{ij}^{-1}(\eta) ds \tag{2.53}$$

For an isotropic material Equation 2.53 may be evaluated to give the Kelvin solution:

$$U^*_{ij}(\vec{x}, \vec{\xi}) = \frac{1}{4\pi\mu|\vec{x} - \vec{\xi}|} \left(\frac{3-4v}{4(1-v)} \delta_{ij} + \frac{1}{4(1-v)} \frac{(x_i - \xi_i)}{|\vec{x} - \vec{\xi})} \frac{(x_j - \xi_j)}{|\vec{x} - \vec{\xi}|} \right)$$

In the general case Equation 2.53 must be evaluated numerically.

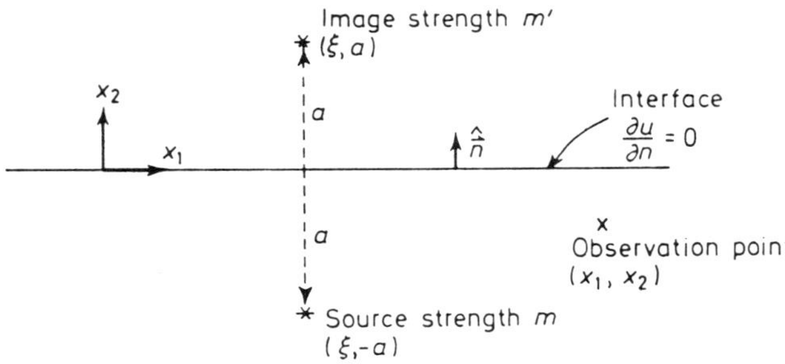

Figure 2.1 Derivation of the fundamental solution in a halfplane

2.4 THE METHOD OF IMAGES

So far we have a method of finding the fundamental solution in empty sapce. In some circumstances our problem region may be confined in some regular way and it may be more convenient to find a fundamental solution specific to a region.

The simplest case is that of a semi-infinite space such as may occur in a foundation or fluids problem. The surface of the soil or fluid may make it more convenient to work with a semi-infinite space fundamental solution. We choose this solution to satisfy the boundary condition on the interface identically; in this way we shall not need to put elements on the surface when using the boundary integral method.

To obtain these solutions it is necessary to consider their physical interpretation. Consider for the moment a two-dimensional space and take coordinates as in Figure 2.1. The interface is along the x_1 axis and the normal to this is measured in the positive x_2 direction. We shall require that $\partial u/\partial n=0$ on this interface.

The fundamental solution is the field due to a point source in infinite space and so the field in the presence of our boundary $x_2=0$ will represent our half space fundamental solution. Consider a source strength m at a point $(\xi_1, -a)$. The field mu^* from this source will be reflected somehow at $x_2=0$, depending on the boundary condition applied there. In order to represent this reflection we place an image source of strength m' behind the interface. The strength of m' required will depend on the kind of interface to be represented. We may then forget the existence of the surface ($x_2=0$) and adjust m' in order to satisfy the boundary condition there. The observed field at any point (x_1, x_2) will then be the superposition of the field from both sources (as we have assumed a linear operator for our governing differential equation).

This field will be given by

$$u(x_1)=mu^*(x_1, x_2; \xi_1, -a)+m'u^*(x_1, x_2; \xi_1, I) \tag{2.54}$$

where $u^*(x_1, x_2; \xi_1, \xi_2)$ is the empty space fundamental solution for a source at (ξ_1, ξ_2) observed at (x_1, x_2). Then applying the boundary condition

$$\frac{\partial u}{\partial n} = \frac{\partial u}{\partial x_2} = 0, \qquad \text{on } x_2 = 0 \tag{2.55}$$

we obtain

$$m \frac{\partial u^*(x_1, 0; \xi_1, -a)}{\partial x_2} + m' \frac{\partial u^*(x_1, 0; \xi_1, a)}{\partial x_2} = 0 \tag{2.56}$$

which for the two-dimensional Laplace equation becomes

$$m \frac{\partial}{\partial x_2} [\ln\{x_1^2 + (x_2 - a)^2\}^{1/2}] + m \frac{\partial}{\partial x_2} [\ln\{x_1^2 + (x_2 + a)^2\}^{1/2}] = 0$$

$$\text{on } x_2 = 0, \qquad \forall x_1 \tag{2.57}$$

or in particular at $x = 0$ gives

$$\frac{m}{-a} + \frac{m'}{a} = 0 \tag{2.58}$$

Hence we choose $m' = m$. The field is then

$$u = \frac{m}{2\pi} \{\ln[x_1^2 + (x_2 - a)^2]^{1/2} + \ln [x_1^2 + (x_2 + a)^2]^{1/2}\} \tag{2.59}$$

$$= \frac{m}{2\pi} \ln \{[x_1^2 + (x_2 - a)^2][x_1^2 + (x_2 + a)^2]\}^{1/2} \tag{2.60}$$

Thus the fundamental solution for the two-dimensional Laplace equation for the lower half plane is (Figure 2.1)

$$u_s^*(x_1, x_2; \xi_1, \xi_2) = \frac{1}{2\pi} \ln \{[(x_1 - \xi_1)^2 + (x_2 - \xi_2)^2] \times$$

$$\times [x_1 - \xi_1)^2 + (x_2 + \xi_2)^2]\}^{1/2} \tag{2.61}$$

Note — we have that u_s^* is the solution of

$$\nabla^2 u_s^* = m\delta(x_1, x_2 - a) + m\delta(x_1, x_2 + a) \tag{2.62}$$

source image

We could equally have solved this equation in the usual way.

For different boundary conditions we obtain a slightly different equation for m, i.e. if $u = 0$ on $x = 0$ then $m = m'$ and

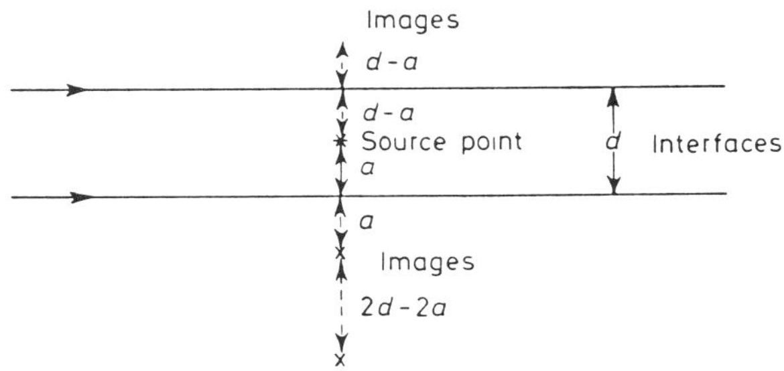

Figure 2.2 Derivation of the fundamental solution in a layer

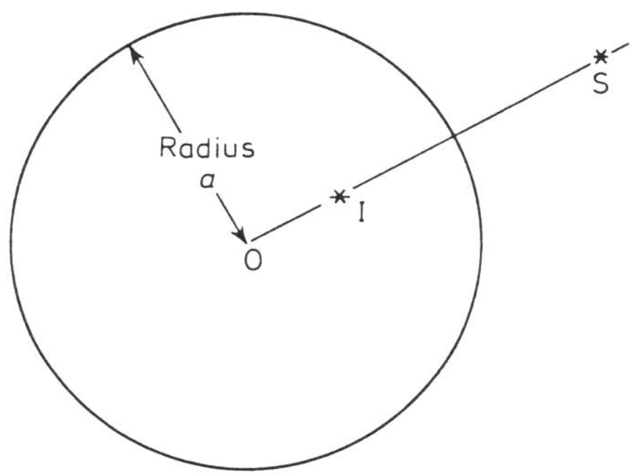

Figure 2.3 Cylindrical geometry: I, image point; S, source

$$u_s^* = \frac{1}{2\pi} \ln \left\{ \frac{x_1^2 + (x_2 - \xi_2)^2}{x_1^2 + (x_2 + \xi_2)^2} \right\}^{1/2} \tag{2.63}$$

in this case.

The other two important cases are shown in Figures 2.2 and 2.3. In Figure 2.2 we have a parallel layered medium such as might occur in a soils problem. Image sources are constructed as shown. For non-parallel layers a numerical treatment might be appropriate.

Figure 2.3 shows a cylindrical cavity or obstruction such as might occur in diffraction or fracture problems. In this case an image is constructed at the inverse point given by

$$\text{OI} + \text{OS} = a^2 \tag{2.64}$$

where a is the radius of the cylinder. The most widely used application of this method so far has been for semi-infinite elastic space problems where the Mindlin solution is appropriate[5].

We shall now deduce this fundamental solution from the Kelvin solution using the method of images. Mindlin considered the problem of finding the field (fundamental solution) due to point forces perpendicular and parallel to

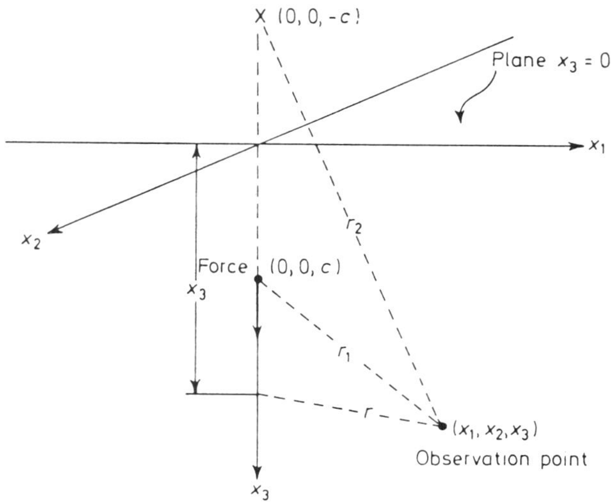

Figure 2.4 Force normal to the boundary in the interior of a semi-infinite solid

the surface of a semi-infinite elastic half space. The boundary condition on this surface was that there was no stress, this gives rise to a system of 'images' of the Kelvin type which in an infinite elastic space would have the effect of creating a situation of zero stress at the position of the interface.

Here we shall consider the case of a point force P normal to the interface at a depth c (see Figure 2.4). Mindlin[5] found that the zero stress condition on $x_3=0$ is produced by a combination of images at $(0, 0, -c)$ of the following types:

(a) a force in the x_3 direction at $(0, 0, -c)$
+(b) a double force in the x_3 direction at $(0, 0, -c)$
+(c) a centre of compression (three mutually perpendicular double forces) at $(0, 0, -c)$
+(d) a line of compression along the z axis from $x_3=-c$ to $x_3=+c$
+(e) a doublet at $(0, 0, -c)$.

Superposing the displacement fields due to the original force and its images we obtain the displacement field for the force normal to the interface for the horizontal directions x_1 and x_2. The corresponding stresses may be obtained by differentiation.

Melan[6] has obtained similar expressions for the two dimensional case and Davies and Banerjee[7] have extended the results to a two layer half space although the result is not original.

2.5 NUMERICAL AND APPROXIMATE FUNDAMENTAL SOLUTIONS

We have seen that the determination of the fundamental solution involves taking transforms of both sides of our original equation with respect to the eigenfunctions φ_n, and later inverting these transforms. These processes can only be performed in closed form for a very limited number of special cases. When we have a governing equation with different coefficients such as with anisotropic media, a numerical approach for the evaluation of our solution becomes essential.

The simplest approach is to consider the region for definition of the fundamental solution to be bounded by say a sphere (or cylinder) of large radius, then the fundamental solution is represented by the sum of a series of eigenfunctions, i.e.

$$u_{\Omega}^*(\vec{x}, \vec{\xi}) = \sum_{n=0}^{\infty} \frac{\hat{\varphi}_n(\vec{\xi})}{\lambda_n} \varphi_n(\vec{x}) \tag{2.65}$$

This function represents the field due to a point source at $\vec{\xi}$ together with 'image' sources generated by the boundary of the region Ω. λ_n are the eigenvalues corresponding to the eigenfunctions φ_n of the differential equation. (*Note* that care is needed in the use of Equation 2.65 for the case of two-dimensional problems as the field from the image sources may be larger than that of the source point, because of the logarithmic behaviour of the fundamental solution. In three dimensions, however, Equation 2.65 may be used.)

The infinite series, Equation 2.65, may be approximated by truncating the series at some appropriate point to give a finite sum.

When it is not appropriate to introduce an artificial boundary (as in 2-D problems), then we must deal directly with some numerical Fourier inversion procedure. Suppose for simplicity we are working in rectangular Cartesian coordinates, then the operator L may be represented as a polynomial combination of the derivatives $\partial/\partial x_i$, $i=1, \ldots, n$' We can write

$$L \equiv P\left(a_1 \frac{\partial}{\partial x_i}, \quad a_0, a_2 \frac{\partial^2}{\partial x_1 \partial x_2}, \ldots\right) \tag{2.66}$$

where $a_0, a_1, a_2, \ldots$ are taken as constants for now, independent of x. Then the equation for the fundamental solution will be

$$L(u^*(x, \xi)) = \delta(x-\xi) \tag{2.67}$$

Transforming this equation a sufficient number of times we obtain an equation which for constant a_0, a_1, a_2 is

$$\frac{1}{(2\pi)^{n/2}} P(a_0, ia_1\kappa_i, \; -a_2\kappa_i\kappa_j, \ldots)\tilde{u}^*(\vec{\kappa}, \vec{\xi}) = 1 \tag{2.68}$$

where n is the number of coordinate directions, hence

$$u^*(\vec{\kappa}, \vec{\xi}) = \frac{1}{P(a_0, ia_1\kappa_k, -a_2\kappa_i\kappa_j, \ldots)} \tag{2.69}$$

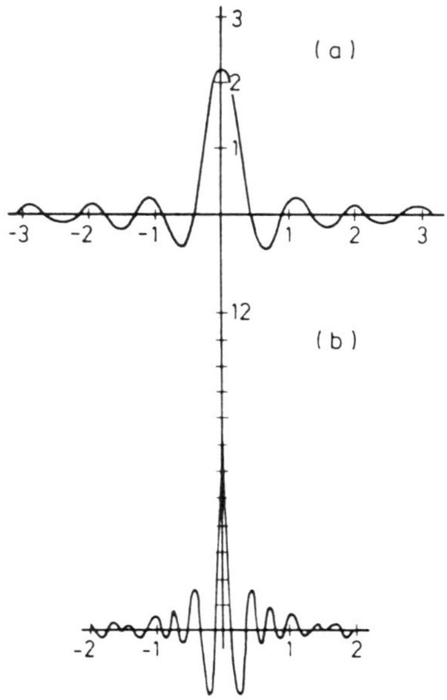

Figure 2.5 Dirac delta function (approximate): (a) integration range = 7, integration increment = 0.1; (b) integration range = 40, integration increment = 0.2

The fundamental solution is then given by the inverse transform:

$$u^*(\vec{x}, \vec{\xi}) = \frac{1}{(2\pi)^{n/2}} \int_{-\infty}^{\infty} \frac{\tilde{u}^*(\vec{\kappa}, \vec{\xi}) e^{-i\vec{\kappa}, \vec{x}}}{P(a_0, ia_1\kappa_i, -a_2\kappa_i\kappa_j, \ldots)} d\vec{\kappa} \tag{2.70}$$

In the case where $a_0, a_1 \ldots$ are functions of space, the transformations must be done numerically, i.e. the integrations involved in the transformations are performed numerically. Fortunately in the BEM we only require $u^*(\vec{x}, \vec{\xi})$ at a finite number of points, corresponding to the element positions at $\vec{x}$ and $\vec{\xi}$. Equation 2.70 will then consist of tables of values of $\tilde{u}^*(\vec{\kappa}, \vec{\xi})$ for various values of the parameter $\vec{\kappa}$ and the source coordinate $\vec{\xi}$. The inverse transform Equation 2.70 then becomes the summation of the integrand for the points where $\tilde{u}^*$ has been calculated.

These numerical integrations will involve a truncation of the range of integration of the integrals which in turn means that an approximation has been made to the Dirac delta function on the right-hand side. Figures 2.5(a) and (b) show the representation of the δ function for various integration ranges.

In many situations an approximate fundamental solution may be used. Garrison[8] exploited the physics of the situation when he considered surface wave diffraction and ratiation problems. By simplification of the free surface boundary conditions, he was able to deduce an approximate fundamental solution which gave good results when compared with the exact formulation. For many situations the actual parameters of the problem may make it possible to use asymptotic forms of fundamental solutions, i.e. widely separated elements in wave problems; in others the fundamental solution may even suggest some boundary condition to be applied at a fictitious boundary[9].

The main use of approximate fundamental solutions is when an exact solution is not known or does not exist. In order to discuss a method of construction and use of these approximate solutions it is necessary to consider some general properties of linear partial differential equations[10].

Consider the general second order linear partial differential equation:-

$$a'u_{xx}+2b'u_{xy}+c'u_{yy}+d'u_x+e'u_y+f'u=g' \tag{2.71}$$

where the coefficients a', b', c', d', e', f' and g' are functions of x and y only, and the subscripts x and y denote differentiation w.r.t. of these variables.

If $b^2-ac>0$ then the equation is elliptic; we shall consider this case. Equation 2.71 may always be reduced to the canonical form (Garabedian[10], Chapter 3)

$$L(u)=u_{xx}+u_{yy}+au_x+bu_y+cu \tag{2.72}$$

If we take u to be the fundamental solution u^* then we observe that the most singular terms involving the highest order of differentiation correspond to Laplace's equation.

$$u^*_{xx}+u^*_{yy}+\delta=0 \tag{2.73}$$

We would expect therefore that the singular part of the fundamental solution is of the form $1/2\pi \log 1/r$ where $r=\{(x-\xi)^2+(y-\eta)^2\}^{1/2}$. (In fact in three dimensions the singular part of the fundamental solutions is of the form $1/4\pi r$.) We can in fact generalise the Laplace solution to give

$$u^*(x, y; \xi, \eta)=\frac{1}{2\pi}A(x, y; \xi, \eta)\log\left(\frac{1}{r}\right)+B(x, y; \xi, \eta) \tag{2.74}$$

$$A(\xi, \eta; \xi, \eta)=1 \tag{2.75}$$

The function $A(x, y; \xi, \eta)$ may be obtained by a process of successive approximations after the equation has been expressed in characteristic coordinates. B may then be obtained from A by solution of an equation of the form

$$L[B(z, z^*; \zeta, \zeta^*)]=L[A(z, z^*; \zeta, \zeta^*)\log r] \tag{2.76}$$

It is not proposed to go into the details of the derivation of A and B here, the reader is referred to Garabedian[10] for a full esposition.

In the context of boundary elements, however, it is suggested that in view of the above the simple solutions $1/2\pi \log 1/r$ (and $1/4\pi r$) may be used as initial guesses. The residual terms associated mainly with $B(x, y; \xi, \eta)$ may then be moved to the right-hand-side and treated as body force terms. Numerical integration over 'cells' throughout the problem region may be used to evaluate these terms assuming an initial form for the solution variable $u=u_0$, say. This procedure may also be used to solve non-linear problems as described below:

Consider the equation

$$L(u)+N(u)=P \tag{2.77}$$

where $L$is a linear operator, N is either a non-linear operator or consists of the terms which make the fundamental solution unknown (or both), and P is a known forcing function. The iterative procedure will consist of writing Equation 2.79 as,

$$L(u)=P-N(u) \tag{2.78}$$

making some realistic guess for the initial form of u, u_0 (or in some cases taking $u_0 \equiv 0$), and putting this in the N terms on the right-hand side, i.e.

$$L(u_1)=P-N(u_0) \tag{2.79}$$

If Equation 2.79 is then solved for u_1 in the usual way we may repeat the process until some criterion for convergence is satisfied. The fundamental solution needed for our boundary element formulation is defined by:

$$L^+(u^*)+\delta_i=0 \tag{2.80}$$

where L^+ is the adjoint operator to L defined below. The resulting matrices for the boundary element formulation will be

$$\mathbf{HU}=\mathbf{GQ}+\mathbf{P}+\mathbf{N} \tag{2.81}$$

where P represents the body force vector and N is the vector obtained from the non-linear terms.

One important feature of this approach is that the influence matrices **H** and **G** remain the same for each iteration. We only need to recalculate the vector **N** and find the new solution vector **U** and **Q** for each iteration.

2.6 USE OF FUNDAMENTAL SOLUTIONS IN BOUNDARY ELEMENTS

In this chapter we shall not use the weighted residuals method referred to in Chapter 1, but will use the Generalised Green's Theorem as it is more consistent with the approach used so far in this chapter, and illustrates the use of alternative formulations more clearly.

For the operator $L(u)$ defined in Equation 2.72, define the adjoint operator on v ($L^+(v)$) such that the combination of terms,

$$v\ L(u) - u\ L^{+}(v), \text{ is a divergence}$$

in fact

$$L^{+}(v) = v_{xx} + v_{yy} - av_{x} - bv_{y} + (c - a_{x} - b_{y})v \tag{2.82}$$

i.e.

$$v\ L(u) - uL^{+}(v) = (vu_{x} - uv_{x} + auv)_{x} + (vu_{y} - uv_{y} + buv)_{y} \tag{2.83}$$

Then with n as the outward normal, for u and v any sufficiently differentiable functions:

$$\int_{\Omega} vL(u) - uL^{+}(v)\mathrm{d}V = \int_{\Gamma}\left[v\frac{\partial u}{\partial n} - u\frac{\partial v}{\partial n} + \left(a\frac{\partial x}{\partial n} + b\frac{\partial y}{\partial n}\right)uv\right]\mathrm{d}S \tag{2.84}$$

then if we choose v to be the fundamental solution $u^{*}(\vec{x}, \vec{\xi})$ defined by:-

$$L^{\dagger}(u^{*}(\vec{x}, \vec{\xi})) + \delta(\vec{x} - \vec{\xi}) = 0 \tag{2.85}$$

then we may use the selective property of the delta function to reduce Equation 2.84 to:

$$cu(\vec{x}) = \int_{\Gamma}\left(u^{*}\frac{\partial u}{\partial n} - u\frac{\partial u^{*}}{\partial n}\right) + \left(a\frac{\partial x}{\partial n} + b\frac{\partial y}{\partial n}\right)uu^{*}\mathrm{d}S(\vec{\xi}) \tag{2.86}$$

We have taken the integrations to be over the variable $\vec{\xi}$, where for smooth boundaries

$$c = 0 \quad \vec{x} \in \Omega$$

$$c = \tfrac{1}{2} \quad \vec{x} \in \Gamma$$

$$c \leqslant 1 \quad \vec{x} \in \Omega$$

Notice that u^{*} is a function of two variables and that Equation 2.86 is an identity for all values of the source or reference point $\bar{x}$. If we choose $\bar{x}$ on the boundary we have now eliminated all integrations over the region hence Equation 2.86 may be used as the starting point for the direct boundary integral formulation, i.e.

$$cu(\vec{x}) = \int_{\Gamma} u^{*}\frac{\partial u}{\partial n}\mathrm{d}S - \int_{\Gamma} u\left[\frac{\partial u^{*}}{\partial n} - \left(a\frac{\partial x}{\partial n} + b\frac{\partial y}{\partial n}\right)u^{*}\right]\mathrm{d}S \tag{2.87}$$

Note a and b may be functions of position.

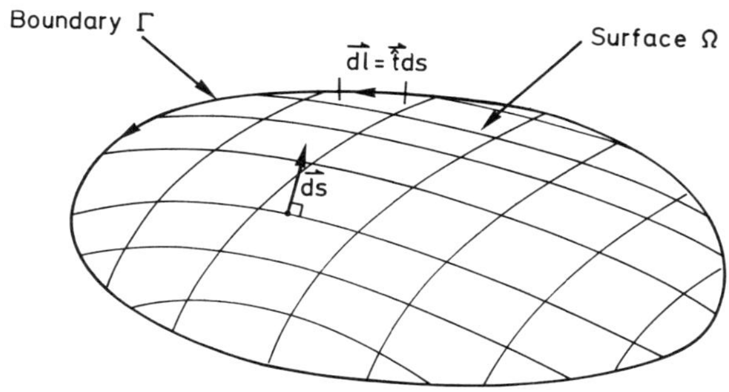

Figure 2.6 Stokes theorem (notation)

The integrals over the boundary Γ involve the normal derivatives $\partial u/\partial n$, if we require the tangential derivatives of a variable directly then Stoke's theorem below may be used[11].

For an open parametrisable surface Ω with boundary curve Γ and $\vec{A}(\vec{x})$ some sufficiently differential integrable vector field then:

$$\int_{\Omega} \vec{\nabla}\vec{A}.\mathrm{d}\vec{S} = \int_{\Gamma} \vec{A}\,.\,\mathrm{d}\vec{l} \tag{2.88}$$

with $\mathrm{d}\vec{l}$ and $\mathrm{d}\vec{S}$ as in Figure 2.6 and $\vec{\nabla}$ the gradient operator $(\partial/\partial x, \partial/\partial y, \partial/\partial z)$.

Then choose

$$\vec{A} = \varphi\vec{\nabla}u^* \tag{2.89}$$

with u^* the fundamental solution of the 2-D Laplace equation and φ some potential representing a vector field u defined by:

$$\vec{u} = \vec{\nabla}\varphi \tag{2.90}$$

Then

$$\int_{\Omega} (\vec{\nabla} \times \varphi\vec{\nabla}u^*).\mathrm{d}\vec{S} = \int_{\Gamma} \vec{\nabla}\varphi \times \vec{\nabla}u^*\,.\,\mathrm{d}S + \int_{\Gamma} \varphi\vec{\nabla} \times \vec{\nabla}u^*\,.\,\mathrm{d}\vec{S} \tag{2.91}$$

and putting $\vec{A} = u^*\vec{\nabla}\varphi$

$$\int_{\Omega} \vec{\nabla} \times (u^*\vec{\nabla}\varphi)\mathrm{d}\vec{S} = \int \vec{\nabla}u^* \times \vec{\nabla}\varphi.\mathrm{d}\vec{S} + \int_{\Gamma} \varphi\vec{\nabla} \times \vec{\nabla}u^*\,.\,\mathrm{d}\vec{S} \tag{2.92}$$

Then using the identity

$$\vec{\nabla} \times \vec{\nabla} \equiv 0 \tag{2.93}$$

and adding Equations 2.91 and 2.92 we obtain:

$$\int_{\Omega} \vec{\nabla} \times \varphi \vec{\nabla} u^* . d\vec{S} + \int_{\Omega} \vec{\nabla} \times u^* \vec{\nabla} \varphi) . d\vec{S}$$

$$= \int_{\Omega} \vec{\nabla} \varphi \times \vec{\nabla} u^* . d\vec{S} - \int_{\Omega} \vec{\nabla} \varphi \times \vec{\nabla} u^* . d\vec{S}$$

$$= \vec{0} = \int_{\Gamma} (\varphi \vec{\nabla} u^* + u^* \vec{\nabla} \varphi) . d\vec{l} \tag{2.94}$$

Now

$$\vec{\nabla} u^* . d\vec{l} = \frac{\partial u^*}{\partial S} \quad \text{and} \quad \vec{\nabla} \varphi . d\vec{l} = \frac{\partial \varphi}{\partial S} \tag{2.95}$$

where $\partial/\partial S$ denotes differentiation round the curve Γ. Then we have the identities

$$\int_{\Gamma} \left\{ \varphi \frac{\partial u^*}{\partial S} + u^* \frac{\partial \varphi}{\partial S} \right\} dS = 0 \tag{2.96}$$

where $\partial \varphi / \partial S$ is the tangential velocity. Thus we obtain the tangential velocities directly without differentiation of the approximate values of φ from the BEM which are in any case inaccurate near the boundary.

An important extension of the formulations mentioned so far is the inclusion of an integration over time in the identities. If u^* is a function of time (τ) then both sides of an equation such as Equation 2.87 may be integrated with respect to this free variable.

Consider the simple case of the diffusion equation

$$\nabla^2 u = \frac{\partial u}{\partial \tau} \tag{2.97}$$

then Equation 2.84 with $a \equiv b \equiv 0$

$$\int_0^T \int_{\Omega} (u^* \nabla^2 u - u \nabla^2 u^*) dV d\tau = \int_0^T \int_{\Gamma} \left(u^* \frac{\partial u}{\partial n} - u \frac{\partial u^*}{\partial n} \right) dS d\tau \tag{2.98}$$

for an impulsive source u^* defined by

$$\nabla^2 u^* + \frac{\partial u^*}{\partial t} = \delta(\vec{x} - \vec{\xi})\delta(t - \tau) \tag{2.99}$$

(note the plus sign in this equation).
Note also

$$\nabla^2 u^* + \frac{\partial u^*}{\partial t} = 0, \qquad \text{for } t > \tau \tag{2.100}$$

Then Equation 2.98 becomes

$$-\int_0^T \int_\Gamma \left(u^* \frac{\partial u}{\partial \tau} + u \frac{\partial u^*}{\partial \tau} \right) \mathrm{d}\Omega \mathrm{d}\tau - u(\vec{x}, t)$$

$$= \int_0^T \int_\Gamma \left(u^* \frac{\partial u}{\partial n} - u \frac{\partial u^*}{\partial n} \right) \mathrm{d}S \mathrm{d}\tau \tag{2.101}$$

Integrating the first term with respect to τ we obtain

$$-\left[\int_\Omega u u^* \mathrm{d}\Omega \right]_{t=0}^{t=T} - u(\bar{x}, t) = \int_0^T \int_\Gamma \left(u^* \frac{\partial u}{\partial n} - u \frac{\partial u^*}{\partial n} \right) \mathrm{d}S \mathrm{d}\tau \tag{2.102}$$

The first term in the above corresponds to the initial conditions (Brebbia[2], pp. 169–171). The integrations over time can in fact be performed analytically leaving only special integrations. The function $u(\bar{x}, t)$ may then be found in the conventional manner.

2.7 THE AVOIDANCE OF SINGULARITIES

One of the main numerical problems associated with the fundamental solution is the evaluation of the diagonal terms of the influence matrices. These diagonal terms often consist of the evaluation of the integral of the fundamental solution over the singularity, typically integrals like

$$\int_{-l}^{l} u(\vec{x}, \vec{\xi}) \mathrm{d}r \tag{2.103}$$

and

$$\int_{-l}^{l} \frac{\partial u(\vec{x}, \vec{\xi})}{\partial n} \, \mathrm{d}r \tag{2.104}$$

are sought, where $r = |\vec{x} - \vec{\xi}|$.

These problems may be simply avoided. Consider Equation 2.86 again with $a \equiv b \equiv 0$, then

$$cu(\vec{x}) = \int_{\Gamma} u^*(\vec{x}, \vec{\xi}) \frac{\partial u(\vec{\xi})}{\partial n} - u(\vec{\xi}) \frac{\partial}{\partial n} u^*(\vec{x}, \xi) \mathrm{d}S(\vec{\xi}) \tag{2.105}$$

Then if we choose $\vec{x}$ to be a reference point $\vec{x}_1$ outside Ω, say

$$\vec{x} = \vec{x}_i$$

then $c = 0$ and Equation 2.105 may be written:

$$\int_{\Omega} u^*(\vec{x}_i, \vec{\xi}) \frac{\partial u(\vec{\xi})}{\partial n} \mathrm{d}S(\vec{\xi}) = \int_{\Gamma} u(\vec{\xi}) \frac{\partial}{\partial n} u^*(\vec{x}_i, \vec{\xi}) \mathrm{d}S(\vec{\xi}) \tag{2.106}$$

or approximately

$$\sum_j \frac{\partial u(\vec{\xi})}{\partial n} \int_{\Gamma_j} u^*(\vec{x}_i, \vec{\xi}) \mathrm{d}S(\vec{\xi}) = \sum_j u(\vec{\xi}) \int_{\Gamma_j} \frac{\partial}{\partial n} u^*(\vec{x}_i, \vec{\xi}) \mathrm{d}S(\vec{\xi}) \tag{2.107}$$

If we discretise Γ into boundary elements then we can now choose n such reference points. We have now constructed n equations for the n unknown u values on the boundary.

We now have no singular integrals to evaluate. One disadvantage of this method is that the influence matrices become less diagonally dominant with the resulting numerical problems as the diagonal terms now correspond to an evaluation of u^* between the ith element and the ith reference point. This difficulty may be overcome by choosing the reference point close to the corresponding element.

Another way of eliminating the integration over our problem region is to choose our weighting function v to be a homogeneous solution for the operator L, i.e.

$$L(v) = 0, \qquad \text{in } \Omega \tag{2.108}$$

Some care is needed here, however, as we need to choose v such that it involves

Table 2.1 SOME FUNDAMENTAL SOLUTIONS

	Equation	*Fundamental solution*
(1) ONE-DIMENSIONAL EQUATIONS		
Laplace	$\frac{d^2u^*}{dx^2}+\delta_0=0$	$u^*=r/2 \quad r=\|x\|$
Helmholtz	$\frac{d^2u^*}{dx^2}+\lambda^2u^*+\delta_0=0$	$u^*=-\frac{1}{2\lambda}\sin(\lambda r)$
Wave equation	$c^2\frac{\partial^2u^*}{\partial x^2}-\frac{\partial^2u^*}{\partial t^2}+\delta_0\delta(t)=0$	$u^*=\frac{1}{2c}H(ct-r)$
		where H is the Heaviside or unit step function
Diffusion equation	$\frac{\partial^2u^*}{\partial x^2}-\frac{1}{k}\frac{\partial u^*}{\partial t}+\delta_0\delta(t)=0$	$u^*=-\frac{H(t)}{\sqrt{4\pi kt}}\exp\left(-\frac{r^2}{4kt}\right)$
Convection/decay equation	$\frac{\partial u^*}{\partial t}+\bar{u}\frac{\partial u^*}{\partial x}+\beta u^*+\delta_0\delta(t)=0$	$u^*=-e^{-\beta r\sqrt{u}}\delta\left(t-\frac{r}{\bar{u}}\right)$
Plate equation	$\left(\frac{\partial^2}{\partial t^2}-\mu^2\nabla^4\right)u^*+\delta_0\delta(t)=0$	$u^*=+\frac{H(t)}{4\pi\mu}S_i\left(\frac{r}{4\mu t}\right)$
	$\nabla^4=(\nabla^2)^2$ in two dimensions	S_i is the integral sine function
		$S_i(u)=-\int_u^\infty \frac{\sin v}{v}dv$
Reduced plate equation	$k_p=\omega/\mu$ $(\nabla^4-k_p^4)u^*+\delta_0=0$	$u^*=-\frac{1}{8ik_p^2}\left(H_0^{(2)}(k_pr)-\frac{2i}{\pi}K_0(k_pr)\right)$
		where K_0 is an elliptic function

Table 2.1 (CONTINUED)

	Equation	*Fundamental solution*
Navier's equation (Kelvin solution)	$\frac{\partial \sigma^*_{jk}}{\partial x_j} + \delta_l = 0$ (point load in direction l)	$u^*_k = U^*_{lk} e_k$ $= \frac{[(3-4\nu)\ln(1/r)\delta_{lk} + r_{,k} r_{,l}]e_l}{8\pi G(1-\nu)}$ Traction in direction k $p^*_k = p^*_{kl} e_l$ $= -\frac{1}{r}\left(\frac{dr}{dn}[(1-2\nu)\delta_{lk} + 2r_{,l} r_{,k}]\right.$ $\left. -(1-2\nu)(n_l r_{,k} - n_k r_{,l})\right)\frac{e_l}{4\pi(1-\nu)}$
Diffusion equation	$\frac{\partial^2 u^*}{\partial x_1^2} + \frac{\partial^2 u^*}{\partial x_2^2} - \frac{1}{K}\frac{\partial u^*}{\partial t} + \delta_0 \delta(t) = 0$	$u^* = -\frac{1}{4\pi k t}\exp\left(-\frac{r^2}{4kt}\right)$
(2) TWO-DIMENSIONAL EQUATIONS		$r = \sqrt{x_1^2 + x_2^2}$
Laplace	$\frac{\partial^2 u^*}{\partial x_1^2} + \frac{\partial^2 u^*}{\partial x_2^2} + \delta_0 = 0$	$u^* = \frac{1}{2\pi}\ln\left(\frac{1}{r}\right)$
Helmholtz	$\frac{\partial^2 u^*}{\partial x_1^2} + \frac{\partial^2 u^*}{\partial x_2^2} + \lambda^2 u^* + \delta_0 = 0$	$u^* = \frac{1}{4i} H_0^{(2)}(\lambda r)$ where H is a Hankel function
D'Arcy	$k_1 \frac{\partial^2 u^*}{\partial x_1^2} + k_2 \frac{\partial^2 u^*}{\partial x_2^2} + \delta_0 = 0$ (orthotropic case)	$u^* = \frac{1}{\sqrt{k_1 k_2}}\frac{1}{2\pi}\ln\left(\frac{1}{r_0}\right)$ where $r_0 = \left(\frac{x_1^2}{k_1} + \frac{x_2^2}{k_2}\right)^{1/2}$

Wave equation	$c^2\left(\frac{\partial^2 u^*}{\partial x_1^2}+\frac{\partial^2 u^*}{\partial x_2^2}\right)-\frac{\partial^2 u^*}{\partial t^2}+\delta_0\delta(t)=0$	$u^*=-\frac{H(ct-r)}{2\pi c(c^2t^2-r^2)}$
	(point load in direction l)	Traction in direction k $p_i^*=p_{ji}^* e_j$ $p_j^*=p_{ij}^* e_i$ $=-\frac{1}{8\pi(1-\nu^2)r^2}\left(\frac{\partial r}{\partial n}[(1-2\nu)\tilde{\delta}_{ij}\right.$ $\left.+3r_{,i}r_{,j}]+(1-2\nu)(n_j r_{,i}-n_i r_{,j})\right]e_j$
Diffusion equation	$\frac{\partial^2 u^*}{\partial x_1^2}+\frac{\partial^2 u^*}{\partial x_2^2}+\frac{\partial^2 u^*}{\partial x_3^2}-\frac{1}{K}\frac{\partial u^*}{\partial t}+\delta_0\delta(t)=0$	$u^*=-\frac{1}{(4\pi kt)^{3/2}}\exp\left(-\frac{r^2}{4kt}\right)$
(3) THREE-DIMENSIONAL EQUATIONS		$r=(x_1^2+x_2^2+x_3^2)^{1/2}$
Laplace	$\frac{\partial^2 u^*}{\partial x_1^2}+\frac{\partial^2 u^*}{\partial x_2^2}+\frac{\partial^2 u^*}{\partial x_3^2}+\delta_0=0$	$u^*=1/4\pi r$
Helmoltz	$\frac{\partial^2 u^*}{\partial x_1^2}+\frac{\partial^2 u^*}{\partial x_2^2}+\frac{\partial^2 u^*}{\partial x_3^2}+\lambda^2 u^*+\delta_0=0$	$u^*=\frac{1}{4\pi r}\exp(-i\lambda r)$
D'Arcy	$k_1\frac{\partial^2 u^*}{\partial x_1^2}+k_2\frac{\partial^2 u^*}{\partial x_2^2}+k_3\frac{\partial^2 u^*}{\partial x_3^2}+\delta_0=0$	$u^*=\frac{1}{\sqrt{k_1k_2k_3}}\frac{1}{4\pi r_0}$ where $r_0=\left\{\frac{x_1^2}{k_1}+\frac{x_2^2}{k_2}+\frac{x_3^2}{k_3}\right\}^{1/2}$
Wave equation	$c^2\nabla^2 u^*-\frac{\partial^2 u^*}{\partial t^2}+\delta_0 d(t)=0$	$u^*=\frac{\delta(t-(r/c))}{4\pi r}$
Navier's equation (isotropic, homogeneous, Kelvin solution)	$\frac{\partial\sigma_{jk}^*}{\partial x_j}+\delta_l=0$	Displacement in direction k $u_k^*=U_{lk}^* e_l$ $u_{lk}^*=\frac{1}{16\pi G(1-\nu)}\left(\frac{3-4\nu}{r}\tilde{\delta}_{lk}+\frac{\partial r}{\partial x_1}\frac{\partial r}{\partial x_2}\right)$

Table 2.1 (CONTINUED)

Equation	Fundamental solution
(4) AXISYMMETRIC PROBLEMS Laplace[13]	$\rho=(x_1^2+x_2^2)^{1/2}$ $z=x_3$ $\theta=\tan^{-1}\left(\frac{x_2}{x_1}\right)$ R = radius of ring load at Z $\gamma=1+[(Z-z)^2+(R-r)^2]/2Rr$ $u^*=\frac{K(k)}{\pi[(r-R)^2+(z-Z)^2]^{1/2}}$ $k=\frac{4rR}{(r+R)^2+(z-Z)^2}$ K(k) is the complete elliptic integral of the first kind $q^*=\frac{\partial u^*}{\partial n}=-\frac{1}{\pi[(r+R)^2+(z-Z)^2]^{1/2}}\left\{\frac{1}{2r}\left[K(k)+\frac{r^2-R^2-(z-Z)^2}{(r-R)^2+(z-Z)^2}E(k)\right]\frac{\partial r}{\partial n}+\left[\frac{z-Z}{(r-R)^2+(z-Z)^2}E(k)\right]\frac{\partial z}{\partial n}\right\}$ E(k) is a complete elliptic integral of the second kind

Navier's equation[12]

$$\left\{\begin{matrix} F_r \\ F_z \end{matrix}\right\} = \left\{\begin{matrix} 1 \\ 1 \end{matrix}\right\} \delta(R-r, Z-z)/2\pi r$$

Displacements

$$U^*_{rr} = \frac{1}{4\pi^2\mu}\left\{\frac{3-4\nu}{4(1-\nu)}\frac{1}{\sqrt{Rr}}Q_{1/2}(\gamma) + \frac{1}{4(1-\nu)}\frac{(Z-z)^2}{Rr}\frac{1}{\sqrt{Rr}}\frac{dQ_{1/2}}{d\gamma}\right\}$$

$$U^*_{zr} = \frac{1}{4\pi^2\mu}\frac{1}{4(1-\nu)}\left\{\frac{Z-z}{r}\frac{1}{\sqrt{Rr}}\left[\tfrac{1}{2}Q_{1/2}(\gamma) - \left(\gamma - \frac{r}{R}\right)\frac{dQ_{1/2}}{d\gamma}\right]\right\}$$

$$U^*_{rz} = -\frac{1}{4\pi^2\mu}\frac{1}{4(1-\nu)}\left\{\frac{Z-z}{r}\frac{1}{\sqrt{Rr}}\left[\tfrac{1}{2}Q_{-1/2}(\gamma) + \left(\gamma - \frac{r}{R}\right)\frac{dQ_{-1/2}}{d\gamma}\right]\right\}$$

$$U^*_{zz} = \frac{1}{4\pi^2\mu}\left\{\frac{3-4\nu}{4(1-\nu)}\frac{1}{\sqrt{Rr}}Q_{-1/2}(\gamma) - \frac{1}{4(1-\nu)}\frac{(Z-z)^2}{Rr}\frac{1}{\sqrt{Rr}}\frac{dQ_{-1/2}}{d\gamma}\right\}$$

where $Q_{1/2}(\gamma)$ and $Q_{-1/2}(\gamma)$ are Legendre functions
Tractions

$$\frac{1}{2\mu}T_{rr} = \left[\frac{1-\nu}{1-2\nu}\frac{\partial u^*_{rr}}{dr} + \frac{\nu}{1-2\nu}\left(\frac{1}{r}u^*_{rr} + \frac{\partial u^*}{\partial z}z_r\right)\right]n_r +$$

$$+\frac{1}{2}\left(\frac{\partial u^*_{rr}}{\partial z} + \frac{\partial u^*_{zr}}{\partial r}\right)n_z$$

$$\frac{1}{2\mu}T_{zr} = \left[\frac{1-\nu}{1-2\nu}\frac{\partial u^*_{zr}}{\partial z} + \frac{\nu}{1-2\nu}\left(\frac{1}{r}u^*_{rr} + \frac{\partial u^*_{rr}}{\partial r}\right)\right]n_z$$

$$+\frac{1}{2}\left(\frac{\partial u^*_{rr}}{\partial z} + \frac{\partial u^*_{zr}}{\partial r}\right)n_r$$

$$\frac{1}{2\mu}T_{rz} = \left[\frac{1-\nu}{1-2\nu}\frac{\partial u^*_{rz}}{\partial r} + \frac{\nu}{1-2\nu}\left(\frac{1}{r}u^*_{rz} + \frac{\partial u^*_{zz}}{\partial z}\right)\right]n_r$$

Table 2.1 (CONTINUED)

Equation	Fundamental solution
	$+\frac{1}{2}\left(\frac{\partial u_{rz}^*}{\partial z}+\frac{\partial u_{zz}^*}{\partial r}\right)n_z$
	$\frac{1}{2\mu}T_{zz}=\left[\frac{1-\nu}{1-2\nu}\frac{\partial u_{zz}^*}{\partial z}+\frac{\nu}{1-2\nu}\left(\frac{1}{r}u_{rz}^*+\frac{\partial u_{rz}^*}{\partial r}\right)\right]n_z$
	$+\frac{1}{2}\left(\frac{\partial u_{rz}^*}{\partial z}+\frac{\partial u_{zr}^*}{\partial r}\right)n_r$

another point $\bar{\xi}$ in order that we may obtain enough equations to solve our problem. In addition if our problem region extends to infinity then we need to ensure that the integrals along the part of the boundary at infinity are bounded for this solution. This condition is automatically satisfied for our choice of the fundamental solution for this function, as can be seen from its physical interpretation.

Having obtained the values of u on the boundary we can re-use Equation 2.105 to obtain values of the function u at any internal point $\bar{x}$ by now putting $\bar{x}\in\Omega$. We must now choose v to be the fundamental solution in order to exploit the selective property of the delta function.

Note that for these internal points we again have no singular integrals to evaluate after the boundary element discretisation.

2.8 EXPLICIT FORMS FOR THE FUNDAMENTAL SOLUTION

In this section we present some explicit forms for the fundamental solution for some commonly occurring problems. These are presented in Table 2.1. We have reverted to the simpler notation for the delta function and have taken our source point to be the origin by a simple coordinate transformation.

REFERENCES

1 Titchmarsh, E. C., *Eigenfunction Expansions Associated with Second-Order Differential Equations (2 Vols)*, Clarendon Press (1958)

2 Brebbia, C. A. and Walker, S., *Boundary Element Techniques in Engineering*, Newnes–Butterworths (1980)

3 Synge, J. L., *The Hypercircle in Mathematical Physics*, Cambridge University Press (1957)

4 Boley, B. A., 'A Method for the Construction of Green's Functions', *Quarterly J. Applied Maths.*, **14**, No. 3 (1956)

5 Mindlin, R. D., 'Force at a Point in the Interior of a Semi-Infinite Solid', *Physics*, **7**, p. 145, May (1936)

6 Melan, E., 'Der Spannungszustand der turch eina Einzelkraft im Innem beanspruchten lialbscheibe', *Zeits Angen Math. Mech.*, **12**, (1932)

7 Davies, T. G. and Banerjee, P. K., 'The Displacement Field Due to a Point Load at the Interface of a Two Layer Elastic Half Space', *Geotechnique*, **28** (1978)

8 Garrison, C. J., 'Hydrodynamic Loading of Large Offshore Structures', *Numerical Methods in Offshore Engineering*, Zienkiewitz et al (Ed.), Wiley (1978)

9 Brebbia, C. A. and Walker, S., 'Simplified Boundary Elements for Radiation Problems', *Applied Math. Modelling*, **2**, June (1978)

10 Garabedian, P. R., *Partial Differential Equations*, John Wiley & Sons (1964)

11 Spiegel, M. R., *Vector Analysis*, Schaum's Outline Series, McGraw Hill (1959)

12 Cruse, T. A., Snow, D. W. and Wilson, R. B., 'Numerical Solutions in Axisymmetric Elasticity', *Computers and Structures* (1977)

13 Wrobel, L. C. and Brebbia, C. A., 'Axisymmetrix Potential Problems', in *New Developments in Boundary Element Methods*, Brebbia, C. A. (Ed.), CML Publications (1980)

BIBLIOGRAPHY

Cruse, T. A. and Rizzo, F. J., 'A Direct Formulation and Numerical Solution of the General Transient Elastodynamic Problem, 1', *J. Math. Anal. and Applications* **22** (1968)

Hearn, G. E., 'Alternative Methods of Evaluating Green's Function in Three-Dimensional Ship-Wave Problems', *J. Ship Research*, **21** (June 1977)

Odabasi, A. Y. and Hearn, G. E., 'Sea Keeping Theories — What is the Choice', *North East Coast Ins. of Eng. & Shipbuilders Trans.* **94** (1978)

Thome, R. C., 'Multipole Expansions in the Theory of Surface Waves', *Camb. Phil. Soc.*, **17** (1978)

Wehausen, J. Y. and Laitone, E. V., 'Surface Waves', *Hankbuch Der Physik*, **9**, Springer Verlag, Berlin (1977)

3

Potential theory

by F. Paris, A. Martin and E. Alarcón

3.1 INTRODUCTION

In this chapter we will introduce the reader to the techniques of the Boundary Element Method applied to simple Laplacian problems. Most classical applications refer to electrostatic and magnetic fields, but the Laplacian operator also governs problems such as Saint-Venant torsion, irrotational flow, fluid flow through porous media and the added fluid mass in fluid-structure interaction problems. This short list, to which it would be possible to add many other physical problems governed by the same equation, is an indication of the importance of the numerical treatment of the Laplacian operator.

Potential theory has pioneered the use of BEM since the papers of Jaswon[1] and Hess[2]. An interesting introduction to the topic is given by Cruse[3].

In the last five years a renaissance of integral methods has been detected. This can be followed in the books by Jaswon and Symm[4] and by Brebbia[5] or Brebbia and Walker[6].

In what follows we shall maintain an elementary level and follow a classical scheme in order to make the content accessible to the reader who has just started to study the technique. The whole emphasis has been put on the so-called 'direct' method because it is the one which appears to offer more advantages.

In the following section we recall the classical concepts of potential theory and establish the basic equations of the method. An alternative formulation using weighted residuals can be seen in chapter 6 on transient problems and in Brebbia[5]. Later on we discuss the discretization philosophy, the implementation of different kinds of elements and the advantages of substructuring which is unavoidable when dealing with heterogeneous materials.

3.2 POTENTIAL THEORY

Here we review some potential theory concepts which are familiar to all engineers. Classical references include the books by Courant and Hilbert[8], Jeffreys and Jeffreys[9], Carslaw and Jaeger[10] and Jaswon and Symm[4].

3.1.1 **Divergence and Green's theorems**

As is well known, the divergence theorem is the most used tool when changing from volume to surface integrals. Text books usually present heuristic proofs and in this sense, we recommend Chapter 4 of Kellog[7].

In short, the theorem establishes that the flux of a vector $\vec{\mathrm{F}}$ through a closed surface Γ equals the integral of its divergence in the domain Ω, i.e.

$$\int_{\Omega} \operatorname{div} \vec{\mathrm{F}} d\Omega = \int_{\Gamma} \vec{\mathrm{F}} . \vec{n} d\Gamma \tag{3.1}$$

where $\vec{F}$ *is a vector field and* $\vec{n}$ *is the exterior normal to* Γ.

Fields for which

$$\operatorname{div} \vec{\mathrm{F}} = 0 \tag{3.2}$$

are called *solenoidal* and, as a consequence of equation 3.1, the flux through every closed surface is null. If, in particular, $\vec{\mathrm{F}}$ is derived as a curl we have

$$\vec{F} = \operatorname{rot} \vec{Q}$$

$$\operatorname{div} \vec{F} = \operatorname{div} \operatorname{rot} \vec{Q} = 0 \tag{3.3}$$

The existence of an *irrotational field implies that*

$$\operatorname{rot} \vec{F} = 0 \tag{3.4}$$

Lamellar fields are derived from a potential U

$$\vec{F} = \operatorname{grad} U \tag{3.5}$$

In particular every *lamellar* field is an *irrotational* field because

$$\operatorname{rot} \operatorname{grad} U = 0 \tag{3.6}$$

Another important mathematical tool are Green's theorems obtained by applying the divergence theorem to a compound lamellar field, i.e

$$\vec{F} = U \operatorname{grad} V \tag{3.7}$$

where U and V are C^2 functions. Using the Hamilton symbolic form

$$\operatorname{div} \vec{F} = \nabla . \vec{F} = \nabla (U . \nabla V) = \nabla U \nabla V + U \nabla^2 V \tag{3.8}$$

where

$$\nabla^2 = \text{div grad} \tag{3.9}$$

is the Laplace's operator. Substituting Equation 3.8 into 3.1

$$\int_\Omega \text{div } \vec{F} d\Omega = \int_\Omega (\nabla U \nabla V + U\nabla^2 V) d\Omega = \int_\Gamma \vec{F}.\vec{n} \, d\Gamma = \int_\Gamma U\nabla V.\vec{n} \, d\Gamma = \int_\Gamma U\frac{\partial V}{\partial n} \, d\Gamma \tag{3.10}$$

and reordering we have,

$$\int_\Omega \nabla U \nabla V \, d\Omega = \int_\Gamma U\frac{\partial V}{\partial n} \, d\Gamma - \int_\Omega U\nabla^2 V \, d\Omega \tag{3.11}$$

This is the *first* Green's theorem.

The second theorem is obtained by noticing the symmetry of the left hand side, interchanging U and V, and subtracting. This gives,

$$\int_\Omega (U\nabla^2 V - V\nabla^2 U) \, d\Omega = \int_\Gamma \left(U\frac{\partial V}{\partial n} - V\frac{\partial U}{\partial n} \right) d\Gamma \tag{3.12}$$

Equation 3.12 is the *second* Green's theorem.

As is well known, every function for which

$$\nabla^2 U = 0 \tag{3.13}$$

is called a 'harmonic' function. If both U and V are harmonic, Equation 3.12 reduces to

$$\int_\Gamma \left(U\frac{\partial V}{\partial n} - V\frac{\partial U}{\partial n} \right) d\Gamma = 0 \tag{3.14}$$

which is the basis for the so-called Trefftz's method.

3.2.2 **Point newtonian potentials. Gauss's theorem**

Gravitational and electric fields follow the law

$$\vec{F} = \gamma \frac{m \, m'}{(r - r')^2} \left(\frac{\vec{r} - \vec{r}'}{r - r'} \right) \tag{3.15}$$

which can also be written as,

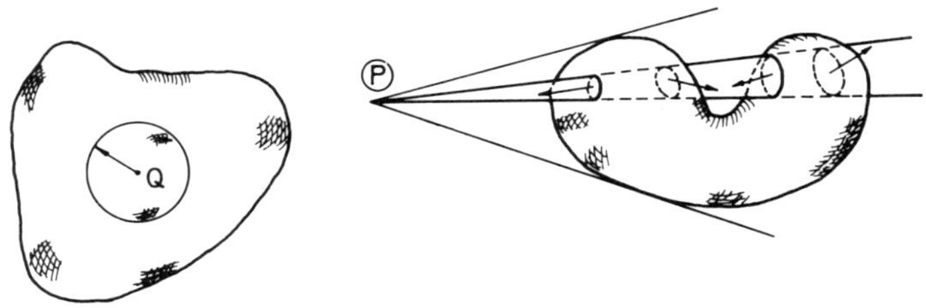

Figure 3.1

$$\vec{E}=\lim_{m'\to 0}\left(\frac{\vec{F}}{m'}\right)=\gamma\frac{m}{(r-r')^2}\left(\frac{\vec{r}-\vec{r}'}{r-r'}\right) \tag{3.16}$$

We assume that m' does not interact with m. m is then an 'active' mass, at the point Q (represented by vector $\vec{r}'$) and the field $\vec{E}$ is measured at point $\vec{P}$ (represented by vector $\vec{r}$). The quotient

$$\frac{\vec{r}-\vec{r}'}{r-r'}$$

represents a unit vector pointing from Q to P. γ is a constant which characterizes the medium under study.

The analysis of the $\vec{E}$ flux through a surface, whether or not it contains the active source (Figure 3.1), produces the well known result of Gauss. If the surface surrounds Q,

$$\int_\Gamma \vec{E}.\vec{n}\,\mathrm{d}\Gamma=\gamma m\int_\Gamma \frac{1}{(r-r')^3}(\vec{r}-\vec{r}').\vec{n}\,\mathrm{d}\Gamma \tag{3.17}$$

But $\dfrac{\vec{r}-\vec{r}'}{r-r'}\vec{n}$ times the area differential is the projection of the surface on a plane orthogonal to $(\vec{r}-\vec{r}')$ and the quotient multiplied by $(r-r')^2$ is the solid angle θ. Hence, if the surface is closed,

$$\int_\Gamma \mathrm{d}\theta=4\pi \tag{3.18}$$

$$\int_\Gamma (\vec{E}\,.\,\vec{n})\,\mathrm{d}\Gamma=4\pi\gamma m \tag{3.19}$$

If the source is outside the surface (Figure 3.1) the angle is null due to the changes of sign of the normal at both sides of the contact line in the cone of influence.

$$\int_{\Gamma} (\vec{E} \,.\, \vec{n}) \, \mathrm{d}\Gamma = 0 \tag{3.20}$$

If, on the other hand, the source is on the surface, at a non-singular point Q,

$$\int_{\Gamma} \mathrm{d}\theta = 2\pi; \quad \int_{\Gamma} (\vec{E} \,.\, \vec{n}) \, \mathrm{d}\Gamma = 2\pi\gamma m \tag{3.21}$$

Hence Gauss's integral

$$\theta = \int_{\Gamma} \frac{1}{(r - r')^3} (\vec{r} - \vec{r}') \,.\, \vec{n} \, \mathrm{d}\Gamma \tag{3.22}$$

presents the following properties characteristic of source distributions

$$\theta_e = 0, \quad (Q \text{ exterior})$$

$$\theta_\Gamma = 2\pi, \quad (Q \text{ on the boundary})$$

$$\theta_i = 4\pi, \quad (Q \text{ interior})$$

It is worth noticing that, due to Equations 3.20 and 3.1

$$\operatorname{div} \vec{E} = 0 \tag{3.23}$$

that is, $\vec{E}$ is solenoidal.

Newtonian fields are *lamellar* (and by Equation 3.6, irrotational), and we note that the function

$$V = \frac{m}{r - r'} \tag{3.24}$$

plays the role of potential and

$$\vec{E} = -\gamma \operatorname{grad} V \tag{3.25}$$

By using Equation 3.23 one obtains

$$\operatorname{div} \vec{E} = -\gamma \nabla^2 V = 0 \tag{3.26}$$

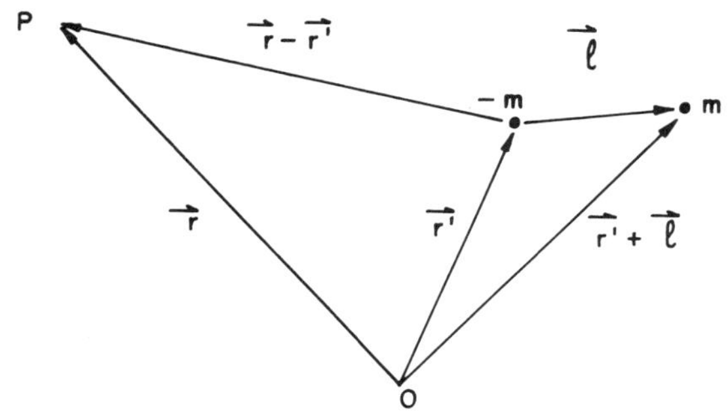

Figure 3.2

Hence the potential function is *harmonic.*

Another interesting field is the one created by two equal sources of different sign, separated by a small distance l (Figure 3.2). Using the principle of superposition, the potential function can be written as

$$V(\vec{r})=m\left(\frac{1}{r-r'-l}-\frac{1}{r-r'}\right) \tag{3.27}$$

We notice that

$$(r-r'-l)^{-1}=[(r-r'-l)^2]^{-1/2}=(r-r')^{-1}\left[1-2\frac{(\vec{r}-\vec{r}').\vec{l}}{(r-r')^2}+\frac{l^2}{r-r'}\right]^{-1/2}$$

Expanding the last expression we find

$$(r-r'-l)^{-1}=(r-r')^{-1}\left[1+\frac{(\vec{r}-\vec{r}').\vec{l}}{(r-r')^2}+\dots\right]$$

After substitution into Equation 3.27 we obtain,

$$V(\vec{r})=\frac{m\vec{l}}{(r-r')^3}(\vec{r}-\vec{r}') \tag{3.28}$$

In the limit when l tends to zero and m tends to p/l

$$\lim m\vec{l}=\vec{p} \tag{3.29}$$

called the *dipolar moment*, and then Equation 3.28 can be written,

$$V(\vec{r})=\frac{1}{(r-r')^3}\vec{p}.(\vec{r}-\vec{r}') \tag{3.30}$$

This potential depends then not only on the magnitude but also on the dipole direction and is positive near its positive end.

3.2.3 Volumetric single and double layer potentials

An immediate generalization of the previous results consists of assuming distributed potentials in place of concentrated ones, i.e. charge densities.

Independently of the physical reality of any continuum model, we can define the following densities.

Volumetric: $\rho(Q)$

$$\text{Surface:} \begin{cases} \text{Single layer, } \sigma(Q) \\ \text{Double layer, } \kappa(Q) \text{ (dipoles)} \end{cases}$$

σ is defined as the limit of ρ distribution when a dimension t of the volume tends to zero (producing a surface) and ρ tends to ∞, so that

$$\lim_{\substack{t \to 0 \\ \rho \to \infty}} (\rho \, d\Omega) = \sigma \, d\Gamma \tag{3.31}$$

κ is obtained as the limit of a double layer build-up of dipoles directed perpendicular to the surface.

If σ tends to ∞ when the distance t of both layers tends to zero, the elemental dipole moment would have a modulus,

$$\kappa = \lim_{\substack{t \to 0 \\ \sigma \to \infty}} (\sigma t) \tag{3.32}$$

and

$$\vec{p} = \kappa \vec{n} \tag{3.33}$$

Since

$$\text{grad} \frac{1}{r - r'} = -\frac{(\vec{r} - \vec{r}')}{(r - r')^3} \tag{3.34}$$

the corresponding definitions would be:

Volume potential

$$V_v = \int_\Gamma \frac{\rho(Q)}{r - r'} \, d\Omega \, (Q) \tag{3.35}$$

Single layer potential

$$V_s = \int_\Gamma \frac{\sigma(Q)}{r-r'} \, \mathrm{d}\Gamma\ (Q) \tag{3.36}$$

Double layer potential (applying Equations 3.30, 3.33 and 3.34)

$$W = -\int_\Gamma \kappa(Q)\vec{n}\,.\,\mathrm{grad}\,\frac{1}{r-r'}\,\mathrm{d}\Gamma(Q) \tag{3.37}$$

Equation 3.37 can be also written as

$$W = -\int_\Gamma \kappa(Q)\frac{\partial\left(\dfrac{1}{r-r'}\right)}{\partial n}\,\mathrm{d}\Gamma(Q) \tag{3.38}$$

An interesting case is obtained when $\kappa = \text{constant}$ because using the solid angle definition of Equation 3.22, Equation 3.38 can be expressed as

$$W = \kappa\int_\Gamma \frac{1}{(r-r')^3}(\vec{r}-\vec{r}')\,.\,\vec{n}\,\mathrm{d}\Gamma = \kappa\Omega\ (\text{constant}) \tag{3.39}$$

The generalization of Gauss's theorem for the volumetric potential case can be written as

$$\int_\Gamma (\vec{E}.\vec{n})\mathrm{d}\Gamma = 4\pi\gamma\int_\Omega \rho\ \mathrm{d}\Omega \tag{3.40}$$

Applying Equation 3.25 we have

$$-\gamma\int_\Gamma (\mathrm{grad}\ V.\vec{n})\ \mathrm{d}\Gamma = 4\pi\gamma\int_\Omega \rho\ \mathrm{d}\Omega \tag{3.41}$$

Substituting Equation 3.1 in the left hand side we obtain

$$-\int_\Omega (\mathrm{div\ grad}\ V + 4\pi\rho)\ \mathrm{d}\Omega = 0 \tag{3.42}$$

This is valid for every domain and produces Poisson's equation,

$$-\nabla^2 V = 4\pi\rho \tag{3.43}$$

one of whose solutions is simply Equation 3.35.

There are two examples of the greatest interest which justifies the non-singularity of Equations 3.35 and 3.36 when $\vec{r}=\vec{r}'$. The first one is the study of the potential generated by a constant spherical distribution. By symmetry we can write

$$\nabla^2 V = \frac{1}{r^2}\frac{\mathrm{d}}{\mathrm{d}r}\left(r^2\frac{\mathrm{d}V}{\mathrm{d}r}\right) \tag{3.44}$$

Equation 3.44 can be easily integrated in this case, i.e.

$$V(r) = 2\pi\rho R^2 - \frac{2\pi}{3}\rho r^2$$

$$\frac{\mathrm{d}V}{\mathrm{d}r} = -\frac{4\pi}{3}\rho r \tag{3.45}$$

where R is the radius of the sphere.

As can be seen V and $\mathrm{d}V/\mathrm{d}r$ are finite and tend to zero with the sphere radius. This suggests the interpretation of Equation 3.35 in the sense of its principle value when $P=Q$, that is, we consider

$$\int_{\Omega} = \int_{\Omega-} + \int_{\mathscr{E}} \tag{3.46}$$

where $\mathscr{E}$ is a sphere of radius ε around the point Q. The first integral of the sum has a continuous integrand. On the other hand the second integral is bounded by that which one could obtain substituting ρ by its maximum value in $\mathscr{E}$ and then, according to Equation 3.45, we should get a value going to zero with ε. Thus V and its first derivate are finite at interior points of a volume distribution.

The second useful example is the potential created by a circular plate with a constant surface distribution. If we have a symmetric problem and call x the normal to the plate

$$\nabla^2 \mathrm{V} = \frac{\partial^2 \mathrm{V}}{\partial x^2} \tag{3.47}$$

After integration, we obtain

$$V = 2\pi\sigma[(x^2+R^2)^{1/2} - x]$$

$$\frac{\mathrm{d}V}{\mathrm{d}x} = 2\pi\sigma\left[\frac{x}{(x^2+R^2)^{1/2}} - 1\right] \tag{3.48}$$

which are null when $R \to 0$ producing for the single layer potential the same kind of results that we have obtained for the volume case.

3.2.4 Discontinuities of the single and double layered potentials

When $x \to 0$, the results shown in Formula 3.48 become

$$V(0) = 2\pi\sigma R$$

$$\frac{\mathrm{d}V}{\mathrm{d}x}(0) = -2\pi\sigma \tag{3.49}$$

If on physical grounds, the distribution is symmetric, continuous and the derivative is not zero, the above results indicates a discontinuity of the function derivative.

Applying Poisson's theorem to a cylinder of height ε with bases parallel to the plate, in the limit the flux would be (Equation 3.41)

$$-\int_{\Gamma_1} \frac{\partial V}{\partial n_1}\,\mathrm{d}\Gamma - \int_{\Gamma_2} \frac{\partial V}{\partial n_2}\,\mathrm{d}\Gamma = \lim\, 4\pi \int \rho\varepsilon\,\mathrm{d}\Gamma \tag{3.50}$$

As $\lim(\rho\varepsilon) = \sigma$, and $\mathrm{d}s_2 = \mathrm{d}s_1 = \mathrm{d}s$, we have

$$\frac{\partial V}{\partial n_1} + \frac{\partial V}{\partial n_2} = -4\pi\sigma \tag{3.51}$$

which is just the result presented in the second of Equations 3.49 after the derivatives at both sides are added (notice that the derivation is done in opposite directions).

The previous analysis suggests a formula to represent the derivative at any interior or exterior point. To do this, a small plate of radius ε can be excluded from the distribution under study. When $\varepsilon \to 0$, $V \to 0$ (Equation 3.49) and so the potential is continuous. On the other hand when we are at an interior point it is neccessary to add $2\pi\sigma$ to the above contribution and to subtract it when at an exterior point. Hence

$$\begin{aligned}
\left(\frac{\partial V}{\partial n}\right)_i &= 2\pi\sigma + \int_\Gamma \sigma \frac{\partial\left(\dfrac{1}{r-r'}\right)}{\partial n}\,\mathrm{d}\Gamma \\
\left(\frac{\partial V}{\partial n}\right)_e &= -2\pi\sigma + \int_\Gamma \sigma \frac{\partial\left(\dfrac{1}{r-r'}\right)}{\partial n}\,\mathrm{d}\Gamma
\end{aligned} \tag{3.52}$$

These relationships were deduced by Fredholm in 1900.

The same kind of reasoning can be used for the double layer potential case. The analogy of Equation 3.38 with the single layer potential derivative shows that W presents a singularity and that the corresponding Fredholm formulae are

$$W_i = -2\pi\kappa - \int_\Gamma \kappa \frac{\partial\left(\dfrac{1}{r-r'}\right)}{\partial n} \, d\Gamma$$

$$W_e = 2\pi\kappa - \int_\Gamma \kappa \frac{\partial\left(\dfrac{1}{r-r'}\right)}{\partial n} \, d\Gamma \tag{3.53}$$

Due to the skew symmetry, the W derivative has a continuous tangent when crossing the plate. Formula 3.53 can be directly found using Equation 3.39. Equation 3.39.

For exterior points $(r-r').n$ is positive and $W_e = 2\pi\kappa$, while for interior points $W_i = -2\pi\kappa$. These proofs have a marked heuristic form. A rigorous treatment is shown in Chapter 6 of Kellog[7].

3.2.5 **Green's formula: boundary equation**

The second Green's theorem produces an interesting formula when $V = 1/(r-r')$ and we try to calculate U at P. In order to apply it the continuity of functions is postulated, and we build a sphere of radius ε around P. Now, since $\nabla V = 0$ Equation 3.12 can be written as

$$-\int_{\Omega-\mathscr{E}} (V\nabla^2 U) \, d\Omega = \int_{\Gamma \cup \partial\mathscr{E}} \left(U\frac{\partial V}{\partial n} - V\frac{\partial U}{\partial n}\right) d\Gamma \tag{3.54}$$

Taking limits

$$\lim_{\varepsilon\to 0} - \int_{\partial\mathscr{E}} \frac{1}{r}\frac{\partial U}{\partial n} d\Gamma = \lim_{\varepsilon\to 0}\left\{-\varepsilon \int_{\partial\mathscr{E}} \frac{\partial U}{\partial r} d\theta\right\} = 0$$

$$\lim_{\varepsilon\to 0} \int_{\partial\mathscr{E}} U \frac{\partial\left(\dfrac{1}{r}\right)}{\partial n} d\Gamma = \lim_{\varepsilon\to 0} \int_{\partial\mathscr{E}} \frac{U}{\varepsilon^2} d\Gamma = \mathit{lim} \int_{\partial\mathscr{E}} U \, d\theta = 4\pi U_p$$

Hence

$$-\int_{\Omega}(V\nabla^2 U)\,\mathrm{d}\Omega=\int_{\Gamma}\left(U\frac{\partial V}{\partial n}-V\frac{\partial U}{\partial n}\right)\mathrm{d}\Gamma+4\pi U_p \tag{3.55}$$

or, reordering

$$-4\pi U_p=\int_{\Omega}\left(\frac{\nabla^2 U}{r-r'}\right)\mathrm{d}\Omega+\int_{\Gamma}U\frac{\partial\left(\frac{1}{r-r'}\right)}{\partial n}\,\mathrm{d}\Gamma-\int_{\Gamma}\left(\frac{1}{r-r'}\frac{\partial U}{\partial n}\right)\mathrm{d}\Gamma \tag{3.56}$$

If

$$\frac{U}{4\pi}=\kappa;\quad \frac{1}{4\pi}\frac{\partial U}{\partial n}=\sigma$$

and our problem is of Poisson type, $-\nabla^2 U=4\pi\rho$, Equation 3.56 produces

$$+U_p=+\int\frac{\rho}{r-r'}\mathrm{d}\Omega+\int_{\Gamma}\frac{\sigma}{r-r'}\,\mathrm{d}\Gamma+\int_{\Gamma}\kappa\frac{\partial\left(\frac{1}{r-r'}\right)}{\partial n}\,\mathrm{d}\Gamma \tag{3.57}$$

Equations 3.56 and 3.57 represent the well known result that every function satisfying Green's 2nd theorem conditions, can be expressed as the sum of a single layer a double layer and a volumetric potential.

The first integral of Equation 3.56 is obviously zero if U is harmonic. It is also worth noticing that the 4π factor becomes 2π when P is on the surface. The final expression is then

$$-C_p U_p=\int_{\Omega}\left(\frac{\nabla^2 U}{r-r'}\right)\mathrm{d}\Omega+\int_{\Gamma}U\frac{\partial\left(\frac{1}{r-r'}\right)}{\partial n}\,\mathrm{d}\Gamma-\int_{\Gamma}\left(\frac{1}{r-r'}\frac{\partial U}{\partial n}\right)\mathrm{d}\Gamma \tag{3.58}$$

3.2.6 **Bidimensional case**

The previous equations will also apply to two-dimensional cases, i.e.

$$\begin{aligned} &C_p=0 \text{ if } P\in\Omega_e \\ &C_p=\pi \text{ if } P\in\partial\Omega \qquad V=\ln\frac{1}{r} \\ &C_p=2\pi \text{ if } P\in\Omega_i \end{aligned} \tag{3.59}$$

and Fredholm's formulae for this case are

$$\left(\frac{\partial V}{\partial n}\right)_i=\pi\sigma+\int_{\Gamma}\sigma\frac{\partial\left(\ln\frac{1}{r-r'}\right)}{\partial n}\,\mathrm{d}\Gamma;\quad \left(\frac{\partial V}{\partial n}\right)_e=-\pi\sigma+\int\sigma\frac{\partial\left(\ln\frac{1}{r-r'}\right)}{\partial n}$$

$$W_i = -\pi\kappa - \int_\Gamma \kappa \frac{\partial\left(\ln \frac{1}{r-r'}\right)}{\partial n} \, d\Gamma$$

$$W_e = \pi\kappa - \int_\Gamma \kappa \frac{\partial\left(\ln \frac{1}{r-r'}\right)}{\partial n} \, d\Gamma \tag{3.60}$$

Green's formula is now

$$-C_p U_p = \int_\Omega \nabla^2 U \ln \frac{1}{r-r'} \, d\Omega + \int_\Gamma U \frac{\partial\left(\ln \frac{1}{r-r'}\right)}{\partial n} \, d\Gamma - \int_\Gamma \ln \frac{1}{r-r'} \frac{\partial U}{\partial n} \, d\Gamma \tag{3.61}$$

$-C_p$ being the angle of the vertex under study.

3.3 NUMERICAL TREATMENT

In this section we shall study the different possibilities in order to apply Equations 3.58 and 3.61 to arbitrarily shaped domains.

Equation 3.11 represents the weak formulation of a potential problem and 3.12 its reciprocity relationship. Equation 3.58 can be applied to the following equations

$$-\nabla^2 \varphi = f \tag{3.62}$$

The solution is sought using the procedures of projective methods; that is, taking n interpolating functions such that, *on the boundary*

$$\varphi \approx \varphi_m = a_j \psi_j \tag{3.63}$$

$$\frac{\partial \varphi}{\partial n} = q \approx q_n = b_j \psi_j^* \tag{3.64}$$

(repeated index means summation)

The computation of the a_j or b_j parameters is done by using Equation 3.58 at as many points of the boundary as needed. The fundamental solution at these points plays the role of weighting functions[5,6]. Note that the first term on the right hand side of Equation 3.52 is known by virtue of Equation 3.62, i.e.

$$\int_\Omega \frac{\nabla^2 \varphi}{r-r'} \, d\Omega = -\int_\Omega \frac{f}{r-r'} \, d\Omega \tag{3.65}$$

Thus

$$-4\pi\varphi_p = -\int_\Omega \frac{f}{r-r'}\,d\Omega + \int_\Gamma \varphi \frac{\partial\left(\frac{1}{r-r'}\right)}{\partial n}\,d\Omega - \int_\Gamma \frac{1}{r-r'}\frac{\partial\varphi}{\partial n}\,d\Omega \tag{3.66}$$

can be written as

$$4\pi\varphi_p + \int_\Gamma \varphi \frac{\partial\left(\frac{1}{r-r'}\right)}{\partial n}\,d\Gamma = b_j \int_\Gamma \psi_j^* \frac{1}{r-r'}\,d\Gamma + \int_\Omega \frac{f}{r-r'}\,d\Omega \tag{3.67}$$

In a Neumann problem, for instance, the right hand side would be known and it would be possible (using Equation 3.63 into 3.67) to write,

$$4\pi\varphi_p + a_j \int_\Gamma \psi_j q^*\,d\Gamma = b_j \int_\Gamma \psi_j^* \frac{1}{r-r'}\,d\Gamma + \int_\Omega \frac{f}{r-r'}\,d\Omega$$

$$q^* = \frac{\partial\left(\frac{1}{r-r'}\right)}{\partial n} \tag{3.68}$$

in as many points as required to solve the problem.

Although it is not necessary (because the fundamental solution is defined everywhere at the boundary) the ψ are chosen as locally based functions for which $a_j = \varphi_j$, $b_j = \partial\varphi/\partial n|_j$ and Equation 3.68 is reduced to a system of equations,

$$\mathbf{KX} = \mathbf{F} \tag{3.69}$$

which can be solved as usual.

3.3.1 Piecewise constant potential case

The simplest hypothesis, to discretize Equation 3.68, is to image a continuous piecewise variation of φ and q over local plane Γ_j modelling the surface Γ. Hence

$$\begin{aligned} \varphi &\approx \varphi_j \delta(\Gamma_j) \\ q &\approx q_j \delta(\Gamma_j) \end{aligned} \tag{3.70}$$

where φ_j and q_j values are assigned to each local plane centroid. In this way

$$4\pi\varphi_p + \varphi_j \int_{\Gamma_j} q^* \, \mathrm{d}\Gamma = q_j \int_{\Gamma_j} \varphi^* \, \mathrm{d}\Gamma + \int_{\Omega} \frac{f}{r-r'} \, \mathrm{d}\Omega$$

$$\varphi^* = \frac{1}{r-r'} \tag{3.71}$$

Note that if p takes the n positions i, the integrals are dependent on the relative positions of observations (i) and source points (j). Then it is possible to define:

$$\hat{a}_{ij} = \int_{\Gamma_j} q^* \, \mathrm{d}\Gamma$$

$$b_{ij} = \int_{\Gamma_j} \varphi^* \, \mathrm{d}\Gamma$$

$$\hat{\hat{a}}_{ij} = 4\pi\varphi_i \delta_{ij}$$

$$a_{ij} = \hat{a}_{ij} + \hat{\hat{a}}_{ij}, \; i,j = 1,2,\ldots n \tag{3.72}$$

Equation 3.71 can now be written

$$a_{ij}\varphi_j = b_{ij}q_j + f_j$$

or

$$\mathbf{A}\boldsymbol{\varphi} = \mathbf{B}\mathbf{q} + \mathbf{f} \tag{3.73}$$

A very important step is the treatment of the integral Equation 3.71. Cruse[3] presents an analytical interpretation and Brebbia[5] gives a set of simple computer programs which are easy to understand.

In a Neumann problem the right-hand side is known, and hence we can solve the problem. In the Dirichlet problem, $\boldsymbol{\varphi}$ is moved to the left-hand side and $\mathbf{q}$ is determined by solving the system of equations.

In a *mixed* problem there will be zones with q known and other with φ known, and this requires a reordering of the system in order to arrive at Equation 3.69. In a Newton problem the boundary conditions are imposed using auxiliary equations as in certain thermal insulation problems in which the boundary conditions imply a relationship between both the normal and tangential derivatives on the boundary. In every case it is important to notice that, due to the construction procedure, **A** and **B** are full asymmetric matrices, so that the computational subroutines have to be carefully planned in order to optimize the computer time.

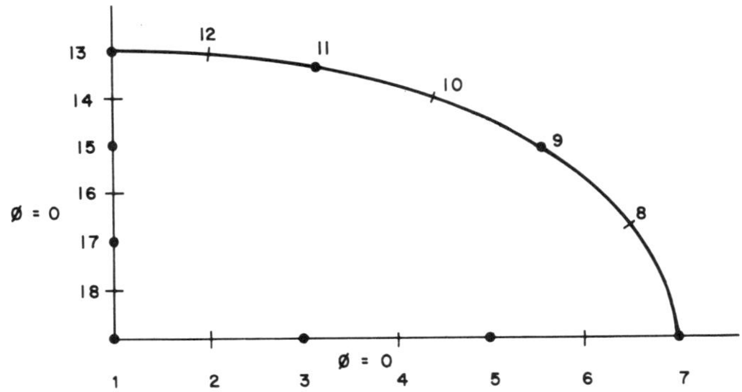

Figure 3.3

3.3.2 Curved elements

The potential, the flux and the geometry can be modelled using the same interpolation functions, in much the same way as is done for isoparametric finite element models. In general a local system of coordinates is established (Figure 3.3) and shape functions $N_1, N_2 \ldots N_m$ are defined, so that

$$\varphi(\xi) = [N_1 N_2 \ldots N_m] \begin{bmatrix} \varphi_1 \\ \varphi_2 \\ \vdots \\ \varphi_m \end{bmatrix}$$

$$q(\xi) = [N_1 N_2 \ldots N_m] \begin{bmatrix} q_1 \\ q_2 \\ \vdots \\ q_m \end{bmatrix} \tag{3.74}$$

To carry out the integrals it is now necessary to evaluate the mapping Jacobian. Numerical integration techniques are generally used and it is important to recall that the time spent in calculating the matrix elements becomes the most important factor in the final cost.

Alarcon *et al.*[11,12] present a linear element for two-dimensional problems. In this case the boundary δD is a closed curve. The nodes are defined by the extreme points of rectilinear elements in which a linear variation of φ and q is assumed,

$$\varphi_\xi = [N_1 N_2] \begin{bmatrix} \varphi_1 \\ \varphi_2 \end{bmatrix}^e; \; q_\xi = \left.\frac{\partial \varphi}{\partial v}\right|_\xi = [N_1 N_2] \begin{bmatrix} q_1 \\ q_2 \end{bmatrix}^e \tag{3.75}$$

and

$$N_1 = -\frac{1}{2}(\xi - 1);\ N_2 = \frac{1}{2}(\xi + 1),\ -1 < \xi < 1$$

The j element runs from node j to node $j+1$ so

$$\int_{\Gamma_j} \varphi \frac{\partial}{\partial v}\left(\ln\frac{1}{r}\right) d\Gamma = \int_{\Gamma_j} [N_1, N_2] \begin{bmatrix} \varphi_j \\ \varphi_{j+1} \end{bmatrix} \frac{\partial}{\partial v}\left(\ln\frac{1}{r}\right) \frac{L_j}{2} d\xi$$

where L_j is the element length. Similarly,

$$\int_{\Gamma_j} \ln\frac{1}{r}\frac{\partial \varphi}{\partial v} d\Gamma = \int_{\Gamma_j} \ln\frac{1}{r}[N_1, N_2] \begin{bmatrix} q_j \\ q_{j+1} \end{bmatrix} \frac{L_j}{2}\, d\xi \tag{3.77}$$

The integral equation now reduces to

$$2c_i\varphi_i + \sum_{j=1}^{N} L_j \left[\int_{\Gamma_j} N_1 \frac{\partial}{\partial v}\left(\ln\frac{1}{r}\right) d\xi;\ \int_{\Gamma_j} N_2 \frac{\partial}{\partial v}\left(\ln\frac{1}{r}\right) d\xi \right] \begin{bmatrix} \varphi_j^j \\ \varphi_{j+1}^j \end{bmatrix} =$$

$$= \sum_{j=1}^{N} L_j \left[\int_{\Gamma_j} N_1 \ln\frac{1}{r}\, d\xi;\ \int_{\Gamma_j} N_2 \ln\frac{1}{r}\, d\xi \right] \begin{bmatrix} q_j^j \\ q_{j+1}^j \end{bmatrix} \tag{3.78}$$

At every boundary point there are three unknowns: the potential φ and the two fluxes before the node (q_B) and after (q_A). In a Dirichlet type corner only a datum is supplied; the use of Equation 3.78 is not sufficient to determine the problem. In a Neumann zone both (q_B) and (q_A) are known and Equation 3.78 can be applied.

In a mixed situation one always knows two of the three unknowns and again Equation 3.78 is appropriate. In this way we can define five types of boundary conditions needed around a node, (q_B,q_A); (q_B,φ); (φ,q_A); (φ,sharp corner); (φ,smooth boundary).

In the last case the obvious choice is $q_B = q_A$ and when the boundary presents an apex, various solutions can be taken: binodes (Brebbia and Dominguez)[13] approximating the direction of the gradient (Alarcon *et al.*)[12], etc. We present an alternative idea which is applicable when φ is continuous at the corner.

Once the system of Equations 3.78 is established, it is possible to write

$$\mathbf{A}\boldsymbol{\varphi} = \mathbf{B}\mathbf{q} \tag{3.79}$$

and, reordering in such a way that all the unknowns are on the left hand side, one can write

$$\mathbf{K}\mathbf{X} = \mathbf{F} \tag{3.80}$$

where **X** is the vector of unknowns. The general law of formation of the matrices $A(N \times N)$ and $B(N \times 2N)$ is

$$\underset{(N \times N)}{\begin{bmatrix} & j & & j+1 & \\ & \vdots & & \vdots & \\ i \; \cdot & \cdot & \cdot & \cdot & \cdot \\ & \vdots & & \vdots & \end{bmatrix}} \underset{(N \times 1)}{\begin{bmatrix} \vdots \\ \cdot \\ \varphi_j \\ \varphi_{j+1} \\ \cdot \\ \vdots \end{bmatrix}} = \underset{(N \times 2N)}{\begin{bmatrix} & 2_j & & 2_{j+1} & \\ & \vdots & & \vdots & \\ \cdot & \cdot & \cdot & \cdot & \cdot \\ & \vdots & & \vdots & \end{bmatrix}} \underset{(2N \times 1)}{\begin{bmatrix} \vdots \\ \cdot \\ q_j^A \\ q_{j+1}^B \\ \cdot \\ \vdots \end{bmatrix}} \tag{3.81}$$

By careful programming one can form the product of every coefficient by its corresponding data and put it in vector F. If the code of the point under consideration indicates the presence of an unknown this is relocated in the appropriate row of **X** and its correlated coefficient in matrix **K**. Repeating this process for every node, one finally finds the system of Equation 3.80 which can be solved in the usual way[14].

Higher order approximations can also be used. Below we show the parabolic approach used by Benitez *et al.*[14] which can be described as follows (Figure 3.4). Shape function in local coordinates are

$$N_1(\xi) = \frac{1}{2}\xi(\xi + 1)$$

$$N_2(\xi) = 1 - \xi^2$$

$$N_3(\xi) = \frac{1}{2}\xi(\xi - 1) \tag{3.81}$$

A typical equation can be written as

$$c_i\varphi_i + \sum_{k=1}^{N} \int_{\Gamma_k} [N_1 N_2 N_3] q^* \begin{bmatrix} \varphi_1 \\ \varphi_2 \\ \varphi_3 \end{bmatrix}^k s\Gamma_k$$

$$= \sum_{k=1}^{N} \int_{\Gamma_k} [N_1 N_2 N_3] \varphi^* \begin{bmatrix} q_1 \\ q_2 \\ q_3 \end{bmatrix}^k d\Gamma_k \tag{3.82}$$

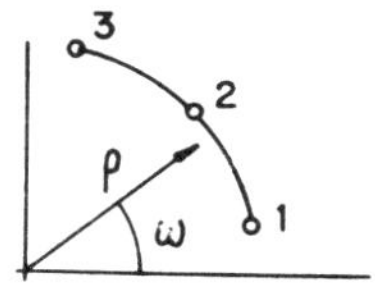

Figure 3.4

or

$$c_i\varphi_i + \sum_{k=1}^{N} [A_1 A_2 A_3] \begin{bmatrix} \varphi_1 \\ \varphi_2 \\ \varphi_3 \end{bmatrix}^k = \sum_{k=1}^{N} [B_1 B_2 B_3] \begin{bmatrix} q_1 \\ q_2 \\ q_3 \end{bmatrix}^k \tag{3.83}$$

Taking the same precautions as with linear elements it is possible to reduce to

$$\mathbf{C}\boldsymbol{\varphi} + \mathbf{A}\boldsymbol{\varphi} = \mathbf{B}\mathbf{q} \tag{3.84}$$

where

$$A_i = \int_{\Gamma_k} N_i q^* \, \mathrm{d}\Gamma_k = -\int_{\Gamma_k} N_i(\xi) \, \mathrm{d}w \tag{3.85}$$

with

$$\mathrm{d}w = \frac{\Sigma N_i^1 y_i \Sigma N_i x_i - \Sigma N_i y_i \Sigma N_i^1 x_i}{(\Sigma N_i x_i)^2 + (\Sigma N_i y_i)^2} \tag{3.86}$$

and

$$B_i = \int_{\Gamma_k} N_i(\xi) \ln \frac{[(\xi a + b)^2 + (\xi c + d)^2]^{1/2}}{\left[\left(\frac{\xi^2}{2} a + \xi b + x_2 - x_1\right)^2 + \left(\frac{\xi^2}{2} c + \xi d + y_2 - y_1\right)^2\right]^{1/2}} \, \mathrm{d}\Gamma$$

$$a = x_1 - 2x_2 + x_3$$

$$b = \frac{1}{2}(x_1 - x_3)$$

$$c = y_1 - 2y_2 + y_3$$

$$d = \frac{1}{2}(y_1 - y_3) \tag{3.87}$$

The analytical integration of Equation 3.85 when the k node is on its own element produces

$$A_1 = -\frac{\beta}{2}(I_1 + I_2)$$

$$A_2 = \beta(I_2 - I_0)$$

$$A_3 = \frac{\beta}{2}(I_1 - I_2)$$

$$I_0 = \frac{1}{\beta}\left[\operatorname{arctg}\frac{\alpha+1}{\beta} - \operatorname{arctg}\frac{\alpha-1}{\beta}\right]$$

$$I_1 = \frac{1}{2}\ln\frac{1+\left(\frac{\alpha+1}{\beta}\right)^2}{1+\left(\frac{\alpha-1}{\beta}\right)^2} + \frac{\alpha}{\beta}\left[\operatorname{arctg}\frac{\alpha-1}{\beta} - \operatorname{arctg}\frac{\alpha+1}{\beta}\right]$$

$$I_2 = 2 + \left(\frac{\alpha^2}{\beta} - \beta\right)\left[\operatorname{arctg}\frac{\alpha+1}{\beta} - \operatorname{arctg}\frac{\alpha-1}{\beta}\right] + \alpha\ln\frac{1+\left(\frac{\alpha-1}{\beta}\right)^2}{1+\left(\frac{\alpha+1}{\beta}\right)^2} \tag{3.88}$$

where α and β depend on the node from which the integration is carried out (1,2,3). The B coefficients are numerically computed by decomposition in two groups of the type

$$B_i = \int_{-1}^{1} f(\xi)\,d\xi + \int_{0}^{1} f(\eta)\ln\frac{1}{\eta}\,d\eta \tag{3.89}$$

Figure 3.3 represents as an example the torsion of a bar with an elliptic cross-section. Following Saint Venant the warping function follows $\nabla^2\varphi = 0$ and the boundary conditions are $\partial\varphi/\partial n = r \cdot t$, r being the position and t the tangent vector. In the case of Figure 3.3 we have solved a mixed problem by taking into account the symmetry conditions. By comparing with the theoretical prediction, we can see that with only three elements in every side the accuracy is good, in spite of the fact that the discretization near the longest axis of the elipse should have been more refined to take account of the high variation in the slope of this zone.

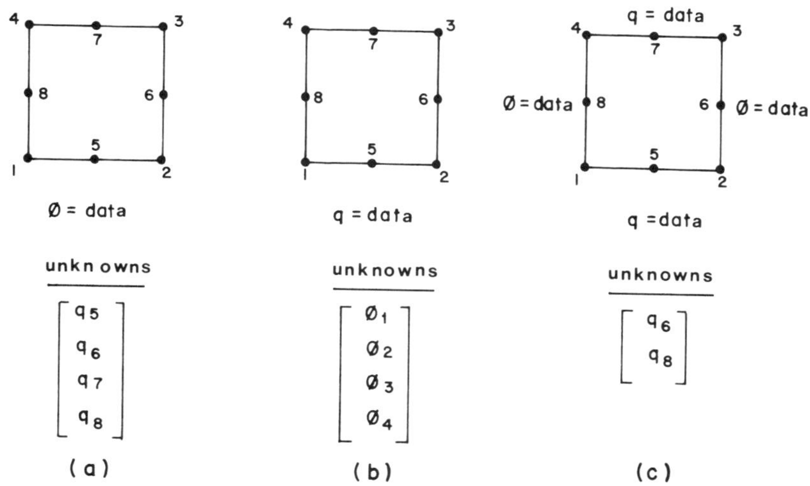

Figure 3.5

3.3.3 Mixed elements

As we said before, the φ and q interpolation can be of different orders, producing the so-called 'mixed' elements.

An atractive case is a combination of the linear potential interpolation and the constant flux one. By having q^e constant, we avoid the discontinuity of the corner problem. We shall use here a plane or linear element as was used by Martin *et al.*[15] for this case.

$$\varphi^k(\xi) = [N_1, N_2] \begin{bmatrix} \varphi_1 \\ \varphi_2 \end{bmatrix}^k$$

$$q^k = q_k$$

$$N_1 = \frac{1}{2}(1-\xi)$$

$$N_2 = \frac{1}{2}(1+\xi) \tag{3.90}$$

The basic governing equation is discretized as follows:

$$C_i \varphi_i + \sum_{k=1}^{N} \int_{\Gamma_k} [N_1, N_2] q^* \begin{bmatrix} \varphi_1 \\ \varphi_2 \end{bmatrix}^k \mathrm{d}\Gamma_k = \sum_{k=1}^{N} q_k \int_{\Gamma_k} \varphi^* \, \mathrm{d}\Gamma_k \tag{3.91}$$

In order to examine the difficulty which appears in the assembling of the equations we shall consider the three cases shown in Figure 3.5. Figure 3.5(a)

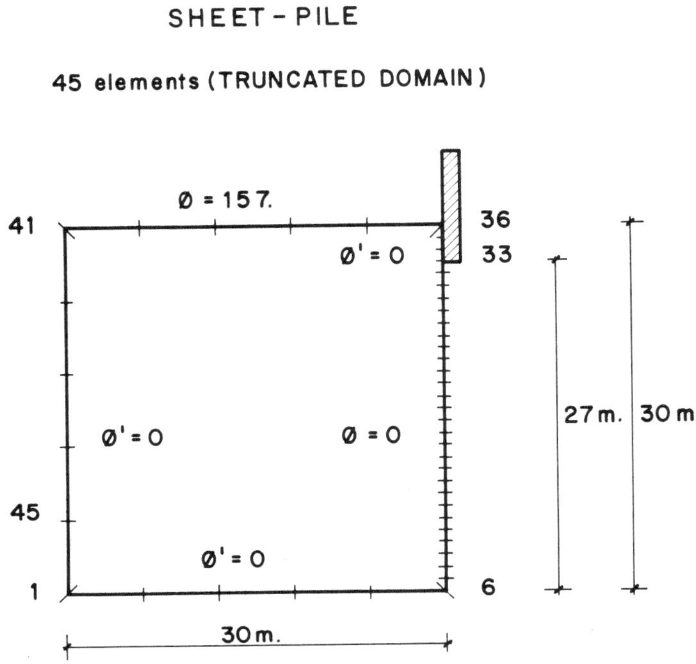

Figure 3.6(a) *Sheet-pile, 45 elements (truncated domain)*

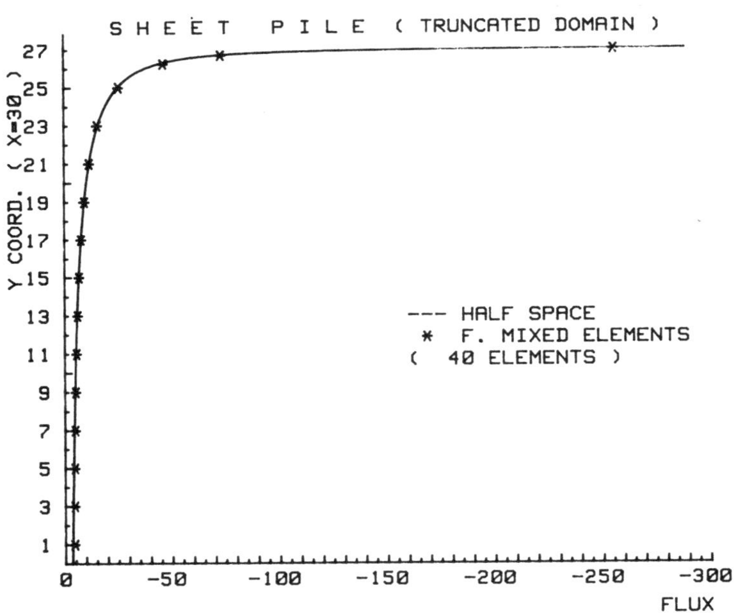

Figure 3.6(b)

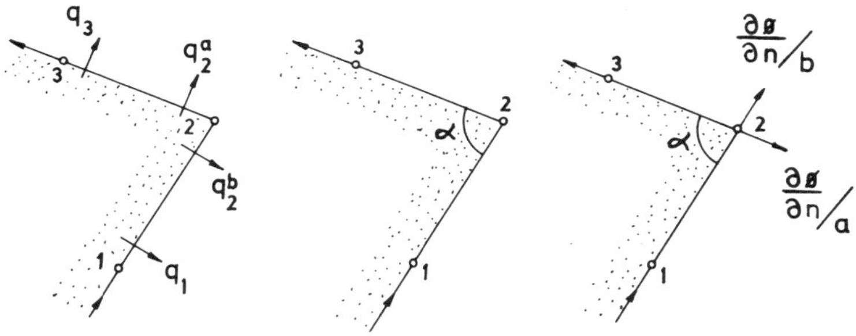

Figure 3.7

represents a Dirichlet problem. It should be enough to apply Equation 3.86 at $i=1,2,3,4$, to obtain the necessary 4 equations. The same happens for Figure 3.5(b) where the boundary data is that corresponding to a Neumann problem. Figure 3.5(c) is a mixed problem where if we try to collocate at points 1,2,3,4 one would obtain more equations than unknowns. However, this is an artificial case because it is only necessary to collocate at 6 and 8 to obtain a well defined problem. To be consistent case (a) should be treated collocating Equation 3.91 at 5,6,7,8 and taking

$$\varphi_5=\frac{1}{2}(\varphi_1+\varphi_2),\ \varphi_6=\frac{1}{2}(\varphi_2+\varphi_3),\ \text{etc.}$$

A computer program, which makes this change automatically has been implemented[15]. The options are the following when arriving at node k.

(1) If $q_k=$datum, $\varphi_k=$unknown the integration is done from end nodes.
(2) If $q_k=$unknown, $\varphi_k=$datum the integration is done from the center of element k.
(3) If $q_k=$datum, $\varphi_k=$datum no equation is generated.

As an example we present the filtration under a sheet-pile, which follows D'Arcy's law, (Gomez-Lera *et al.*[16]) Figure 3.6. In order to see the flux evolution along the sheet-pile direction we have considered a skew-symmetric problem and compared the results with the halfspace solution. As can be seen the accuracy is good. It is worth noticing the good simulation of the strong singularity which appears at the end of the sheet pile.

3.3.4 Special elements

In some cases the overdetermination created by the discretization hypothesis can be used to reduce the number of equations of a problem, similar to that which was done with constant elements. A typical case is the corner of a plane-problem when we know Dirichlet's conditions (Figure 3.7).

If we know the evolution law of φ along the boundary we can compute the derivative along the contour before and after the node; if we only have point values on a linear elements this gives

$$\left.\frac{\partial\varphi}{\partial S}\right|_b\approx\frac{\varphi_1-\varphi_2}{L_{12}}\qquad\qquad \left.\frac{\partial\varphi}{\partial S}\right|_a\approx\frac{\varphi_3-\varphi_2}{L_{32}}\qquad(3.92)$$

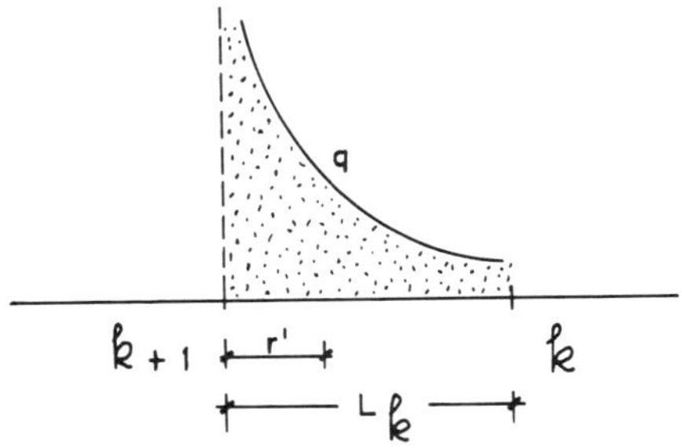

Figure 3.8 Singular element

From these values we can compute the values at the corner. Projecting on the normals before and after node 2 allow us to know the two values of fluxes. The element is now the corner 1–2–3 and the unknowns are the fluxes q_1 and q_3.

Different solutions can be used to get good results in the presence of singularities; mesh refinement round the singular point, asymptotic analysis by extrapolation of results near the singular node and introduction of special shape functions including the singularity. The last solution can be very easily illustrated. Imagine the case of Figure 3.8 in which the singularity is of the type $i/\sqrt{r'}$. In this case instead of interpolating q we can use

$$q'_{DK} = \sqrt{r'}\, q_{DK} \tag{3.93}$$

$$q'_{DK} = [N_1 N_2]\begin{bmatrix} q'_K \\ q'_{K+1} \end{bmatrix} \tag{3.94}$$

but

$$q'_K = L_K\, q_K \tag{3.95}$$

Introducing Equation 3.93 in 3.94 gives

$$q_{DK} = \begin{bmatrix} \dfrac{N_1}{\sqrt{r'}}; & \dfrac{N_2}{\sqrt{r'}} \end{bmatrix}\begin{bmatrix} \sqrt{L_K}\, q_K \\ q'_{K+1} \end{bmatrix} = \begin{bmatrix} N_1\sqrt{\dfrac{L_K}{r'}}; & \dfrac{N_2}{\sqrt{r'}} \end{bmatrix}\begin{bmatrix} q_K \\ q'_{K+1} \end{bmatrix} \tag{3.96}$$

q'_{K+1} plays the role of a 'concentration factor' and is the variable used by the programm to avoid an 'infinite' quantity.

The reasoning can be repeated using another kind of interpolation function. The integrations to be carried out are

$$\begin{aligned} B_1 &= \int_{\Gamma_k} \frac{1}{2\pi}\ln\left(\frac{1}{r}\right) N_1 \sqrt{\frac{L_K}{r'}}\; d\Gamma_k \\ B_2 &= \int_{\Gamma_k} \frac{1}{2\pi}\ln\left(\frac{1}{r}\right) N_2 \frac{1}{\sqrt{r'}}\; d\Gamma_k \end{aligned} \tag{3.97}$$

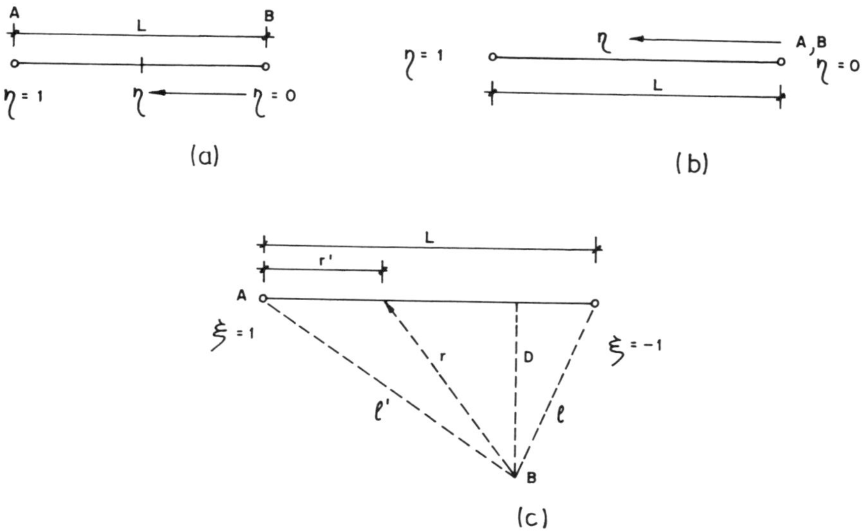

Figure 3.9 Three different cases for singular element integration: A, singular point, B integration point

Two different cases can be present depending on the position of the point of observation with respect to the singular element:

(1) The observation point and the singularity are in the same element.

(a) The observation point does not coincide with the singular one (Figure 3.4(a)). Then

$$B_1 \leqslant \int_0^1 \frac{1-\eta}{\sqrt{L(1-\eta)}} \ln\left(\frac{1}{L_\eta}\right) L \, \mathrm{d}\eta = -\sqrt{L}\left\{\frac{2}{3}\ln(L) - 0.853645\right\}$$

$$B_2 = \int_0^1 \frac{\eta}{\sqrt{L(1-\eta)}} \ln\left(\frac{1}{L_\eta}\right) L \, \mathrm{d}\eta = \sqrt{L}\left\{-\frac{4}{3}\ln(L) + 0.370132\right\} \quad (3.98)$$

(b) Both points coincide (Figure 3.9(b)). In this case

$$B_1 = \int_0^1 \frac{1-\eta}{\sqrt{L_\eta}} \ln\left(\frac{1}{L_\eta}\right) L \, \mathrm{d}\eta = -\sqrt{L}\left\{\frac{4}{3}\ln(L) - 2.514794\right\}$$

$$B_2 = \int_0^1 \frac{\eta}{\sqrt{L_\eta}} \ln\left(\frac{1}{L_\eta}\right) L \, \mathrm{d}\eta = -\sqrt{L}\left\{\frac{2}{3}\ln(L) - 0.447764\right\} \quad (3.99)$$

The intermediate integrals which appear under the form

$$I=\int_0^1 f(\eta)\ \ln\frac{1}{\eta}\ \mathrm{d}\eta \tag{3.100}$$

can be evaluated numerically by a Gauss formulat with 4 integration points.

(2) Observation and singularity points are not in the same element (Figure 3.9(c)). An integration by parts eliminates the singular term in the integrand. The integrations needed are of the type

$$B_1=\int_{-1}^{1}\frac{1}{\sqrt{r'}}N_1(\xi)\ \ln\left(\frac{1}{r}\right)\frac{L}{2}\ \mathrm{d}\xi=$$

$$=\sqrt{\frac{L}{2}}\left[-2\sqrt{2}\ \ln l-\int_{-1}^{1}(1-\xi)^{1/2}\ \ln\frac{1}{r}\ \mathrm{d}\xi-\int_{-1}^{1}(1-\xi)^{3/2}\frac{L}{2r^2}\sqrt{r^2-D^2}\ \right]\mathrm{d}\xi \tag{3.101}$$

$$B_2=\int_{-1}^{1}\frac{1}{\sqrt{r'}}N_2(\xi)\ \ln\left(\frac{1}{r}\right)\frac{L}{2}\ \mathrm{d}\xi=\frac{L}{2}\int_{-1}^{1}(1-\xi)^{1/2}\left[\ln\left(\frac{1}{r}\right)-\frac{1+\xi}{2r^2}Lr^2-D^2\right]\mathrm{d}\xi$$

As an example we present again the sheet-pile problem (Figure 3.10) with the factor $r^{-1/2}$ round point A.

Results are shown in Figure 3.10(c).

3.4 NON-HOMOGENEOUS MEDIA

The previous development has assumed that the factor γ (characteristic of the material at which the field was established) was constant. In many cases the medium is non-homogeneous and it is necessary to divide it into regions and then to superimpose the equations deduced for every region.

3.4.1 **Subregions**

To fix our ideas, imagine a domain D with boundary Γ and permeability k in which there are two inclusions with properties $(\bar{D},\bar{\Gamma},\bar{k})$ and $(\bar{\bar{D}},\bar{\bar{\Gamma}},\bar{\bar{k}})$ and where it

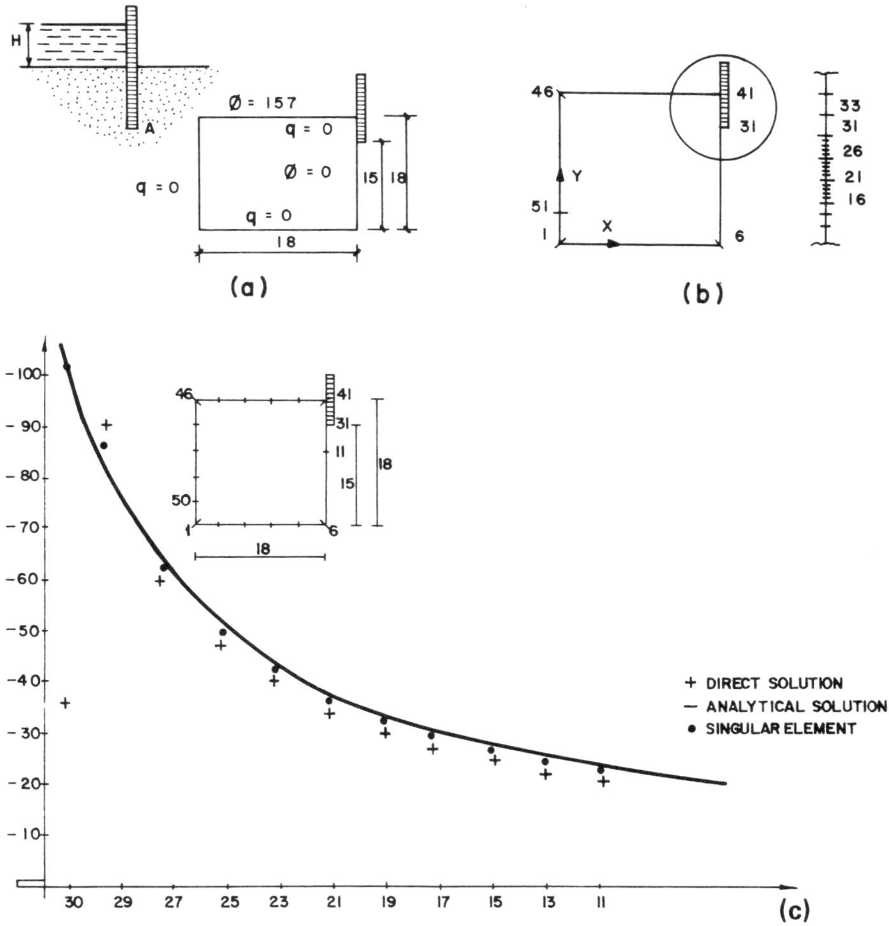

Figure 3.10

is neccesary to study the problems

$$
\begin{aligned}
k\nabla^2 h &= 0 \\
\bar{k}\nabla^2 h &= 0 \qquad (3.102) \\
\bar{\bar{k}}\nabla^2 h &= 0
\end{aligned}
$$

If we imagine that the corresponding discretizations have N, $\bar{n}$, $\bar{\bar{n}}$ linear elements, the partitioned equations would be

$$
[A'\bar{A}_1'\bar{A}_1''] \begin{bmatrix} \varphi \\ \bar{\varphi}_1 \\ \bar{\bar{\varphi}}_1 \end{bmatrix} = [B\bar{B}_1\bar{\bar{B}}_1] \begin{bmatrix} q \\ \bar{q}_1 \\ \bar{\bar{q}}_1 \end{bmatrix} \qquad (3.103)
$$

$N+\bar{n}+\bar{\bar{n}}=$total number of elements, where

$$\varphi=kh;\ \bar{\varphi}_1=\bar{k}h;\ \bar{\bar{\varphi}}_1=\bar{\bar{k}}h \tag{3.104}$$

We can also write,

$$\begin{aligned} \bar{A}'_2\bar{\varphi}^2&=\bar{B}_2\bar{q}^2\\ \bar{\bar{A}}'_3\bar{\bar{\varphi}}^3&=\bar{\bar{B}}_3\bar{\bar{q}}^3 \end{aligned} \tag{3.105}$$

Notice that the continuity equations are

$$\begin{aligned} h_1&=\frac{\varphi_1}{k_1}=\frac{\varphi_2}{k_2}=h_2\\ q_1&=-q_2 \end{aligned} \tag{3.106}$$

which in our particular case means that,

$$\begin{aligned} \frac{\bar{\varphi}^1}{k}&=\frac{\bar{\varphi}^2}{\bar{k}}=\bar{\varphi}\\ \frac{\bar{\bar{\varphi}}^1}{k}&=\frac{\bar{\bar{\varphi}}^2}{\bar{\bar{k}}}=\bar{\bar{\varphi}}\\ \bar{q}^1&=-\bar{q}^2=\bar{q}\\ \bar{\bar{q}}^1&=-\bar{\bar{q}}^2=\bar{\bar{q}} \end{aligned} \tag{3.107}$$

Assembling the equations with these conditions duly incorporated, one obtains,

$$\begin{bmatrix} A & \bar{A}_1 & \bar{\bar{A}}_1\\ 0 & \bar{A}_2 & 0\\ 0 & 0 & \bar{\bar{A}}_3 \end{bmatrix}\begin{bmatrix}\varphi\\ \bar{\varphi}\\ \bar{\bar{\varphi}}\end{bmatrix}=\begin{bmatrix} B_1 & \bar{B}_1 & \bar{\bar{B}}_1\\ 0 & -\bar{B}_2 & 0\\ 0 & 0 & \bar{\bar{B}}_3 \end{bmatrix}\begin{bmatrix}q\\ \bar{q}\\ \bar{\bar{q}}\end{bmatrix} \tag{3.108}$$

For Neumann's problem, for instance, we need to solve

$$\begin{bmatrix} A & \bar{A}_1 & -\bar{B}_1 & \bar{\bar{A}}_1 & -\bar{\bar{B}}_1\\ 0 & \bar{A}_2 & \bar{B}_2 & 0 & 0\\ 0 & 0 & 0 & \bar{\bar{A}}_3 & -\bar{\bar{B}}_3 \end{bmatrix}\begin{bmatrix}\varphi\\ \bar{\varphi}\\ \bar{q}\\ \bar{\bar{\varphi}}\\ \bar{\bar{q}}\end{bmatrix}=\begin{bmatrix}B_1q\\ 0\\ 0\end{bmatrix} \tag{3.109}$$

Imposing conditions of the kind

$$\begin{aligned} \bar{q}^b&=\bar{q}^a\\ \bar{\bar{q}}^b&=\bar{\bar{q}}^a \end{aligned} \tag{3.110}$$

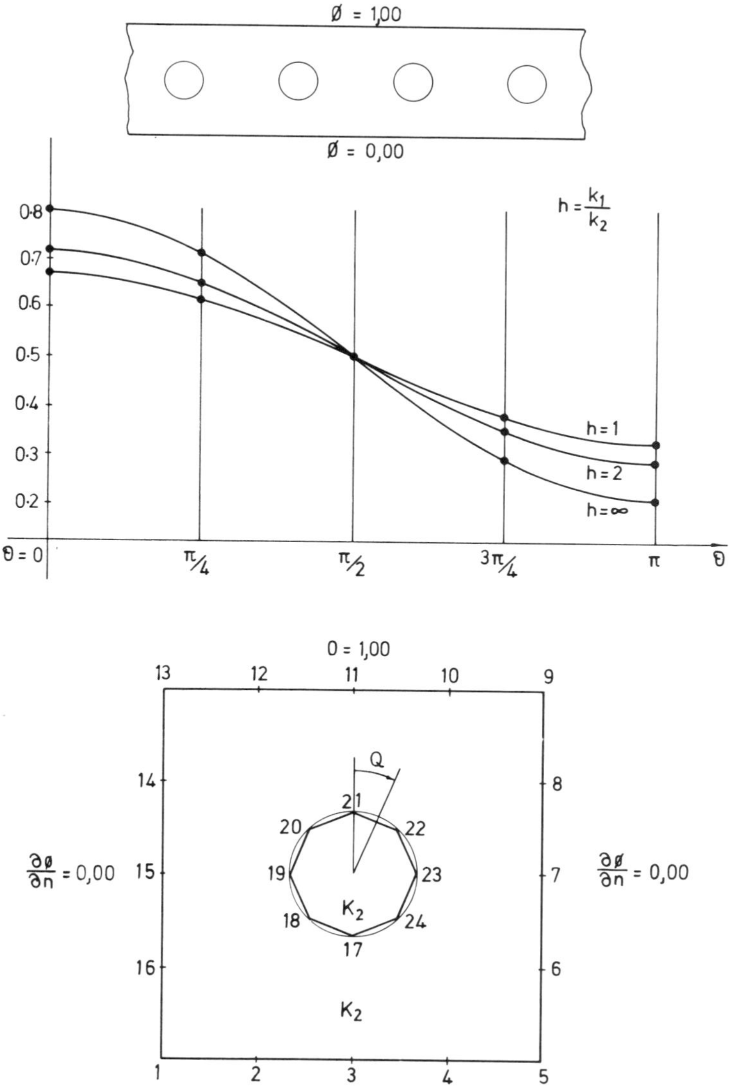

Figure 3.11

it is possible to solve a system of $N+2(\bar{n}+\bar{\bar{n}})$ equations and obtain the fluxes and potentials on the interface.

As an example, we present the heat propagation problem through a plate with a periodic array of circular inclusions (Figure 3.11). Using symmetry we need only discretize one square. We represent the potentials in a quarter of the interface as functions of different relative permeabilities.

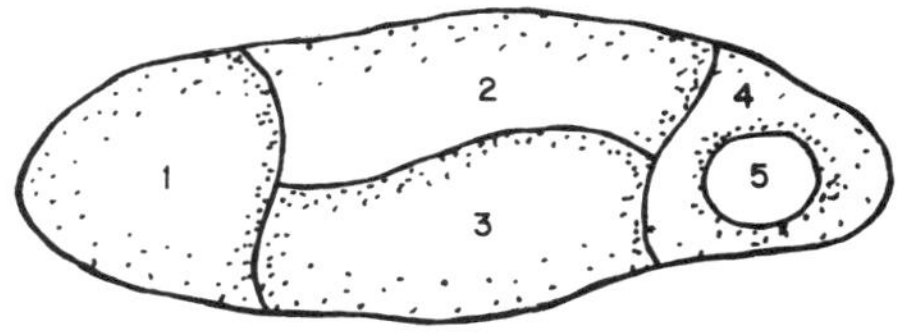

Figure 3.12

3.4.2 Ordering the system of equations

The careful storing of the matrix governing the system of equations is very important for heterogeneous media and for the treatment of large problems, because on it hinges the efficiency of the method and the size of the required storage devices. Below we discuss different ways of ordering the matrix coefficients.

Consider the case shown in Figure 3.12 where there are 5 different subregions. As a first solution, the integral equations have been ordered by subregions and the unknowns by subregions and interfaces as they appear when the numbering progresses. It is easy to see that the storage of the integration coefficients is very compact and, subsequently, the total core requirement and the pointers neccesary to determine the significant blocks are considerably reduced. Nevertheless, with this procedure, we have no previous information about the quantities that are occupying the main diagonal, and there is even the possibility of finding zeros if a subregion has no contact with the one numbered next. This makes the system unsuitable in many cases.

In Figure 3.13 a solution scheme is presented with the following symbols:

$S^{(i)}$: Integral equations associated with subregion i.

S^{i}: Integral equations associated with nodes only included in region i.

S^{ij}: Integral equations associated with nodes belonging to the interface ij.

φ^{i}/q^{i}: potential/flux at nodes belonging only to subregion i.

φ^{ij}/q^{ij}: potential/flux at nodes belonging to ij interface.

To guarantee a non-zero main diagonal it is possible to use the second organization, in which the integral equations are assembled by subregions and then by interfaces numbered in order. The unknowns are ordered acording to the number of integral equations on every interface, which guarantees the existence of elements on the main diagonal.

Note that in certain cases one can have several values for the flux at a given node, although the potential is always uniquely defined. For these cases we will have several equations at the node, each associated with a flux value.

For these problems the storage size increases if a direct method of solution is used (because it is necessary to store blocks of zeros). In this case, in every different configuration column, there are only two separate blocks different from zero, and this is why it would be necessary to store only four pointers for every column (Figure 3.14).

The other possibility is to fix the unknowns in the same order as in the first case but, in place of writing the integral equations by subregions, doing this alternately in order to guarantee the presence of non-zero elements in the main diagonal. The result can be seen in Figure 3.15. The worst feature of this

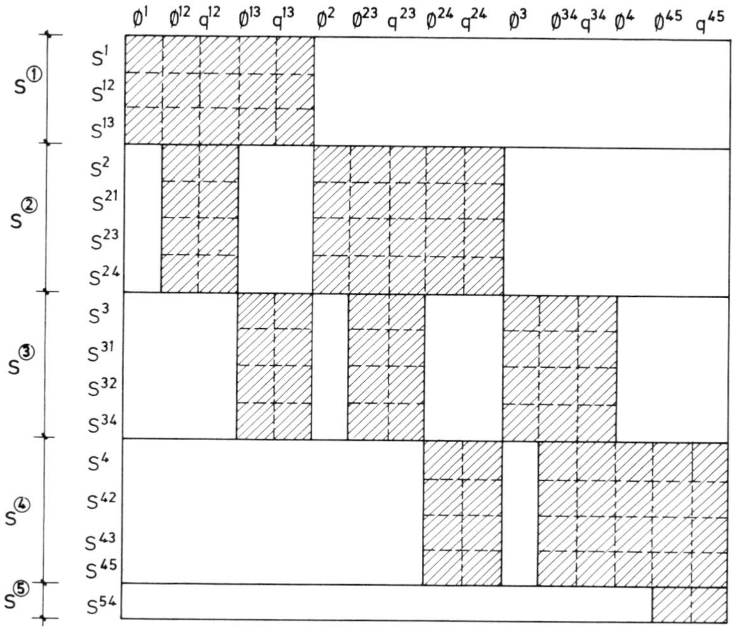

Figure 3.13

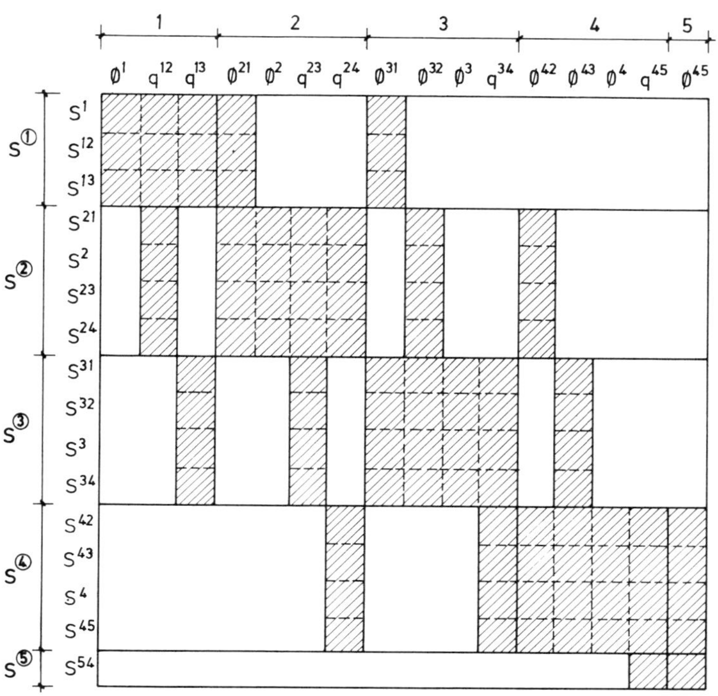

Figure 3.14

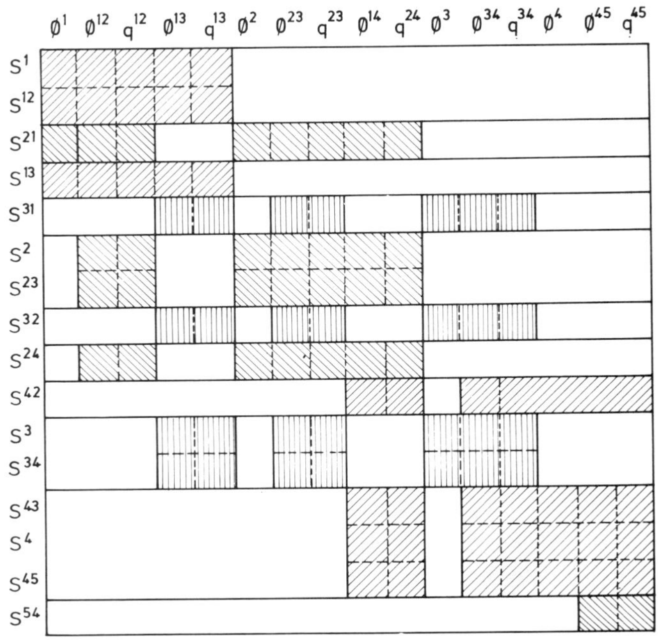

Figure 3.15

solution is the large amount of pointers which are necessary to detect the zero blocks.

From the above discussion it seems reasonable to say that the second solution is the best. The organization for subregion i is presented in detail in Figure 3.16.

3.5 MISCELLANEOUS PROBLEMS

In what follows we shall comment on the possibility of studying certain specific problems using the boundary element method. Among the several references to these topics it is worthwhile to cite Ligget[17] for free surface problems, Brebbia and Chang[19] for anisotropic media, Brebbia and Wrobel[18] and Tomlin[20] for indirect methods. The transient problem is also particularly important but for this the reader is referred to Chapter 6.

3.5.1 **Free boundary**

One of the more interesting problems of seepage through porous media is that in which, the conditions for flux and potential are specified on a certain part of the boundary, but the boundary geometry is not known.

A typical example is the earth dam of Figure 3.17. The boundary-conditions are:

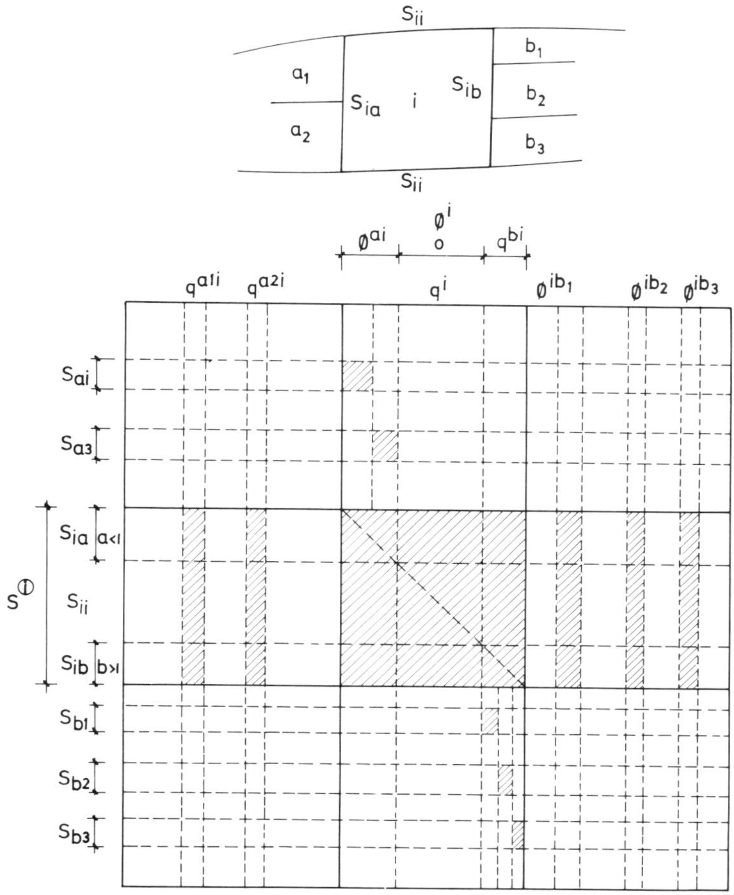

Figure 3.16

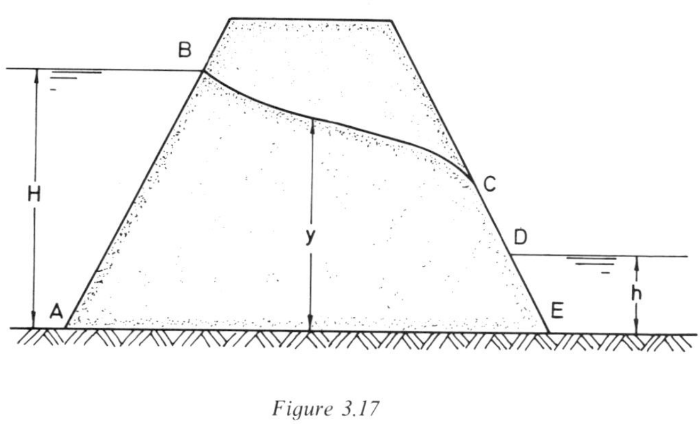

Figure 3.17

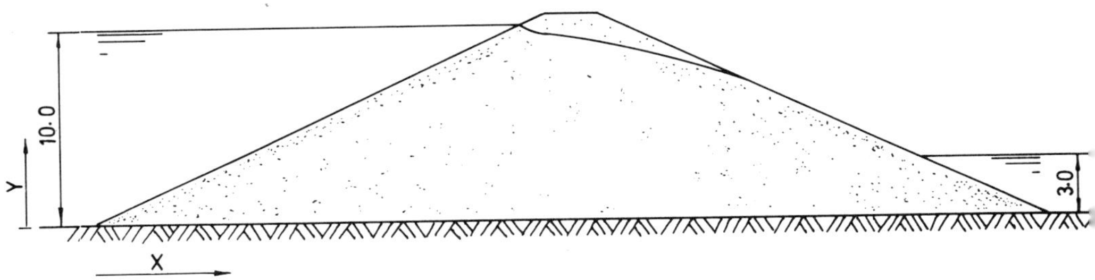

Figure 3.18 *Accepted line (AL) versus obtained results (OR)*

$$
\begin{aligned}
&\varphi = H, \text{ on AB} \\
&\left.\begin{aligned}\varphi &= y \\ \frac{\partial \varphi}{\partial n} &= 0\end{aligned}\right\} \text{on BC} \\
&\varphi = y, \text{ on CD} \\
&\varphi = h, \text{ on DE} \\
&\frac{\partial \varphi}{\partial n} = 0, \text{ on AE}
\end{aligned}
\qquad (3.111)
$$

B is a fixed point, but C is unknown and the same happens with the head 'y' on BC.

To solve the problem, one can use an iterative solution, starting with an arbitrary geometry and correcting it according to the results. For instance, it is possible to start with $\varphi = y^0$ and to get the corresponding fluxes q^0. Imagining that the fluxes are velocities, it is possible to establish a new boundary

$$y_i^1 = y_i^0 + q_i^0 \Delta t \qquad (3.112)$$

where Δt is adequately chosen (see Desai[21]). In this way, after a few iterations, the final profile is obtained.

In the finite element method (Martins *et al.*[22]) the procedure is rather complicated because it is necessary to modify the whole mesh or at least to distort it in the iteration zone.

In this sense, the BEM presents a striking simplicity. In the example of Figure 3.18 only five iterations were needed to get a good solution.

3.5.2 **Anisotropic media**

When treating anisotropic solids, the problem becomes rather more complicated. The field vector now has the form (Carslaw and Jaeger[10])

$$\vec{E} = -\vec{K} \text{ grad } \varphi \qquad (3.113)$$

K is not the constant γ used at the very beginning of the chapter as representative of the medium properties but a full matrix containing the conductivity coefficients k_{ij}, i.e. K is a second-order tensor. The field equations are now

$$\text{div}\ \vec{E}=0 \tag{3.114}$$

that is

$$\text{div}\ (\vec{K}\ \text{grad}\ \varphi)=0 \tag{3.115}$$

or in indicial form

$$(k_{ij}\ \varphi_{,j})_{,i}=k_{ij}\ \varphi_{,ji}=0 \tag{3.116}$$

Expanding we can write,

$$\begin{gathered}k_{11}\frac{\partial^2\varphi}{\partial x^2}+k_{22}\frac{\partial^2\varphi}{\partial y^2}+k_{33}\frac{\partial^2\varphi}{\partial z^2}+(k_{23}+k_{32})\frac{\partial^2\varphi}{\partial y\partial z}+(k_{31}+k_{13})\frac{\partial^2\varphi}{\partial z\partial x}+\\+(k_{12}+k_{21})\frac{\partial^2\varphi}{\partial x\partial y}=0\end{gathered} \tag{3.117}$$

which can always be referred to a new principal axis system, i.e.

$$K_1\frac{\partial^2\varphi}{\partial\xi^2}+K_2\frac{\partial^2\varphi}{\partial\eta^2}+K_3\frac{\partial^2\varphi}{\partial\zeta^2}=0 \tag{3.118}$$

To treat this equation we can define a new system such that

$$\xi_1=\xi\sqrt{(k/K_1)};\ \eta_1=\eta\sqrt{(k/K_2)};\ \zeta_1=\zeta\sqrt{(k/K_3)} \tag{3.119}$$

where k is, in principle, arbitrary and then recover the form

$$k\left(\frac{\partial^2\varphi}{\partial\xi_1^2}+\frac{\partial^2\varphi}{\partial\eta_1^2}+\frac{\partial^2\varphi}{\partial\zeta_1^2}\right)=0 \tag{3.120}$$

which has to be applied to the distorted geometry.

One special case is when the solid is orthotropic. In this case, the governing equation is

$$K_1\frac{\partial^2\varphi}{\partial x^2}+K_2\frac{\partial^2\varphi}{\partial y^2}+K_3\frac{\partial^2\varphi}{\partial z^2}=0 \tag{3.121}$$

and the three orthogonal axis are in the same direction as the conductivities. The same transformation as before can be made, but in order to use the fundamental solution and the whole philosophy of BEM discretization, it is necessary to take

$$k=\sqrt{(K_1,K_2,K_3)} \tag{3.122}$$

The fundamental solution is now

$$\varphi^*=\frac{1}{4\pi r}\cdot\frac{1}{\sqrt{(K_1,K_2,K_3)}}$$

and

$$r=\sqrt{\left(\frac{x^2}{K_1}+\frac{y^2}{K_2}+\frac{z^2}{K_3}\right)} \tag{3.123}$$

The method has been used by several authors, Brebbia and Chang[20], for instance, have presented the example shown in Figure 3.19 for which the anisotropic domain was divided into subregions.

3.5.3 Indirect methods

Several authors (Hess and Smith[2], Tomlin[20], Jaswon and Symm[4]) have used an alternative representation of the solution based on the possibility of looking for a single kind of layer and adjusting it to the specified boundary conditions. This is called an 'indirect' method. It is nothing more that the discretization of potentials. For instance, the single layer potential was defined as

$$V(P)=\int_\Gamma \frac{\sigma}{r}\,\mathrm{d}\Gamma \tag{3.124}$$

where P can be considered in the boundary and V is continuous. Differentiating in the direction of the outer normal, the Fredholm formula produces

$$\frac{\partial V}{\partial v}=-2\pi\sigma+\int_\Gamma \sigma\frac{\partial\frac{1}{r}}{\partial v}\mathrm{d}\Gamma \tag{3.125}$$

Dividing the boundary into a series of n elements, in every one of which either V or $\partial V/\partial v$ is known, one can establish a system of equations using the previous formulae such that for the case of constant elements

$$V_i=\sum_{j=1}^{n}\sigma_j\int_{\Gamma_j}\frac{1}{r}\,\mathrm{d}\Gamma$$

$$q_i=\frac{\partial V_i}{\partial v}=\sum_{j=1}^{n}\sigma_j\int_{\Gamma_j}\frac{\partial\left(\frac{1}{r}\right)}{\partial v}\mathrm{d}\Gamma-2\pi\sigma_i \tag{3.126}$$

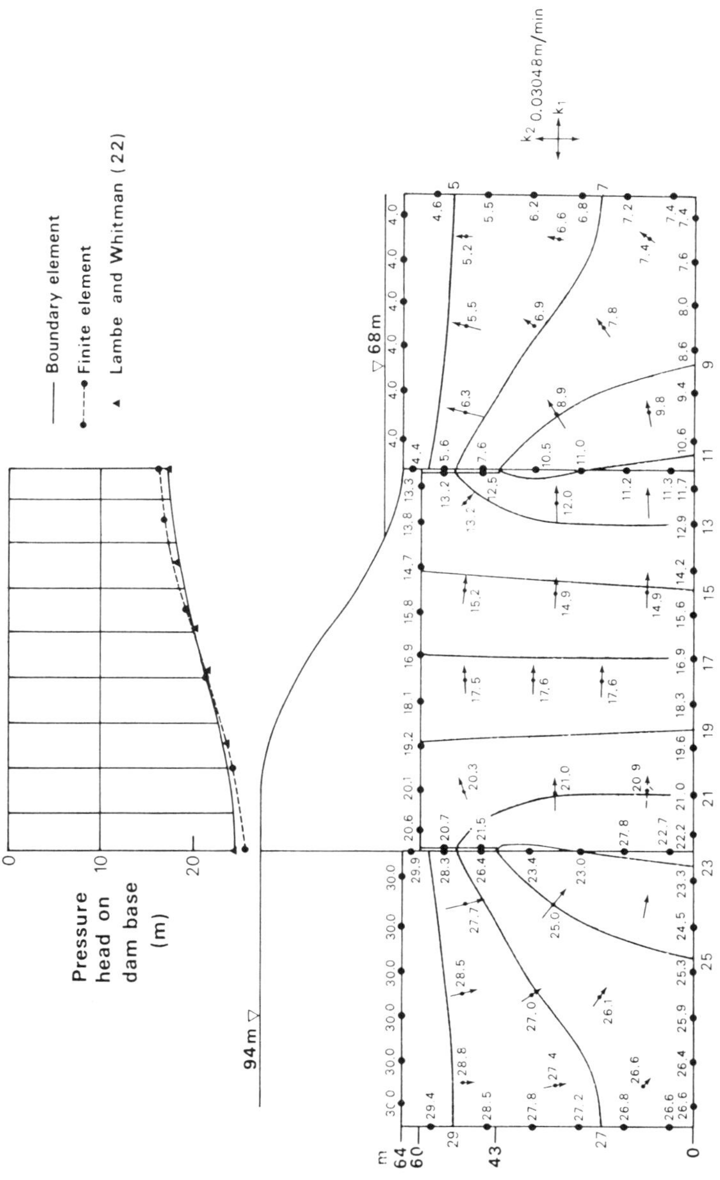

Figure 3.19 Pressure distribution and flow under two-piled dam

Once the n values σ_i are found the first equation can be used to obtain the values of interest, hence the name 'indirect method'.

Another alternative is to use the dipole definition, and using the Fredholm formulae, establish

$$W_i = 2\pi k + \int_{\Gamma} k \frac{\partial\left(\frac{1}{r}\right)}{\partial v} \mathrm{d}\Gamma$$

$$\frac{\partial W_i}{\partial v} = \int_{\Gamma} k \frac{\partial^2\left(\frac{1}{r}\right)}{\partial v^2} \mathrm{d}\Gamma \tag{3.127}$$

Once again, the discretization of the boundary will produce the necessary equations for solving the problem. Because the Fredholm integrals of the second kind are better behaved than those of the first kind, the dipole approach is preferred for solving Dirichlet problems and the source approach for Neumann ones. Further developments can be found in Jaswon and Symm[4].

REFERENCES

1 Jaswon, M. A., 'Integral Equation Methods in Potential-Theory I.' *Proc. Royal Soc.* (a), p. 275, (1963)

2 Hess, J. L. and Smith, A. M. O., 'Calculation of Potential-Flow about Arbitrary Bodies.' In *Progress in Aeronautical Sciences*, Vol. 8. D. Kuchemann (Ed.), Pergamon Press, (1967)

3 Cruse, T. A., *Mathematical Foundations of the Boundary-Integral Equation Method in Solid Mechanics*, Pratt and Whitney Aircraft Group. Speical Scientific Report, (1977)

4 Jaswon, M. A. and Symm, G. I., *Integral Equation Methods in Potential Theory and Elastostatics*, Academic Press, (1977)

5 Brebbia, C. A., *The Boundary Element Method for Engineers*, Pentech Press, (1978)

6 Brebbia, C. A. and Walker, S., *Boundary Element Techniques in Engineering*, Newnes-Butterworth, (1979)

7 Kellog, O. D., *Foundations of Potential Theory*, Springer, (1929)

8 Courant, R. and Hilbert, D., *Methods of Mathematical Physics*, Interscience (1953)

9 Jeffreys, H. and Jeffreys, B. S., *Methods of Mathematical Physics*, Cambridge U.P. (1972)

10 Carslaw, H. S. and Jaeger, J. C., *Conduction of Heat in Solids*, 2nd ed, Clarendon Press, Oxford (1959)

11 Alarcon, E., Marin, A. and Paris, F., 'Some Minor Problems with BEM' in *Applied Numerical Modelling*, E. Alarcon and C. A. Brebbia (Ed.), Pentech Press (1979)

12 Alarcon, E., Martin, A. and Paris, F., 'Boundary Elements in Potential and Elasticity Theory,' *Computers and Structures*, **10**, 1979.

13 Brebbia, C. A. and Dominguez, J., 'Boundary Element Methods for Potential Problems,' Applied Mathematical Modelling, **2**, No. 7 (1977)

14 Benitez, F. G. and Alarcon, E., 'Simulacion numerica por el metodo de los elementos de controno. Utilizacion de los elementos isoparametricos parabolicos,' in *1st Simposium sobre Modelado y Simulacion*; Comite Español de la IFAC (1980)

15 Martin, A., Rodriguez, I. and Alarcon, E., 'Mixed Elements,' in *Recent Advances in BEM*, C. A. Brebbia, (Ed.), CML Pub. (1980)

16 Gomez-Lera, S., Corona, J. A. and Alarcon, E., 'The BIEM Applied to Flow through Porous Media,' *Proc. 3rd Int. Conf. on Finite Elements in Water Resources*, Mississippi University Press (1980)

17 Liggett, J. A., 'Location of Free Surface in Porous Media,' *ASCE, Hy.*, **4**, April (1977)
18 Brebbia, C. A. and Wrobel, L., 'The Boundary Element Methods for Fluid Flow,' *Proc. 3rd Int. Conf. on Finite Elements in Water Resources*, Mississippi University Press (1980)
19 Brebbia, C. A. and Chang, O. V., 'Boundary Elements Applied to Seepage Problems in Zoned Anisotropic Solid,' *Advances Eng. Software*, **1** (1979)
20 Tomlin, G. R., *Numerical Analysis of Continuum Problems in Zoned Anisotropic Media*, Ph.D. Thesis, Southampton University (1973)
21 Desai, C. S. and Christian, J. T., *Numerical Methods in Geotechnical Engineering*, McGraw (1977)
22 Martins, J. B., Matos, A. C. and Bianchi, A., *Free Surface Flow in Porous Media by FEM*, Report ENG., No. 101, Universidade do MINHO (1978)

4

Elastostatics

by F. Hartmann

4.1 INTRODUCTION

This chapter is a summary of the boundary element method in classical elasticity and plate bending theory.

Boundary element methods rest essentially on the fact that a function — at least a function we encounter in elasticity — that satisfies in a n-dimensional region a homogeneous differential equation, is already determined by its 'trace' on a $(n-1)$-dimensional manifold. So it is, that a displacement field in a body with zero body forces is uniquely determined by its value on the surface and its traction on the surface. Similar properties hold for the solutions of other differential equations in elasticity.

The surface, or more generally the boundary, becomes the dominant domain in boundary value problems. It suffices to determine the 'trace' on the boundary to solve a boundary value problem. The word 'trace' comprises at this point not only the value of a function on the boundary,† but also further quantities, as normal derivative, etc. The single functions which constitute the trace satisfy an integral relation on the boundary. This relation becomes, due to the fact that the boundary values are elements of the trace, an integral equation for the unknown components of the trace and thus serves to determine the trace completely. Once the trace is determined, quantities in the interior of the region can then be calculated by influence functions. These influence functions, which also appear in the integral equations are volume or surface integrals. Their integrand is a weakly or strongly singular kernel times a bounded function. The influence functions exhibit a behaviour which is quite uncommon in ordinary functions and therefore call for a careful study of their properties.

In the following sections we will discuss this boundary integral equation approach in more details. The discussion concentrates on problems in classical elastomechanics and plate bending problems, but closes with a (long) list of those problems in elasticity which are amenable to the same treatment.

Section 4.2 serves to introduce the basic equations of classical elasticity and the Kelvin solutions, which are the fundamental solutions of the Cauchy-Navier equations.

† Which is its common meaning. We will use the word in this sense later on.

In Section 4.3 we study the properties of the influence functions — or elastic potentials as we call them — which are the basic ingredients of the direct and indirect method and formulate the integral equations of the indirect method and give an interpretation of the physical nature of the layers of the elastic potentials.

Section 4.4 treats the direct method starting with Betti's theorem and the Somigliana identity which serves to derive the integral equations. Some theorems concerning the theoretical background are discussed before finally practical problems as the calculation of stresses on the boundary and the discretization and numerical handling of the integral equations are treated.

In Section 4.5 we discuss the application of the direct method to axisymmetric problems. The integral equations reduce in such problems to equations on the curve C that generates the surface of the axisymmetric body and are in parts uncoupled, as radial (r) and axial (z) quantities do not influence tangential (φ) quantities. The section closes with an extension of the method to 'arbitrary' boundary value problems in axisymmetric bodies.

In Section 4.6 we mention problems on elastic halfspaces which can be handled with special elastic potentials (Kupradze) and study some boundary value problems which can be solved by using continuation properties of 'exterior' and 'interior' solutions and so we can avoid the solution of integral equations.

Section 4.7 treats plate bending problems starting with the basic equations and then turns to biharmonic potentials which are the basic tools for the indirect approach to boundary value problems in plate bending. The direct method hinges here on the Rayleigh-Green identity which we introduce to derive integral representations for the deflection w and its normal derivative. These equations are finally used to derive integral equations in the sense of the direct method.

Section 4.8 closes this chapter with a list of problems in elasticity for which fundamental solutions are known and therefore can be treated with integral equation methods.

4.2 BASIC CONCEPTS

We start with the introduction of the equations of classical elasticity which we will present preferably in matrix and vector notation. If we use the indicial notation of tensor analysis, the subscripts have the usual range 1,2,3 in three dimensions (3-D) and 1, 2 in two dimensions (2-D), summation over repeated subscripts is implied and it is δ_{ij} Kronecker's delta. To designate partial differentiation we use the symbols

$$u_{i,j} = u_{i,x_j} = \frac{\partial u_i}{\partial x_j}.$$

The first symbol will only be used if it is evident with respect to which coordinate partial differentiation is done.

4.2.1 Fundamental solutions

The classical theory is governed by the Cauchy-Navier differential equation

$$\mathbf{L}=\left[\mu\Delta+\frac{\mu}{1-2v}\text{ grad div}\right]\mathbf{u}(\mathbf{x})=-\mathbf{b}(\mathbf{x}) \tag{4.1}$$

where $\mathbf{u}(\mathbf{x})=[u_1, u_2, u_3]^T$ is the displacement vector of the body B and $\mathbf{b}(\mathbf{x}) =[b_1, b_2, b_3]^T$ is the body force, and μ and v designate the shear modulus and Poisson's ratio

$$\mu>0 \qquad 0\leqslant v<\frac{1}{2} \tag{4.2}$$

The strain-displacement relations in the linear theory are

$$\varepsilon_{ij}=\frac{1}{2}[u_{i,j}+u_{j,i}] \tag{4.3}$$

and the stress–strain relations are

$$\sigma_{ij}=2\mu\ \varepsilon_{ij}+\frac{2\mu v}{1-2v}\delta_{ij}\ \varepsilon_{kk} \tag{4.4}$$

The terms σ_{ij} and ε_{ij} are the elements of the strain tensor **E** and the stress tensor **S**. The product

$$\mathbf{S}\cdot\mathbf{n}=\mathbf{t} \tag{4.5}$$

where $\mathbf{n}=[n_1, n_2, n_3]^T$ is the outward normal on the surface S, is called the traction at the point $\mathbf{x}\in$S. The traction $\mathbf{t}(\mathbf{x})=[t_1,t_2,t_3]^T$ is a function of **u** and **n** which is seen immediately if we introduce the 'traction operator'

$$\boldsymbol{\tau}(\cdot)=2\mu\frac{\partial}{\partial n}+\frac{2\mu}{(1-2v)}\mathbf{n}\text{ div}+\mu\ (\mathbf{n}\wedge\text{rot}) \tag{4.6}$$

which applied to **u** yields **t**.

$$\boldsymbol{\tau}(\mathbf{u})=\mathbf{S}\cdot\mathbf{n}=\mathbf{t} \tag{4.7}$$

This operator also appears in the Betti formula and plays the same role as the normal derivative in the theory of Laplace potentials.

A subscript e or i on the operator $\boldsymbol{\tau}$ will, if necessary, designate that the normal **n** on the surface, which appears in the coefficients of $\boldsymbol{\tau}$, points into the exterior region B_e or the interior region B_i.

The displacement fields due to single loads $\mathbf{F}^{(i)}=e_i$ $i=1,2,3$ at the point $\mathbf{x}$ of the elastic continuum B_∞ are the three fundamental (or Kelvin) solutions $\mathbf{g}^{(i)}$ $(\mathbf{y},\mathbf{x})$ $i=1, 2, 3$, which satisfy formally the equation

$$\mathbf{L}\ \mathbf{g}^{(i)}=-\delta(|\mathbf{x}-\mathbf{y}|)\mathbf{e}_i \tag{4.8}$$

They are the three columns of the Kelvin (or Somigliana) matrix

$$\mathbf{U}(\mathbf{y},\mathbf{x})=[\mathbf{g}^{(1)},\ \mathbf{g}^{(2)},\ \mathbf{g}^{(3)}] \tag{4.9}$$

which has the elements

$$\mathrm{U}_{ij}=\frac{1}{16\pi\mu(1-\nu)r}[(3-4\nu)\delta_{ij}+r,_{x_i}\cdot r,_{x_j}] \tag{4.10}$$

and where

$$r=|\mathbf{y}-\mathbf{x}|=\left[\sum_{i=1}^{3}(y_i-x_i)^2\right]^{1/2}$$

Operating on the columns $\mathbf{g}^{(i)}(\mathbf{y},\mathbf{x})$ with the differential operator $\boldsymbol{\tau}$ once with respect to $\mathbf{y}$ and once with respect to $\mathbf{x}$, we obtain the matrices

$$\boldsymbol{\tau}_y[\mathrm{U}]=[\boldsymbol{\tau}_y(\mathbf{g}^{(1)}),\ \boldsymbol{\tau}_y(\mathbf{g}^{(2)}),\ \boldsymbol{\tau}_y(\mathbf{g}^{(3)})]=\mathbf{T}(\mathbf{y},\mathbf{x})^T \tag{4.11}$$

$$\boldsymbol{\tau}_x[\mathrm{U}]=[\boldsymbol{\tau}_x(\mathrm{g}^{(1)}),\ \boldsymbol{\tau}_x(\mathbf{g}^{(2)}),\ \boldsymbol{\tau}_x(\mathbf{g}^{(3)})]=\mathbf{T}^*(\mathbf{x},\mathbf{y}) \tag{4.12}$$

with the elements

$$T_{ij}=-\frac{1}{8\pi(1-\nu)r^2}\left[\frac{\partial r}{\partial n_{\mathbf{y}}}\{(1-2\nu)\delta_{ij}+3r,_{y_i}\ r,_{y_j}\}-\right.$$
$$\left.-(1-2\nu)\{r,_{y_i}\ n_j(\mathbf{y})-r,_{y_j}n_i(\mathbf{y})\}\right] \tag{4.13}$$

$$T^*_{ij}=-\frac{1}{8\pi(1-\nu)r^2}\left[\frac{\partial r}{\partial n_{\mathbf{x}}}\{(1-2\nu)\delta_{ij}+3r,_{x_i}r,_{x_j}\}-\right.$$
$$\left.-(1-2\nu)\{r,_{x_j}n_i(\mathbf{x})-r,_{x_i}n_j(\mathbf{x})\}\right] \tag{4.14}$$

(It is $\dfrac{\partial r}{\partial n_{\mathbf{x}}}:=\mathrm{grad}_{\mathbf{x}}^T r\cdot\mathbf{n}(\mathbf{x})$, $\dfrac{\partial r}{\partial n_{\mathbf{y}}}:=\mathrm{grad}_{\mathbf{y}}^T r\cdot\mathbf{n}(\mathbf{y})$)

$\mathbf{n}(\mathbf{x})$ resp. $\mathbf{n}(\mathbf{y})$ means, that the normal runs with $\mathbf{x}$ resp. $\mathbf{y}$. In $\mathbf{T}$ it runs with $\mathbf{y}$, in $\mathbf{T}^*$ with $\mathbf{x}$, which we indicate by setting the variable in the first place in the ordered pairs $(\mathbf{y},\mathbf{x})$ resp. $(\mathbf{x},\mathbf{y})$ which follow the symbols **T** and **T***. As **T** and **T*** depend on the normal **n** the subscripts e and i will be used in the sense defined above, wherever necessary. Note that we designate by $\mathbf{T}(\mathbf{y},\mathbf{x})$, (as everybody does), the transpose of the matrix $\boldsymbol{\tau}_{\mathbf{y}}[\mathrm{U}]$. This is expedient, when it comes to the derivation of the Somigliana identity from three scalar equations.

The formulae of the basic equations for two-dimensional elasticity are equal to those in three-dimensional space, but the fundamental solutions are not equal. The same matrices as above have in two-dimensional space the following elements (plane strain):

$$\mathrm{U}_{ij}=\frac{1}{8\pi\mu(1-\nu)}\left[(3-4\nu)\ \ln\frac{1}{r}\delta_{ij}+r_{,x_i}r_{,x_j}\right] \tag{4.15}$$

$$T_{ij}=-\frac{1}{4\pi(1-\nu)r}\left[\frac{\partial r}{\partial n_{\mathbf{y}}}\{(1-2\nu)\delta_{ij}+2r_{,y_i}r_{,y_j}\}-\right.$$
$$\left.(1-2\nu)\{r_{,y_i}n_j(\mathbf{y})-r_{,y_j}n_i(\mathbf{y})\}\right] \tag{4.16}$$

$$T^*_{ij}=-\frac{1}{4\pi(1-\nu)r}\left[\frac{\partial r}{\partial n_{\mathbf{x}}}\{(1-2\nu)\delta_{ij}+2r_{,x_i}r_{,x_j}\}-\right.$$
$$\left.(1-2\nu)\{r_{,x_j}n_i(\mathbf{x})-r_{,x_i}n_j(\mathbf{x})\}\right] \tag{4.17}$$

4.2.2 **Regular regions**

Surfaces or more general boundaries play an important part in boundary element methods and so we better give a precise definition of what we like to call a regular boundary. We start to this end with the concept of a regular region.

We call a plane region $B\subset R^2$ a regular region, if (1) B or its complement $CB=R^2/\bar{B}$ is bounded and (2) the boundary S consists of closed piecewise smooth ($x_i(s)\in C^1$, $i=1{,}2$) curves Γ_j. $S=\cup\Gamma_j$. A three-dimensional region $B\in R^3$ is called a regular region, if (1) B or its complement $CB=R^3/\bar{B}$ is bounded and (2) the boundary S consists of closed surfaces S_j, $S=\cup S_j$, which themselves are for any j the union of finite many smooth surface elements $\dot{G}_{ij}$, $S_j=\cup\dot{G}_{ij}$, that are parts of surfaces whose curvature is continuous, i.e. $\dot{G}_{ij}\in C^2$.

The boundaries of regular regions are called regular surfaces or regular curves.

A halfspace and halfplane are not regular regions but any other ‘real life’ region belongs to this class and especially the boundaries, which are generated by continuous polynomial shape functions, are regular boundaries.

The region bounded by S is called the interior region B_i and the region exterior to it B_e, it contains the point at infinity.

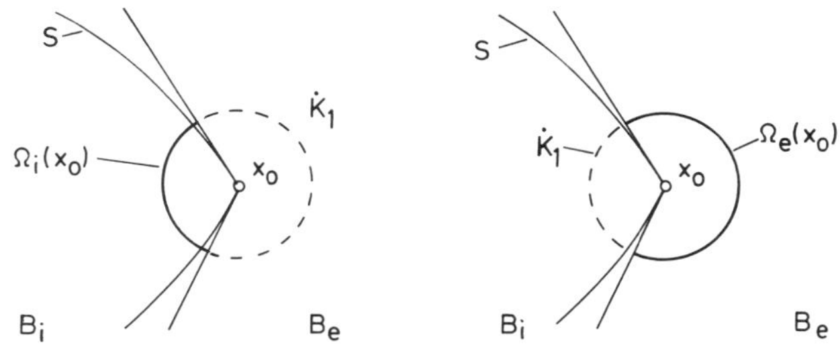

Figure 4.1 Sectional view of a boundary point x on S

A surface S has two sides S_i and S_e, which are the boundary of B_i and B_e respectively (Figure 4.1). The half tangents of a boundary point $\mathbf{x}$ on S form the mantle of a tangent pyramid, that has its vertex at $\mathbf{x}$ and which intersects the unit sphere $\dot{K}_1$ placed around $\mathbf{x}$ and so decomposes it in two surfaces $\Omega_i(\mathbf{x})$ and $\Omega_e(\mathbf{x})$. The index i (e) indicates the surface, that lies for most of its part in B_i (B_e). We call these the characteristic surfaces of the boundary point $\mathbf{x}$.

Let $\varphi_i(\mathbf{x})$ and $\varphi_e(\mathbf{x})$ denote the interior and exterior solid angle of $\mathbf{x}$. It is

$$\varphi_i(\mathbf{x})+\varphi_e(\mathbf{x})=\int_{\Omega_i(\mathbf{x})}|\mathrm{d}s|+\int_{\Omega_e(\mathbf{x})}|\mathrm{d}s|=4\pi \tag{4.18}$$

$$\Omega_i(\mathbf{x})\cup\Omega_e(\mathbf{x})=\dot{K}_1 \text{ (unit sphere)} \tag{4.19}$$

In two dimensions the situation is similar. The two half-tangents of a point $\mathbf{x}$ on the curve S (see Figure 4.1 which might serve too for illustrating the situation in 2-D) cut the unit circle $\dot{K}_1$ placed around $\mathbf{x}$ in two points, which divide $\dot{K}_1$ in two connected parts: $\Omega_i(\mathbf{x})$ and $\Omega_e(\mathbf{x})$, where $e(i)$ denotes the arc, that lies for most of its part in B_e (B_i). If φ_i and φ_e denote interior and exterior angle of the point $\mathbf{x}$, we have

$$\varphi_i(\mathbf{x})+\varphi_e(\mathbf{x})=\int_{\Omega_i(\mathbf{x})}|\mathrm{d}s|+\int_{\Omega_e(\mathbf{x})}|\mathrm{d}s|=2\pi \tag{4.20}$$

$$\Omega_i(\mathbf{x})\cup\Omega_e(\mathbf{x})=\dot{K}_1 \text{ (unit circle)} \tag{4.21}$$

Suppose S is a regular surface (regular curve). We introduce two characteristic functions

$$\mathbf{C}_i(\mathbf{x})=\begin{cases}\mathbf{I} & \mathbf{x}\in B_i\\ \mathbf{C}_i(\mathbf{x}) & \mathbf{x}\in S\\ \mathbf{O} & \mathbf{x}\in B_e\end{cases} \qquad \mathbf{C}_e(\mathbf{x})=\begin{cases}\mathbf{O} & \mathbf{x}\in B_i\\ \dot{\mathbf{C}}_e(\mathbf{x}) & \mathbf{x}\in S\\ \mathbf{I} & \mathbf{x}\in B_e\end{cases} \tag{4.22}$$

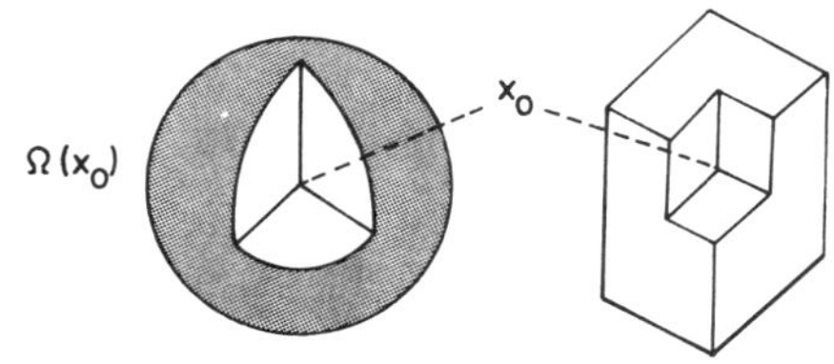

Figure 4.2 Re-entrant corner of a body and its characteristic surface (shaded part) on the unit sphere $\dot{K}_1$

with

$$\dot{\mathbf{C}}_i(\mathbf{x}) = \int_{\Omega_i(\mathbf{x})} \mathbf{T}(\mathbf{y},\mathbf{x})ds_y \tag{4.23}$$

$$\dot{\mathbf{C}}_e(\mathbf{x}) = \int_{\Omega_e(\mathbf{x})} \mathbf{T}(\mathbf{y},\mathbf{x})ds_y \tag{4.24}$$

In Equations 4.23 and 4.24 the normal **n** points to the center **x** of the sphere (This is easily understood if one remembers how the Somigliana identity has been derived.)

At any point **y** on $\dot{K}_1$ we have with $r = |\mathbf{y} - \mathbf{x}|$

$$\text{grad}_y\, r = -\mathbf{n} \tag{4.25}$$

and therefore it is

$$\frac{\partial \mathbf{r}}{\partial \mathbf{n}} = -1, \quad \text{grad}_y\, r \wedge \mathbf{n} = \mathbf{0} \tag{4.26}$$

and so we have in 3-D,

$$T_{ij} = \frac{1}{8\pi(1-\nu)}\{(1-2\nu)\delta_{ij} + 3r,_{y_i}r,_{y_j}\} \quad \mathbf{y} \in \dot{K}_1(\mathbf{x}) \tag{4.27}$$

and in 2-D,

$$T_{ij} = \frac{1}{4\pi(1-\nu)}\{(1-2\nu)\delta_{ij} + 2r,_{y_i}r,_{y_j}\} \quad \mathbf{y} \in \dot{K}_1(\mathbf{x}) \tag{4.28}$$

In 2-D, $\dot{\mathbf{C}}(\mathbf{x})$ is a curve integral. The integration has to be done clockwise on $\dot{K}_1$ (then the region B lies to the left if we progress on $\Omega(\mathbf{x})$).

We present three examples for $\dot{\mathbf{C}}(\mathbf{x})$ in 3-D (Figure 4.3).

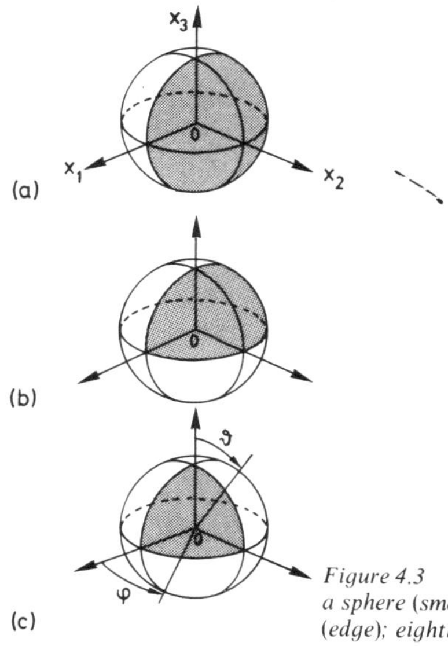

Figure 4.3 *Integration over shaded part:* (*a*) *half of a sphere* (*smooth boundary*); (*b*) *quarter of a sphere* (*edge*); *eighth of a sphere* (*corner*)

(1) Half of a sphere (smooth boundary)

$$\dot{\mathbf{C}} = \frac{1}{8\pi}\begin{bmatrix} 4\pi & 0 & 0 \\ 0 & 4\pi & 0 \\ 0 & 0 & 4\pi \end{bmatrix}$$

(2) Quarter of a sphere (edge)

$$\dot{\mathbf{C}} = \frac{1}{8\pi}\begin{bmatrix} 2\pi & 0 & 0 \\ 0 & 2\pi & \dfrac{2}{1-\nu} \\ 0 & \dfrac{2}{1-\nu} & 2\pi \end{bmatrix}$$

(3) Eighth of a sphere (corner)

$$\dot{\mathbf{C}} = \frac{1}{8\pi}\begin{bmatrix} \pi & \dfrac{1}{1-\nu} & \dfrac{1}{1-\nu} \\ \dfrac{1}{1-\nu} & \pi & \dfrac{1}{1-\nu} \\ \dfrac{1}{1-\nu} & \dfrac{1}{1-\nu} & \pi \end{bmatrix}$$

Contrary to the situation in 3-D, where the calculation of $\dot{\mathbf{C}}(\mathbf{x})$ is in general

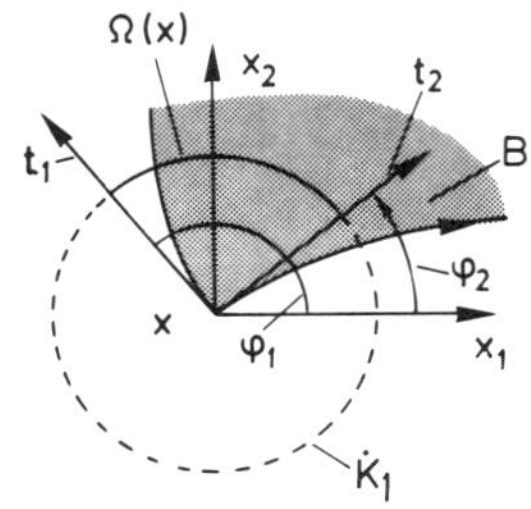

Figure 4.4 $\Omega(x)$ *is the arc between* t_1 *and* t_2

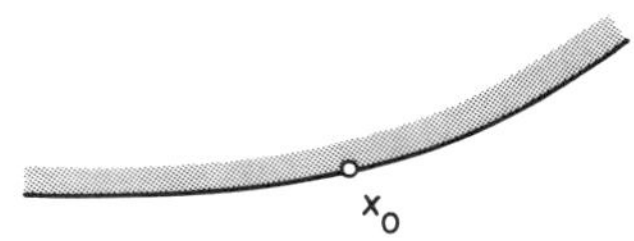

Figure 4.5

only possible if the boundary of $\Omega(\mathbf{x})$ on $\dot{K}_1$ consists of lines $\varphi = c.$, $v = c.$, the calculation in 2-D is possible at any point. To derive a general expression we introduce some notations. At the boundary point $\mathbf{x} \in S$ exist two half-tangents, $\mathbf{t}_1$ and $\mathbf{t}_2$, which are the first resp. second tangent touched on running through $\mathbf{x}$ (Figure 4.4). Let φ_1 denote the angle between the tangent $\mathbf{t}_1$ and the axis x_1, measured in the positive sense. With $\Delta\varphi$, the length of the characteristic arc (it is the angle between $\mathbf{t}_1$ and $\mathbf{t}_2$), we define the angle $\varphi_2 = \varphi_1 - \Delta\varphi$

$$\dot{\mathbf{C}}(\mathbf{x}) = \int_{\Omega(\mathbf{x})} \mathbf{T}(\mathbf{y},\mathbf{x})\, ds_y =$$

$$= \frac{1}{4\pi(1-v)} \begin{bmatrix} 2(1-v)(\varphi_1 - \varphi_2) + \frac{1}{2}[\sin 2\varphi_1 - \sin 2\varphi_2], & \sin^2\varphi_1 - \sin^2\varphi_2 \\ \text{sym.} & 2(1-v)(\varphi_1 - \varphi_2) - \frac{1}{2}[\sin^2\varphi_1 - \sin^2\varphi_2] \end{bmatrix} \quad (4.29)$$

We present three examples:

(1) $\mathbf{x}$ is a smooth boundary point (Figure 4.5)

$$\dot{\mathbf{C}}(\mathbf{x}) = \frac{1}{2}\begin{bmatrix} 1 & 0 \\ 0 & 1 \end{bmatrix} = \frac{1}{2}\mathbf{I}$$

(2) $\mathbf{x}$ lies in a corner of B (Figure 4.6),

$$\dot{\mathbf{C}}(\mathbf{x}) = \begin{bmatrix} \frac{1}{4} & -\frac{1}{4\pi(1-v)} \\ \text{sym.} & \frac{1}{4} \end{bmatrix}$$

$$\dot{\mathbf{C}}(\mathbf{x}) = \begin{bmatrix} \frac{1}{4} & \frac{1}{4\pi(1-v)} \\ \text{sym.} & \frac{1}{4} \end{bmatrix}$$

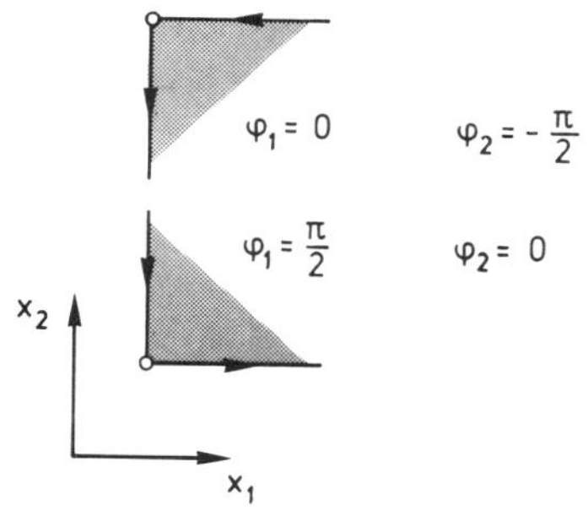

Figure 4.6

The last two examples are to illustrate that $\dot{\mathbf{C}}(\mathbf{x})$ not only depends on the size of $\Omega(\mathbf{x})$ but also on the orientation of $\Omega(\mathbf{x})$ with respect to the global coordinate system whose axes are parallel with the axes of the system at the boundary point $\mathbf{x}$ and has the same orientation.

4.2.3 **Boundary value problems**

We give a short list of some boundary value problems of elasticity in terms of the displacement vector $\mathbf{u}(\mathbf{x})$. Because the differential equation is equal for all problems

$$\mathbf{L}\mathbf{u} = -\mathbf{b} \tag{4.30}$$

where $\mathbf{b}(\mathbf{x})$ is the vector of body forces, we list only the different boundary conditions.

A subscript i or e on $\boldsymbol{\tau}$ denotes that the normal points into B_i or B_e. $\lim\limits_i$ and $\lim\limits_e$ denote interior and exterior limits respectively. A tilde designates a prescribed function.

$$
\begin{aligned}
&\text{1st interior problem } [D_i, \tilde{\mathbf{u}}], &&\lim_i \mathbf{u} = \tilde{\mathbf{u}} = [\tilde{u}_1, \tilde{u}_2, \tilde{u}_3]^T \\
&\text{1st exterior problem } [D_e, \tilde{\mathbf{u}}], &&\lim_e \mathbf{u} = \tilde{\mathbf{u}} = [\tilde{u}_1, \tilde{u}_2, \tilde{u}_3]^T \\
&\text{2nd interior problem } [T_i, \tilde{\mathbf{t}}], &&\lim_i \boldsymbol{\tau}_e(\mathbf{u}) = \tilde{\mathbf{t}} = [\tilde{t}_1, \tilde{t}_2, \tilde{t}_3]^T \\
&\text{2nd exterior problem } [T_e, \tilde{\mathbf{t}}], &&\lim_e \boldsymbol{\tau}_i(\mathbf{u}) = \tilde{\mathbf{t}} = [\tilde{t}_1, \tilde{t}_2, \tilde{t}_3]^T
\end{aligned}
\tag{4.31}
$$

3rd interior problem,

$$
\lim_i \begin{cases} \boldsymbol{\tau}_e(\mathbf{u})^T \cdot \mathbf{a}_{(1)} = \tilde{t}_1 \\ \boldsymbol{\tau}_e(\mathbf{u})^T \cdot \mathbf{a}_{(2)} = \tilde{t}_2 \\ \mathbf{u}^T \cdot \mathbf{a}_{(3)} = \tilde{u}_3 \end{cases} \tag{4.32}
$$

where $\mathbf{a}_{(1)}, \mathbf{a}_{(2)}, \mathbf{a}_{(3)}$ are the three basis vectors of the surface S at $\mathbf{x}$. The vectors $\mathbf{a}_{(1)}$ and $\mathbf{a}_{(2)}$ span the tangent plane at $\mathbf{x}$ and it is $\mathbf{a}_{(3)}=\mathbf{n}$, the unit outward normal.

4th interior problem,

$$\lim_{i} \begin{cases} \mathbf{u}^{T}\cdot \mathbf{a}_{(1)}=\tilde{u}_1 \\ \mathbf{u}^{T}\cdot \mathbf{a}_{(2)}=\tilde{u}_2 \\ \boldsymbol{\tau}_e(\mathbf{u})^{T}\cdot \mathbf{a}_{(3)}=\tilde{t}_3 \end{cases} \tag{4.33}$$

5th interior mixed problem $[M_i,\tilde{\mathbf{t}},\tilde{\mathbf{u}}]$,

$$\lim_{i} \boldsymbol{\tau}_e(\mathbf{u})=\tilde{\mathbf{t}} \text{ on } S_1, \quad \lim_{i} \mathbf{u}=\tilde{\mathbf{u}} \text{ on } S_2 \tag{4.34}$$

where $S=S_1 \cup S_2$ is the boundary of B. The formulation of the last three exterior problems is analogous to the interior problem. While the solutions of well posed interior problems differ at most by rigid body movements, exterior problems have unique solutions, if we impose the following radiation conditions

$$u_i(\mathbf{x})=0\left(\frac{1}{|\mathbf{x}|}\right) \qquad \mathbf{u}_{i,x_j}=0\left(\frac{1}{|\mathbf{x}|^2}\right) \tag{4.35}$$

for $|\mathbf{x}|\rightarrow\infty$. If $\mathbf{u}(\mathbf{x})$ is a regular solution, see Section 4.4, and satisfies these conditions in B_e, then the Somigliana identity, Theorem 4.2 holds. A second radiation condition exists for exterior body forces which we state in the following context: if there exists a constant C, such that

$$|\mathbf{b}(\mathbf{y})|<\frac{C}{|\mathbf{x}-\mathbf{y}|^3} \text{ for } |\mathbf{y}|\rightarrow\infty \tag{4.36}$$

where $\mathbf{x}$ is any arbitrary point in B_e, then the function

$$\mathbf{P}(\mathbf{x})=\int_{B_e} \mathbf{U}(\mathbf{y},\mathbf{x})\mathbf{b}(\mathbf{y})\, \mathrm{d}v_{\mathbf{y}} \tag{4.37}$$

exists at any point $\mathbf{x}\in B_e$. The conditions will be refered to as radiation conditions (A), Equation 4.35, and (B), Equation 4.36.

Besides radiation conditions we will often impose smoothness conditions on the functions. The symbol $C^n(A)$ denotes the class of functions whose n-th partial derivatives are continuous in the domain A, i.e. are elements of $C^0(A)$ $=C(A)$. The symbol $C^{0,\gamma}(A)$ denotes the class of Hölder-continuous functions and it satisfies the inclusions:

$$C^1(A) \subset C^{0,\gamma}(A) \subset C(A)$$

Many proofs in potential theory do not hold for functions which are only continuous, they must be at least Hölder-continuous, that is why this class came into being. Note that any polynomial is Hölder-continuous.

4.2.4 A maximum principle

It is well known, that scalar valued solutions of elliptic equations of 2nd order, e.g. harmonic functions, satisfy in a closed region B a maximum principle. Fichera[1] proved that such a principle also holds for regular displacement fields, i.e. solutions of the equation $\mathbf{Lu} = \mathbf{0}$.

Theorem 2.1

Suppose: (1) $\mathbf{u}(\mathbf{x}) \in C^2(B) \cap C^1(\bar{B})$, (2) $\mathbf{Lu} = \mathbf{0}$ and (3) the surface S is regular, then there exists a constant H, such that

$$\max_{\mathbf{x} \in B} |\mathbf{u}(\mathbf{x})| < H \cdot \max_{\mathbf{x} \in S} |\mathbf{u}(\mathbf{x})| \tag{4.38}$$

holds. The constant H depends only on B and the moduli μ and ν can be calculated.

A consequence of this theorem is that the biggest error in a solution with boundary elements (suppose $H \leqslant 1$) occurs on the boundary if the body force $\mathbf{b}$ is zero. This is easily seen as follows: suppose $\mathbf{u}$ satisfies the equation $\mathbf{Lu} = \mathbf{0}$ and is the exact solution of the boundary value problem and $\bar{\mathbf{u}}$ is an approximate solution constructed with potentials of the first and second kind, then $\mathbf{u} - \bar{\mathbf{u}}$ too is a homogeneous solution, $\mathbf{L}(\mathbf{u} - \bar{\mathbf{u}}) = \mathbf{0}$. Therefore Theorem 2.1 applies, i.e.

$$\max_{\mathbf{x} \in B} |\mathbf{u} - \bar{\mathbf{u}}| < H \max_{\mathbf{x} \in S} |\mathbf{u} - \bar{\mathbf{u}}| \tag{4.39}$$

For other properties of regular solutions, we refer to Gurtin[2] and especially Adler[3].

4.3 ELASTIC POTENTIALS

In this section we introduce elastic potentials and give a description of the indirect method. This method is not as versatile as the direct method and suffers some drawbacks, but its introduction gives us a chance to study the properties of the kernels $\mathbf{U}(\mathbf{y},\mathbf{x})$, $\mathbf{T}(\mathbf{y},\mathbf{x})$ and $\mathbf{T}^*(\mathbf{x},\mathbf{y})$, and the potentials they generate. These potentials will be encountered again in the equations of the direct method.

4.3.1 Properties of elastic potentials

Suppose S is a regular surface and $\mathbf{v}(\mathbf{y})=[v_1,v_2,v_3]^T$ and $\mathbf{w}(\mathbf{y})=[w_1,w_2,w_3]^T$ are two smooth functions on S. The function $\mathbf{V}(\mathbf{x})=[V_1,V_2,V_3]^T$

$$\mathbf{V}(\mathbf{x})=\oint_S \mathbf{U}(\mathbf{y},\mathbf{x})\mathbf{v}(\mathbf{y})\ ds_y \tag{4.40}$$

is called a potential of the 1st kind (single-layer potential) and the function $\mathbf{W}(\dot{\mathbf{x}})=[W_1,W_2,W_3]^T$

$$\mathbf{W}(\mathbf{x})=\oint_S \mathbf{T}_e(\mathbf{y},\mathbf{x})\mathbf{w}(\mathbf{y})\ ds_y \tag{4.41}$$

a potential of the 2nd kind (double-layer potential). Let $\mathbf{b}(\mathbf{y})=[b_1,b_2,b_3]^T$ be a smooth function defined in B. If $B=B_e$, then $\mathbf{b}(\mathbf{y})$ is also supposed to satisfy the radiation condition (B), see Section 4.2.3. The function $\mathbf{P}(\mathbf{x})=[P_1,P_2,P_3]^T$

$$\mathbf{P}(\mathbf{x})=\int_B \mathbf{U}(\mathbf{y},\mathbf{x})\mathbf{b}(\mathbf{y})\ dv_y \tag{4.42}$$

is called a volume potential.

The most interesting feature of these potentials is that they satisfy the differential Equation 4.1. We have

$$\mathbf{LV}(\mathbf{x})=\mathbf{0},\quad \mathbf{LW}(\mathbf{x})=\mathbf{0},\quad \mathbf{x}\in R^3/S \tag{4.43}$$

$$\mathbf{LP}(\mathbf{x})=-\mathbf{b}(\mathbf{x})\quad \mathbf{x}\in B \tag{4.44}$$

Due to the last equation we say that the volume potential has a reproducing kernel with respect to the operator $\mathbf{L}$. If the body force $\mathbf{b}(\mathbf{y})$ is the gradient of a scalar valued function, which itself is the solution of a Poisson equation then the volume integral can be reduced, as the next theorem tells us, to a surface integral.

Theorem 3.1

Suppose there exists a scalar valued function $\psi(\mathbf{x})$ in the region B with surface S, such that $\Delta\psi=k_0$, k_0 a constant, and $\mathbf{b}=[b_1,b_2,b_3]^T=\nabla\psi$ $=[\psi_{,x_1},\psi_{,x_2},\psi_{,x_3}]$ then the function

$$\mathbf{P}(\mathbf{x})=\int_B \mathbf{U}(\mathbf{y},\mathbf{x})\mathbf{b}(\mathbf{y})\ dv_y \tag{4.45}$$

can be written as

$$\mathbf{P}(\mathbf{x}) = a \cdot \oint_S [\psi(\mathbf{y}) \cdot \mathbf{k}(\mathbf{y},\mathbf{x}) - \frac{\partial \psi}{\partial n_{\mathbf{y}}} \nabla_{\mathbf{y}} r + k_0 r \; \mathbf{n}(\mathbf{y})] \; \mathrm{d}s_{\mathbf{y}} \tag{4.46}$$

Here $\mathbf{k}(\mathbf{y},\mathbf{x}) = [k_1, k_2, k_3]^T$ designates the vector

$$k_i = \left[\frac{\partial}{\partial n_{\mathbf{y}}} (r_{,y_i}) - \frac{1}{a} U_{ij}(\mathbf{y},\mathbf{x}) n_j(\mathbf{y}) \right] \tag{4.47}$$

with the constant $a = \dfrac{1-2v}{16\pi\mu(1-v)}$ and the functions $\nabla_{\mathbf{y}} r = [r_{,y_1}, r_{,y_2}, r_{,y_3}]^T$ and $\mathbf{n}(\mathbf{y}) = [n_1, n_2, n_3]^T$.

For a proof we refer to Stippes and Rizzo[4]. A similar reduction to a curve integral in 2-D is also possible. Before proceeding we would like to mention that a problem with non-zero body forces $\mathbf{b} \neq \mathbf{0}$ can be reduced to a homogeneous problem, $\mathbf{L}\mathbf{u} = \mathbf{0}$, if we write

$$\mathbf{u}(\mathbf{x}) = \mathbf{P}(\mathbf{x}) + \bar{\mathbf{u}}(\mathbf{x}) = \int_B \mathbf{U}(\mathbf{y},\mathbf{x}) \mathbf{b}(\mathbf{y}) \; \mathrm{d}v_{\mathbf{y}} + \bar{\mathbf{u}}(\mathbf{x}) \tag{4.48}$$

with a yet unknown function $\bar{\mathbf{u}}$, which satisfies the equation $\mathbf{L}\bar{\mathbf{u}} = \mathbf{0}$ and whose boundary values match the boundary values of $\mathbf{P}$, i.e., the sum $\mathbf{u} = \mathbf{P} + \bar{\mathbf{u}}$ satisfies the boundary conditions for $\mathbf{u}$. We have seen above that the potentials $\mathbf{V}(\mathbf{x})$ and $\mathbf{W}(\mathbf{x})$ satisfy the equations $\mathbf{L}\mathbf{V} = \mathbf{0}$, $\mathbf{L}\mathbf{W} = \mathbf{0}$ and that this property does not depend on the choice of the layers. This choice becomes important if we want to satisfy the boundary conditions, which lead, due to the special character of the potentials, to integral equations of the first or second kind on the boundary. The following theorem will help to formulate these integral equations.

Theorem 3.2

Suppose S is a regular surface and $\mathbf{x}_0$ an arbitrary point on S, which is the limit of a sequence of points $\mathbf{x}$, which converge to $\mathbf{x}_0$ from the inside or the outside. The following statements are true:

(1) If $\mathbf{v}(\mathbf{y}) \in L_\infty(S)$, (e.g. a piecewise continuous function) then $\mathbf{V}(\mathbf{x})$ is continuous everywhere.

(2) If $\mathbf{w}(\mathbf{y}) \in C^{0,\gamma}(S)$, then

$$\lim_i \mathbf{W}(\mathbf{x}) = \mathbf{W}(\mathbf{x}_0) - \dot{\mathbf{C}}_e(\mathbf{x}_0) \mathbf{w}(\mathbf{x}_0) \tag{4.49}$$

$$\lim_e \mathbf{W}(\mathbf{x}) = \mathbf{W}(\mathbf{x}_0) + \dot{\mathbf{C}}_i(\mathbf{x}_0) \mathbf{w}(\mathbf{x}_0) \tag{4.50}$$

See Equations 4.23 and 4.24 for $\dot{\mathbf{C}}_i(\mathbf{x})$, $\dot{\mathbf{C}}_e(\mathbf{x})$.

(3) The function on the right-hand side of Equation 4.49 and 4.50 is a continuous function on S.

(4) If $\mathbf{w}(\mathbf{y}) \in C^{1,\gamma}(S)$, then

$$\lim_i \boldsymbol{\tau}_e(\mathbf{W})(\mathbf{x}) = \lim_e \boldsymbol{\tau}_e(\mathbf{W})(\mathbf{x}) \tag{4.51}$$

at any smooth point on S.

(5) if $\mathbf{x}_0$ is a smooth boundary point (the normal is continuous at $\mathbf{x}_0$) and $\mathbf{v}(\mathbf{y}) \in C^{0,\gamma}(S)$, then

$$\lim_i \boldsymbol{\tau}_e(\mathbf{V})(\mathbf{x}) = \boldsymbol{\tau}_e(\mathbf{V})(\mathbf{x}_0) + \frac{1}{2}\mathbf{v}(\mathbf{x}_0) \tag{4.52}$$

$$\lim_e \boldsymbol{\tau}_e(\mathbf{V})(\mathbf{x}) = \boldsymbol{\tau}_e(\mathbf{V})(\mathbf{x}_0) - \frac{1}{2}\mathbf{v}(\mathbf{x}_0) \tag{4.53}$$

with

$$\boldsymbol{\tau}_e(\mathbf{V})(\mathbf{x}_0) = \oint_S \mathbf{T}_e^*(\mathbf{x}_0, \mathbf{y})\mathbf{v}(\mathbf{y})\, ds_y \tag{4.54}$$

(6) If $\mathbf{x}_0$ is a non-smooth boundary point, i.e. the normal is not continuous in $\mathbf{x}_0$ and the layer $\mathbf{v}(\mathbf{y})$ has no zero at $\mathbf{x}_0$, i.e. $\mathbf{v}(\mathbf{x}_0) \neq \mathbf{0}$, then the function $\boldsymbol{\tau}(\mathbf{V})(\mathbf{x})$, $\mathbf{x} \in S$, has in general at $\mathbf{x}_0$ a logarithmic singularity:

$$\boldsymbol{\tau}(\mathbf{V})(\mathbf{x}) = 0(\ln |\mathbf{x} - \mathbf{x}_0|^2) \tag{4.55}$$

The restriction 'in general' exempts non-smooth points whose neighbourhood on S satisfies a certain symmetry condition usually not encountered in edges and vertices of standard problems. With Equation 4.55 it follows that the function on the right hand-side of Equation 4.52 and 4.53 becomes (if $\mathbf{v}(\mathbf{y})$ is bounded) singular, if $\mathbf{x} \in S$ approaches a non-smooth point on S. For a proof of theorem 3.2 we refer to Hartmann[5], Chapter 5.

Figures 4.7 and 4.8 try to visualize the behaviour of the potentials near the surface S and in smooth points of S. The potential of the first kind, $\mathbf{V}(\mathbf{x})$, is shown in Figure 4.7 and the potential of the second kind, $\mathbf{W}(\mathbf{x})$, in Figure 4.8.

4.3.2 The indirect method

After these preparatory results we turn to the actual solution of boundary value problems with elastic potentials in the sense of the so called indirect method.

Suppose we handle the problems as follows:

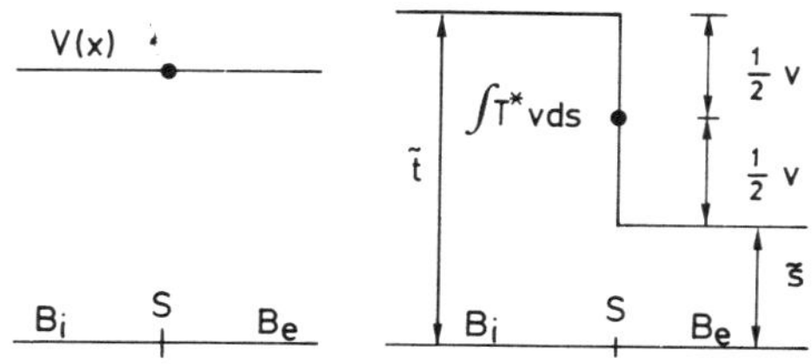

Figure 4.7 Potential of the first kind:

$$\lim_i \tau'_e(V)(x) = \frac{1}{2}v + \int T^* v ds = \tilde{t}$$

$$\lim_i V(x) = \lim_e V(x)$$

$$\lim_e \tau_e(V)(x) = -\frac{1}{2}v + \int T^* v ds = \tilde{s}$$

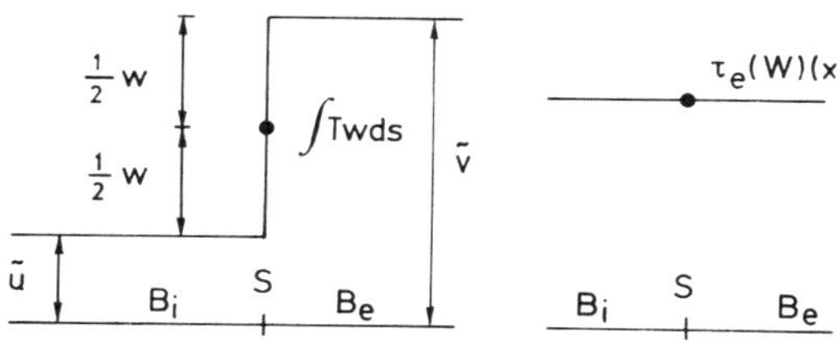

Figure 4.8 Potential of the second kind:

$$\lim_i W(x) = -\frac{1}{2}w + \int Twds = \tilde{u}$$

$$\lim_i \tau_e(W) = \lim_e \tau_e(W)$$

$$\lim_e W(x) = +\frac{1}{2}w + \int Twds = \tilde{v}$$

Problem (interior and exterior)	Solution
1st $[D,\hat{\mathbf{u}}]$	$\mathbf{W}(\mathbf{x})$ Potential of the 2nd kind
2nd $[T,\hat{\mathbf{t}}]$	$\mathbf{V}(\mathbf{x})$ Potential of the 1st kind
Mixed $[M,\hat{\mathbf{t}},\hat{\mathbf{u}}]$	$\mathbf{V}(\mathbf{x})$ Potential of the 1st kind

then the boundary conditions lead, if we consider the jumps occurring on S, see Theorem 3.2, to the following integral equations for the layers $\mathbf{w}(\mathbf{y})$ and $\mathbf{v}(\mathbf{y})$.

Problem $[D_i,\hat{\mathbf{u}}]$

$$-\dot{\mathbf{C}}_e(\mathbf{x})\mathbf{w}(\mathbf{x}) + \oint_S \mathbf{T}_e(\mathbf{y},\mathbf{x})\mathbf{w}(\mathbf{y})\, ds_y = \hat{\mathbf{u}}(\mathbf{x}) \qquad (4.56)$$

Problem $[D_e, \tilde{\mathbf{u}}]$

$$\dot{\mathbf{C}}_i(\mathbf{x})\mathbf{w}(\mathbf{x}) + \oint_S \mathbf{T}_e(\mathbf{y},\mathbf{x})\mathbf{w}(\mathbf{y})\, ds_y = \tilde{\mathbf{u}}(\mathbf{x}) \tag{4.57}$$

Problem $[T_i, \tilde{\mathbf{t}}]$ (at a smooth point $\mathbf{x}$ on S)

$$\frac{1}{2}\mathbf{v}(\mathbf{x}) + \oint_S \mathbf{T}_e^*(\mathbf{x},\mathbf{y})\mathbf{v}(\mathbf{y})\, ds_y = \tilde{\mathbf{t}}(\mathbf{x}) \tag{4.58}$$

Problem $[T_e, \tilde{\mathbf{t}}]$ (at a smooth point $\mathbf{x}$ on S)

$$-\frac{1}{2}\mathbf{v}(\mathbf{x}) + \oint_S \mathbf{T}_e^*(\mathbf{x},\mathbf{y})\mathbf{v}(\mathbf{y})\, ds_y = -\tilde{\mathbf{t}}(\mathbf{x}) \tag{4.59}$$

Mixed interior problem $[M_i, \tilde{\mathbf{t}}, \tilde{\mathbf{u}}]$

$$\oint_S \mathbf{U}(\mathbf{y},\mathbf{x})\mathbf{v}(\mathbf{y})\, ds_y = \tilde{\mathbf{u}}(\mathbf{x}), \qquad \mathbf{x}\in S_1 \tag{4.60a}$$

$$\frac{1}{2}\mathbf{v}(\mathbf{x}) + \oint_S \mathbf{T}_e^*(\mathbf{x},\mathbf{y})\mathbf{v}(\mathbf{y})\, ds_y = \tilde{\mathbf{t}}(\mathbf{x}), \qquad \mathbf{x}\in S_2 \qquad S = S_1 \cup S_2 \tag{4.60b}$$

These equations are (except Equation 4.60(a), which is weakly singular) strongly singular integral equations on S owing to the nature of the kernels $\mathbf{T}$ and $\mathbf{T}^*$ and the integrals are therefore to be understood in the sense of a Cauchy principal value. While $\mathbf{W}(\mathbf{x})$ exists on S at non-smooth points, the function $\boldsymbol{\tau}(\mathbf{V})(\mathbf{x})$ does not exist (in general) at such points, see Theorem 3.2 and therefore the Equations 4.58 and 4.59 hold only at smooth points. This is a severe drawback of the indirect method, as far as the problem $[T, \tilde{\mathbf{t}}]$ is concerned.

We do not want to miss the chance to present the results of Mikhlin[6] concerning the integral equations of the problems $[D, \tilde{\mathbf{u}}]$ and $[\mathbf{T}, \tilde{\mathbf{t}}]$ on smooth regular surfaces. They are an example of what one would like to have in the more general case of piecewise smooth surfaces.

Suppose now that S is smooth, then it is $\dot{\mathbf{C}}(\mathbf{x}) = 1/2\,\mathbf{I}$ for all $\mathbf{x}\in S$. The integral operators $(\lambda\mathbf{I} + \mathbf{K})$ and $(\lambda\mathbf{I} + \mathbf{K}^*)$ on $C^{0,\gamma}(S)$, which appear in Equations 4.56–4.59:

$$(\lambda\mathbf{I} + \mathbf{K})\mathbf{w} = \lambda\mathbf{w} + \int_S \mathbf{T}\mathbf{w}\, d\mathbf{s}_y \tag{4.61}$$

$$(\lambda \mathbf{I} + \mathbf{K}^*)\mathbf{v} = \lambda \mathbf{v} + \int_S \mathbf{T}^* \mathbf{v} \, ds_y \tag{4.62}$$

are adjoint in the sense of the scalar product $(\mathbf{f},\mathbf{g})$ on $C^{0,\gamma}(S)$

$$(\mathbf{f},\mathbf{g}) := \int_S \mathbf{f}(\mathbf{y})\mathbf{g}(\mathbf{y}) \, ds \tag{4.63}$$

if λ assumes the values $\lambda = \pm 1/2$ and satisfy the Fredholm alternative, which implies that the equations

$$(\lambda \mathbf{I} + \mathbf{K})\mathbf{w} = \hat{\mathbf{u}} \tag{4.64}$$

$$(\lambda \mathbf{I} + \mathbf{K}^*)\mathbf{v} = \hat{\mathbf{t}} \tag{4.65}$$

have solutions, if and only if, $(\hat{\mathbf{u}},\mathbf{s}^*) = (\hat{\mathbf{t}},\mathbf{s}) = 0$ for all $\mathbf{s},\mathbf{s}^*$, where $\mathbf{s}$ and $\mathbf{s}^*$ are the eigen-solutions of Equation 4.64 and 4.65 respectively.

In the case $\lambda = -1/2$ there are no eigensolutions beside $\mathbf{s} = \mathbf{s}^* = \mathbf{0}$. The integral Equations 4.56 and 4.59 have in the class $C^{0,\gamma}(S)$ a solution for every $\hat{\mathbf{u}},\hat{\mathbf{t}} \in C^{0,\gamma}(S)$.

In the case $\lambda = +1/2$ there are six linear independent eigensolutions $\mathbf{s}$ and $\mathbf{s}^*$. The eigensolutions $\mathbf{s}(\mathbf{x})$ of Equations 4.57 are the rigid body movements, i.e. any vector that has the form

$$\mathbf{s}(\mathbf{x}) = \mathbf{a} + \mathbf{b} \wedge \mathbf{x} \tag{4.66}$$

where $\mathbf{a}$, $\mathbf{b}$ are constant vectors. The six vectors

$$\mathbf{e}_i, \quad i = 1,2,3, \quad \mathbf{e}_i \wedge \mathbf{x}, \; i = 1,2,3 \tag{4.67}$$

are a complete set of linear independent rigid body movements. The eigensolutions $\mathbf{s}^*$ cannot be classified so easily, but we know at least that if $\mathbf{s}^*$ is an eigensolution of equation (4.58), then its $\mathbf{U}$-transform is a rigid body movement.

$$\oint_S \mathbf{U}(\mathbf{y},\mathbf{x})\mathbf{s}^*(\mathbf{y}) \, ds_y = \mathbf{a} + \mathbf{b} \wedge \mathbf{x} \tag{4.68}$$

So the condition for the problem $[T_i,\hat{\mathbf{t}}]$ becomes

$$(\hat{\mathbf{t}},\mathbf{s}) = 0 \text{ for all } \mathbf{s} = \mathbf{a} + \mathbf{b} \wedge \mathbf{x} \tag{4.69}$$

This is no real restriction, because these are just the global equilibrium conditions for $\hat{\mathbf{t}}$ on S.

The condition for problem $[D_e,\tilde{\mathbf{u}}]$ becomes

$$(\tilde{\mathbf{u}},\mathbf{s}^*)=0 \text{ for all } \mathbf{s}^* \tag{4.70}$$

which is a true restriction. If it is satisfied, the traction of the solution $\mathbf{W}(\mathbf{x})$ satisfies the global equilibrium condition on S, but this is an unnecessary restriction from a physical point of view.

This restriction can be avoided by solving $[D_e,\tilde{\mathbf{u}}]$ with a sum of two potentials, i.e. a potential of the 1st and 2nd kind, $\mathbf{V}(\mathbf{x})+\mathbf{W}(\mathbf{x})$. This approach leads to an integral equation, which has in $C^{0,\gamma}(S)$ for any $\tilde{\mathbf{u}}\in C^{0,\gamma}(S)$, exactly one solution, see Kupradze[7], Chapter VI.

Instead of using the potential $\mathbf{W}(\mathbf{x})$ for the problem $[D_i,\tilde{\mathbf{u}}]$ we could also use the potential $\mathbf{V}(\mathbf{x})$ which leads to an integral equation of the first kind (see Jaswon and Symm[8], p. 92 for a discussion of the equation). The solution of the problem $[T,\tilde{\mathbf{t}}]$ with a potential of the second kind, $\mathbf{W}(\mathbf{x})$, cannot be recommended. The kernel of $\boldsymbol{\tau}(\mathbf{W})$ has the order $0(r^{-3})$ for $r\to 0$, which imposes stronger restrictions on S and the layer $\mathbf{v}(\mathbf{y})$ for lim $\boldsymbol{\tau}(\mathbf{W})$ to exist, not to mention the numerical problems we would encounter.

The integral equation of the mixed problem, Equation 4.60, is not (in general) subject to the foregoing theory. It should be studied in the context of general results on mixed problems in elasticity. We refer to Kupradze[9] and especially Fichera[10] for results concerning the existence of solutions to mixed problems.

If the surface S is not smooth, i.e. only piecewise smooth, an analogous theory for the integral equations does not exist, as far as we know. An important contribution which aims in this direction is the work of Pieranita Rizzonelli[11]. She solves the first boundary value problem, $[D,\tilde{\mathbf{u}}]$, in interior and exterior domains with piecewise smooth surfaces by using the approach, as we would call it, of the direct method and proves with Hilbert-space-methods that the integral equations encountered have unique solutions.

That the hope for such a theory might not be totally in vain shows us the field of Laplace potentials where a theory for the integral equations on piecewise smooth boundaries exists, see Wendland[12] and Burago, Maz'ya and Sapozwikova[13].

While the **T**-transform of a layer $\mathbf{w}(\mathbf{y})$ is bounded at non-smooth points on the surface, the $\mathbf{T}^*$-transform of a bounded layer $\mathbf{v}(\mathbf{y})$ becomes singular at such points prohibiting any sensible collocation procedure. This could be overcome by using a Galerkin procedure to solve the integral equations, which contain the kernel $\mathbf{T}^*$, but our next example will show that also the layer $\mathbf{v}(\mathbf{y})$ can become singular at non-smooth points.

4.3.3 The physical nature of the layers

It is well known, see Rieder[14], Jaswon and Symm[8], Chapters 6, 7, that the potentials $\mathbf{V}(\mathbf{x})$ and $\mathbf{W}(\mathbf{x})$ represent the displacement field of the elastic continuum due to a discontinuity in tractions or displacements of magnitude $\mathbf{v}(\mathbf{y})$ or $\mathbf{w}(\mathbf{y})$ on the surface S. The name, indirect method, stems from the fact that these layers, which appear in the potentials, seem to have no direct physical meaning with respect to the boundary value problem they help to solve. This can be overcome if we introduce the concept of a complementary problem (CP).

The complement of a bounded body $B = B_i$ is the exterior region B_e, whose boundary S_e coincides with the boundary S_i of B_i. (We use three names for the same thing $S = S_i = S_e$.)

We call the first exterior boundary value problem $[D_e, \tilde{\mathbf{v}}]$

$$\mathbf{L}\mathbf{v} = \mathbf{0}, \ \mathbf{x} \in B_e, \ \lim_e \mathbf{v} = \tilde{\mathbf{v}}, \ \mathbf{x} \in S \tag{4.71}$$

a CP of the problem $[D_i, \tilde{\mathbf{u}}]$, if

$$\lim_i \boldsymbol{\tau}_e(\mathbf{u}) = \lim_e \boldsymbol{\tau}_e(\mathbf{v}) \tag{4.72}$$

holds on S. We call the second exterior boundary value problem $[T_e, \tilde{\mathbf{s}}]$,

$$\mathbf{L}\mathbf{w} = \mathbf{0}, \ \mathbf{x} \in B_e, \ \lim_e \boldsymbol{\tau}_i(\mathbf{w}) = \tilde{\mathbf{s}} \tag{4.73}$$

a CP of the problem $[T_i, \tilde{\mathbf{t}}]$, if

$$\lim_i \mathbf{u} = \lim_e \mathbf{w} \tag{4.74}$$

holds on S.

Studying Figures 4.7 and 4.8 we easily conclude the following theorem.

Theorem 3.3

A CP of the problem $[D_i, \tilde{\mathbf{u}}]$, $\mathbf{u} \in C^{1,\gamma}(S)$ is the problem $[D_e, \tilde{\mathbf{u}} + \mathbf{w}]$, where $\mathbf{w}(\mathbf{y}) \in C^{1,\gamma}(S)$ is the layer of the potential of the 2nd kind, which solves the problem $[D_i, \tilde{\mathbf{u}}]$ (Figure 4.8).

A CP of the problem $[T_i, \tilde{\mathbf{t}}]$, $\tilde{\mathbf{t}} \in C^{0,\gamma}(S)$ is the problem $[T_e, \mathbf{v} - \tilde{\mathbf{t}}]$, where $\mathbf{v}(\mathbf{y}) \in C^{0,\gamma}(S)$ is the layer of the potential of the 1st kind, which solves the problem $[T_i, \tilde{\mathbf{t}}]$ (Figure 4.7).

If we consider the boundary conditions, Equations 4.72 and 4.74, that connect the interior problem and its exterior complementary problem, we can state the contents of Theorem 3.3 in the following form:

The layer $\mathbf{v}(\mathbf{y})$ of the potential of the first kind that solves in the bounded body $B = B_i$ the second bvp $[T_i, \tilde{\mathbf{t}}]$ is, up to a bounded term ($\tilde{\mathbf{t}}$), the traction vector on the boundary S_e of the complement B_e if we impose on this boundary the same displacements as on the boundary S_i of the body B.

The layer $\mathbf{w}(\mathbf{y})$ of the potential of the second kind that solves in the bounded body B the first bvp $[D_i, \tilde{\mathbf{u}}]$ is up to a bounded term ($\tilde{\mathbf{u}}$) the displacement vector on S_e, the boundary of the complement, if this has the same traction vector as

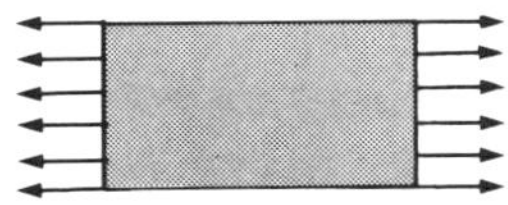

Figure 4.9

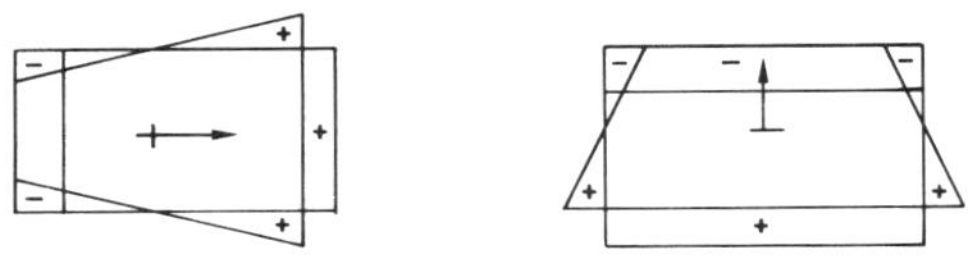

Figure 4.10

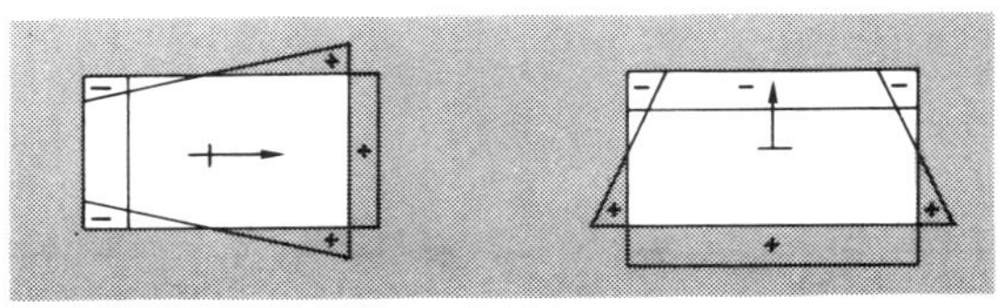

Figure 4.11

the boundary S_i of the body, i.e. $\lim_i \boldsymbol{\tau}_e(\mathbf{u}) = \lim_e \boldsymbol{\tau}_e(\mathbf{v})$ holds on S.

So the layers $\mathbf{v}(\mathbf{y})$ become physical quantities which are closely connected with the original problem and we can use physical insight to judge the behaviour of the layers near critical points.

The problem of a plate under tension, see Figure 4.9, is used to illustrate the theory, although the boundary S and the layers (we use piecewise constant functions together with a collocation procedure) violate the regularity conditions, which would be in the plane analogous to those in three dimensions, see Theorem 3.2.

We wanted to see how the layer $\mathbf{v}(\mathbf{y})$ behaves near corners and exploit the chance to calculate $\mathbf{v}(\mathbf{y})$ in two different ways. Figure 4.10 shows the displacement $\tilde{u}_1$ and $\tilde{u}_2$ of the exact solution of the problem in Figure 4.9 on the edge of the plate. According to Theorem 3.3 the layer $\mathbf{v}(\mathbf{y})$ of the potential $\mathbf{V}(\mathbf{x})$ which solves the plate problem $[T_i,\hat{\mathbf{t}}]$, is up to the term $\hat{\mathbf{t}}$ the traction vector $\tilde{\mathbf{s}} = \mathbf{v} - \hat{\mathbf{t}}$ on the boundary S_e of the complement, if we impose on S_e the displacements $\tilde{u}_1$ and $\tilde{u}_2$ as in Figure 4.11.

First we solved the problem $[T_i,\hat{\mathbf{t}}]$ with a potential of the first kind, $\mathbf{V}(\mathbf{x})$. This leads to the integral Equation 4.58 for the layer $\mathbf{v}(\mathbf{y})$ on S, which was solved numerically with an approximation $\mathbf{v}_h(\mathbf{y})$ for $\mathbf{v}(\mathbf{y})$. Next we turned to the complementary problem which is $[T_e,\mathbf{v}-\hat{\mathbf{t}}]$ and has the displacement $\tilde{\mathbf{u}}$ on S_e, where $\tilde{\mathbf{u}}$ is the trace (= value) of the exact interior solution $\mathbf{u}(\mathbf{x})$ of $[T_i,\hat{\mathbf{t}}]$ on $S = S_i = S_e$.

Suppose there exists a regular solution $\mathbf{v}(\mathbf{x})$ of $[T_e,\mathbf{v}-\hat{\mathbf{t}}]$, then we would have the Somigliana identity

$$\dot{\mathbf{C}}_e(\mathbf{x})\tilde{\mathbf{u}}(\mathbf{x}) = \oint_S [\mathbf{U}(\mathbf{y},\mathbf{x})\tilde{\mathbf{s}}(\mathbf{y}) - \mathbf{T}_i(\mathbf{y},\mathbf{x})\tilde{\mathbf{u}}(\mathbf{y})]\, ds_y \tag{4.75}$$

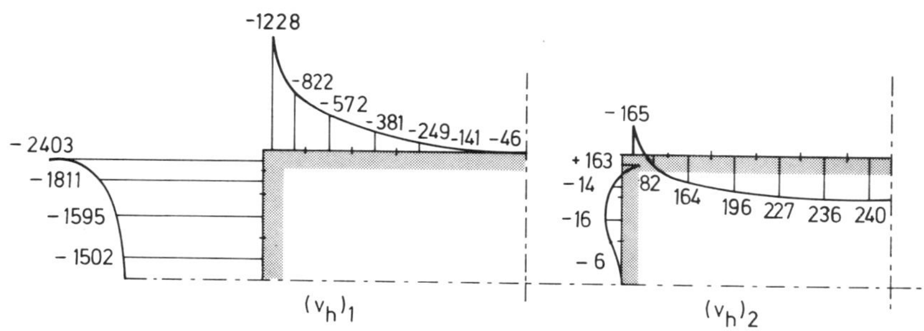

Figure 4.12

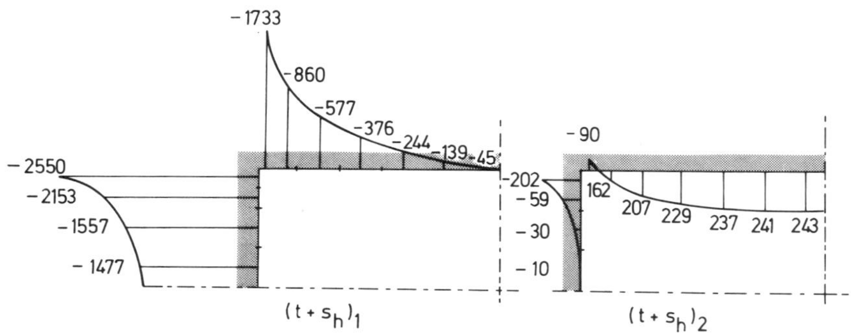

Figure 4.13

with $\hat{\mathbf{s}} = \lim_{e} \boldsymbol{\tau}_i(\mathbf{y})(\mathbf{x})$ and $\hat{\mathbf{u}} = \lim_{e} \mathbf{v}(\mathbf{x})$, see Section 4.4. Rearranging Equation (4.75) it becomes an integral equation

$$\oint_S \mathbf{U}(\mathbf{y},\mathbf{x})\hat{\mathbf{s}}(\mathbf{y})\ \mathrm{d}s_y = \dot{\mathbf{C}}_e(\mathbf{x})\hat{\mathbf{u}}(\mathbf{x}) + \oint_S \mathbf{T}_i(\mathbf{y},\mathbf{x})\hat{\mathbf{u}}(\mathbf{y})\ \mathrm{d}s_y \tag{4.76}$$

for the 'unknown' traction $\hat{\mathbf{s}}(\mathbf{y})$ of the problem $[T_e,\mathbf{v}-\hat{\mathbf{t}}]=[T_e,\hat{\mathbf{s}}]$ on S. The right-hand side of this integral equation depends on the trace $\hat{\mathbf{u}}$ of the interior solution.

The integral equation was solved numerically with an approximation $\mathbf{s}_h(\mathbf{x})$. The exact functions $\mathbf{v}$ and $\mathbf{s}$ satisfy the equation $\mathbf{v}-\hat{\mathbf{t}}=\mathbf{s}$ and so we should have appoximately

$$\hat{\mathbf{t}} + \mathbf{s}_h = \mathbf{v}_h \tag{4.77}$$

The right-hand side of Equation 4.77, $\mathbf{v}_h$, is presented in Figure 4.12. The left-hand side, $\hat{\mathbf{t}}+\mathbf{s}_h$, in Figure 4.13. Due to the double symmetry of the problem, it is sufficient to list the values on the boundary of one quarter of the plate.

The unknown and known functions S were approximated by step functions, i.e. constant vectors on intervals ('boundary elements') of a fixed length on S. N

of those elements represent the whole boundary. Numerical quadrature has been done, if possible, with a four point Gauss formula. The integral equations were solved by collocation, the collocation points were the middles of the elements. Near the corners the standard element was replaced by two elements, each half its length.

The numerical parameters of the plate were: Tension $p_0 = 1000$ kN/cm, length 12 cm, breadth 6 cm, $\mu = 9\,545$ kN/cm^2, $\nu = 0.1$. The exact solution of $[T_i, \mathbf{\tilde{t}}]$ is $\mathbf{u}(\mathbf{x}) = [0.04762\, x_1,\ -0.004762\, x_2]^T$, where $\mathbf{x} = (x_1, x_2)^T = \mathbf{0}$ is the point in the middle of the plate.

The results are approximately equal. A better coincidence should be expected with a refined numerical scheme, which somehow tries to cope with the singularities which occur in the layer $\mathbf{v}(\mathbf{y})$ and the term $\boldsymbol{\tau}(\mathbf{V})(\mathbf{x})$, $\mathbf{x} \in S$, i.e. the integral in the integral Equation 4.58 of the indirect method.

To cope with the singularity of the layer $\mathbf{v}(\mathbf{y})$ we could introduce an approximation $\mathbf{v}(\mathbf{y})$ which behaves near a corner $\mathbf{x}_C$ like $|\mathbf{y} - \mathbf{x}_C|^{-\alpha}$, $0 < \alpha < 1$. This would be a natural choice and should improve the solution of the integral equation of the direct method. But turning to the integral equation of the indirect method we still have to cope with the singularity of the term $\boldsymbol{\tau}(\mathbf{V})(\mathbf{x}_C)$. Furthermore, the layer $\mathbf{v}(\mathbf{y})$ in this potential itself is singular at $\mathbf{x}_C$. In the integral Equation 4.58 the layer appears not only under the integral sign but is also the free term. It is obvious that these two singularities $\mathbf{v}(\mathbf{x}_C)$ and $\boldsymbol{\tau}(\mathbf{V})(\mathbf{x}_C)$ must cancel each other out in such a way that their sum in the corner equals the bounded traction $\mathbf{\tilde{t}}(\mathbf{x}_C)$. All this is easily understood but not so easily it seems is the numerical realization of this insight.

4.3.4 Further remarks on the indirect method

Beside the potentials $\mathbf{V}(\mathbf{x})$, $\mathbf{W}(\mathbf{x})$ and $\mathbf{P}(\mathbf{x})$ there exist other potentials. Kupradze[9] (see p. 328 Equations 9.15 and 9.16), has constructed for example special potentials $\mathbf{P}_{(3)}$, $\mathbf{P}_{(4)}$ for the 3rd and 4th problem with kernels $\mathbf{K}_{(3)}$ and $\mathbf{K}_{(4)}$ which exhibit, if we apply to these the boundary 'operator' (=condition) $\mathbf{b}_{(i)}(\cdot)$, $\mathbf{i} = 3, 4$, of the 3rd or 4th problem a jump

$$\lim \mathbf{b}_{(i)}(\mathbf{P}_{(i)}) = \frac{1}{2}\mathbf{v}(\mathbf{x}) + \oint_S \mathbf{K}_{(i)}(\mathbf{y},\mathbf{x})\mathbf{v}(\mathbf{y})\, ds_y \qquad i = 3,4 \tag{4.78}$$

which leads again to an integral equation of the 2nd kind for the layers of the 3rd and 4th problem.

An approach to elastic potentials and integral equation methods, that strongly favours physical reasoning, is the work of Rieder[14] and Heise[15]. For an illustration of their approach we refer to the many papers of Heise[16,17,18] (we list only a few), who is also familiar with the contributions of Antes[19], Borgwardt[20], Pahnke[21] and Massonet[22] and gives in references 15 and 23 a review of different approaches in plane elasticity with elastic potentials.

The indirect method follows in 2-D the same line as in 3-D, except that 2-D potentials do not satisfy a priori radiation conditions as 3-D potentials do, but this does not alter the approach essentially. We refer to Kupradze[7], Chapter

VIII, for a detailed discussion. In the plane some authors have used an auxiliary curve S' outside of the region B of the plate. The potentials are generated on S' and lead therefore naturally to integral equations of the first kind with regular kernels on the original boundary S. For results and discussions we refer to Oliveira[24], Friemann[25], Heise[16].

Perhaps we should mention here too Kupradze's method of 'functional equations', which is an earlier version of the direct method, but with a strange solution procedure. The Somigliana identity is a relation between the trace and the traction, which must not only hold on S, but also in the complement of the plate B where $\mathbf{C}(\mathbf{x})=\mathbf{0}$. Kupradze draws (in the plane) an auxiliary curve S' outside of B, in the complement, which circumscribes the original boundary S and finds two functions $\hat{\mathbf{u}}$ and $\hat{\mathbf{t}}$ on S, such that the Somigliana identity holds at least on S'. (It should hold in the whole of the complement). The motivation for this solution procedure seems to be the wish to avoid the numerical solution of the strongly singular integral equations on S.

For a detailed description we refer to Kupradze[7], Chapter X, p. 231, and for a discussion of this method we refer to Christiansen[26].

4.4 THE NEW APPROACH

The layer approach, as it is sometimes called, or the indirect method discussed in Section 4.3 is the classical approach of potential theory to boundary value problems and its application to the bvp of elasticity followed in the wake of a parallel approach in the theory of Laplace potentials. Simultaneously with the development of the direct method for the bvp of the Laplace equation, the new direct formulation was applied to the bvp of elasticity by Rizzo† and published[27] together with encouraging numerical results and later used intensively by Cruse[28,29,30] to solve engineering problems. Their works had a great influence on the use of integral equation methods in elastomechanics. Though basically integral equation methods appeared earlier in the works of Muskhelisvili and Kupradze and elsewhere engineers never became too familiar with these methods and the new approach whose starting point is the Somigliana identity has first turned their eyes to integral equation methods.

From a mathematical point of view the direct method is naturally also a potential method, simply because it uses the sum of two (or three) potentials, but the Somigliana identity offers the chance to interpret the layers which appear in the potentials, see Theorem 4.3, as the trace and the traction of the solution of the boundary value problem itself and so the engineer can use judgment based on physical insight to estimate the behaviour of the layers near critical points. After what has been said about the layers of the indirect method at the end of Section 4.3, a similar property is also shared by the layers of the indirect method, but the new method offers more: Equally important is that the 2nd bvp $[T,\hat{\mathbf{t}}]$ leads in domains with piecewise smooth boundaries to a regular integral equation, i.e. the integral which appears in the integral

† An earlier version is Kupradze's method of 'functional equations'[9], which differs from Rizzo's approach by its solution procedure.

equation exists at non-smooth points. A further advantage is, that the layter $\tilde{\mathbf{t}}$, the traction, which can be discontinuous or even singular in practical problems, enters the integral equation not as a free term as the continuous displacement $\tilde{\mathbf{u}}$ does but only its $\mathbf{U}$-transform enters the integral equation. Due to Theorem 3.2 are the $\mathbf{U}$-transforms of piecewise continuous layers continuous functions on S, i.e. the traction $\tilde{\mathbf{t}}$ appears in the integral equations of the direct method only after having been sufficiently 'smoothed'.

4.4.1 The Somigliana identity

Before introducing the Somigliana identity we would like to state Betti's theorem which is the analogue of Green's second identity for the Laplace operator and plays an important part in the proof of the Somigliana identity.

Theorem 4.1

If (1) $B \subset R^n$, $n=2,3$ is a regular region with boundary S and (2) $\mathbf{u}(\mathbf{x})$, $\mathbf{v}(\mathbf{x}) \in C^1(\bar{B}) \cap C^2(B)$ and satisfy the radiation condition (A) (see Section 4.2.3) if $B = B_e$, then

$$\int_B [\mathbf{u}^T \cdot \mathbf{L}\mathbf{v} - \mathbf{v}^T \cdot \mathbf{L}\mathbf{u}] \, \mathrm{d}v = \oint_S [\mathbf{u}^T \cdot \boldsymbol{\tau}(\mathbf{v}) - \mathbf{v}^T \cdot \boldsymbol{\tau}(\mathbf{u})] \, \mathrm{d}s \qquad (4.79)$$

This is Betti's theorem, which rests on Gauss' divergence theorem (integration by parts) and the nature of the coefficients of the operator $\mathbf{L}$. We call a solution of $\mathbf{L}\mathbf{u} = -\mathbf{b}$ a regular solution, if it satisfies conditon (2).

Suppose $\mathbf{u}(\mathbf{x})$ is a regular solution of $\mathbf{L}\mathbf{u} = -\mathbf{b}$ in B. Applying Betti's theorem consecutively to the pairs $\mathbf{u}(\mathbf{y})$, $\mathbf{g}^{(i)}(\mathbf{y},\mathbf{x})$ $i=1,2,3$ in the region $B/K_\varepsilon(\mathbf{x})$, where $K_\varepsilon(\mathbf{x})$ is a ball with radius ε and center $\mathbf{x}$, combining them in an appropriate matrix notation, and letting ε tend to zero, we would derive the Somigliana identity. We want to state this identity in an extended form, i.e. there exist, besides body forces $\mathbf{b}$, single loads $\mathbf{F}^{(i)}$ at interior points and $\mathbf{u}(\mathbf{x})$ behaves near these points like the Kelvin solutions.

Theorem 4.2. *Somigliana identity*

Suppose (1) $\mathbf{b}(\mathbf{x}) \in C(B)$, (2) there exist n single loads $\mathbf{F}^{(i)} = [F_1^{(i)}, F_2^{(i)}, F_3^{(i)}]^T$ at interior points $\mathbf{y}^{(i)} = [y_1^{(i)}, y_2^{(i)}, y_3^{(i)}]^T i = 1,2\ldots n$, (3) $\mathbf{u}(\mathbf{x}) \in C^2(B')$ where $B' = B/Y$and $Y = \{\mathbf{y}^{(1)}, \mathbf{y}^{(2)}, \ldots, \mathbf{y}^{(n)}\}$, (4) $\mathbf{L}\mathbf{u} = -\mathbf{b}$ in B' and

$$\mathbf{u}(\mathbf{x}) = 0\left(\frac{1}{|\mathbf{x} - \mathbf{y}^{(i)}|}\right), \qquad \mathbf{S}(\mathbf{x}) = 0\left(\frac{1}{|\mathbf{x} - \mathbf{y}^{(i)}|^2}\right)$$

$$\lim_{\rho \to 0} \oint_{\dot{K}_\rho} \mathbf{S} \cdot \mathbf{n} \, \mathrm{d}s = \mathbf{F}^{(i)}, \qquad \text{for } i = 1,2\ldots n$$

where $\dot{K}_\rho$ is a sphere with center $\mathbf{y}^{(i)}$ and radius ρ then

$$\mathbf{C}_i(\mathbf{x})\mathbf{u}(\mathbf{x}) = \oint_S [\mathbf{U}(\mathbf{y},\mathbf{x})\boldsymbol{\tau}_e(\mathbf{u})(\mathbf{y}) - \mathbf{T}_e(\mathbf{y},\mathbf{x})\mathbf{u}(\mathbf{y})]\ ds_y + \int_{B_i} \mathbf{U}(\mathbf{y},\mathbf{x})\mathbf{b}(\mathbf{y})\ dv_y + \sum_{i=1}^{n} \mathbf{U}(\mathbf{y}^{(i)},\mathbf{x})\cdot\mathbf{F}^{(i)} \tag{4.80}$$

or if $B = B_e$,

$$\mathbf{C}_e(\mathbf{x})\mathbf{u}(\mathbf{x}) = \oint_S [\mathbf{U}(\mathbf{y},\mathbf{x})\boldsymbol{\tau}_i(\mathbf{u})(\mathbf{y}) - \mathbf{T}_i(\mathbf{y},\mathbf{x})\mathbf{u}(\mathbf{y})]\ ds_y + \int_{B_e} \mathbf{U}(\mathbf{y},\mathbf{x})\mathbf{b}(\mathbf{y})\ dv_y + \sum_{i=1}^{n} \mathbf{U}(\mathbf{y}^{(i)},\mathbf{x})\cdot\mathbf{F}^{(i)} \tag{4.81}$$

with $\mathbf{C}_i(\mathbf{x})$ and $\mathbf{C}_e(\mathbf{x})$ as in Equation 4.22.

For the proof of this identity (without single loads) we refer to Hartmann[5], Chapter 4. The extension with single loads at interior points is due to Turteltaub and Sternberg[31].

The Kelvin matrices $\mathbf{U}(\mathbf{y}^{(i)}, \mathbf{x})$ become singular with sources points $\mathbf{x} = \mathbf{y}^{(i)}$. The Equations 4.80 and 4.81 are, therefore, not defined at these points. In what follows we will refer to the case without single loads.

4.4.2 The direct method

The Somigliana identity is the starting point and motivation of the direct method, but actually it uses not the Somigliana identity but a theorem which is rather the logical converse of the Somgiliana identity and which we would like to state in the following form:

Theorem 4.3

If S is the smooth surface of a regular region and $\hat{\mathbf{t}} \in C^{0,\gamma}(S)$, $\hat{\mathbf{u}} \in C^{1,\gamma}(S)$, $\mathbf{b} \in C^2(B)$ satisfy on S the integral relation

$$\dot{\mathbf{C}}_i(\mathbf{x})\hat{\mathbf{u}}(\mathbf{x}) = \oint_S [\mathbf{U}(\mathbf{y},\mathbf{x})\hat{\mathbf{t}}(\mathbf{y}) - \mathbf{T}_e(\mathbf{y},\mathbf{x})\hat{\mathbf{u}}(\mathbf{y})]\ ds_y + \int_{B_i} \mathbf{U}(\mathbf{y},\mathbf{x})\mathbf{b}(\mathbf{y})\ dv_y \tag{4.82}$$

then this is sufficient for the function

$$\mathbf{u}(\mathbf{x}) = \oint_S [\mathbf{U}(\mathbf{y},\mathbf{x})\tilde{\mathbf{t}}(\mathbf{y}) - \mathbf{T}_e(\mathbf{y},\mathbf{x})\tilde{\mathbf{u}}(\mathbf{y})]\ ds_y + \int_{B_i} \mathbf{U}(\mathbf{y},\mathbf{x})\mathbf{b}(\mathbf{y})\ dv_y,\ \mathbf{x} \in B_i \quad (4.83)$$

to assume the boundary values

$$\lim_i \mathbf{u} = \tilde{\mathbf{u}} \quad (4.84)$$

$$\lim_i \boldsymbol{\tau}_e(\mathbf{u}) = \tilde{\mathbf{t}} \quad (4.85)$$

If S is only piecewise smooth, the last equation does not hold in this strict, pointwise sense anymore, but in the weaker sense

$$\lim_{v \to \infty} \oint_{S_v} \mathbf{U}(\mathbf{y},\mathbf{x}_v)\boldsymbol{\tau}_e(\mathbf{u})(\mathbf{y})\ ds_y = \oint_S \mathbf{U}(\mathbf{y},\mathbf{x})\tilde{\mathbf{t}}(\mathbf{y})\ ds_y \quad (4.86)$$

where S_v are surfaces, which converge to S from the inside such that there is a one to one relation between the points $\mathbf{x}_v$ and $\mathbf{x}$ on S_v and S, and $\boldsymbol{\tau}_e(\mathbf{u})(\mathbf{y})$ is the traction of Equation 4.83 on S_v. For a proof of this theorem we refer to Hartman[5], Theorem 8.1, 8.2. The formulation of Theorem 4.3 for exterior regions $B = B_e$ is obtained by changing the subscripts from i to e and vice versa in any symbol, which appears above.

Theorem 4.3 states the idea of the direct method: Find two functions $\tilde{\mathbf{u}}$ and $\tilde{\mathbf{t}}$ on S which satisfy the integral relation and construct with those functions a solution in the form of Equation 4.83. As $\tilde{\mathbf{u}}$ and $\tilde{\mathbf{t}}$ are physical quantities on the boundary S and therefore one of them (or any combination of components of $\tilde{\mathbf{u}}$ and $\tilde{\mathbf{t}}$) is given at any point of S, the integral relation becomes an integral equation for the other unknown quantity.

The problem $[D_i,\tilde{\mathbf{u}}]$ with $\mathbf{Lu} = -\mathbf{b}$ in B_i leads to the integral equation

$$\oint_S \mathbf{U}(\mathbf{y},\mathbf{x})\tilde{\mathbf{t}}(\mathbf{y})\ ds_y = \dot{\mathbf{C}}_i(\mathbf{x})\tilde{\mathbf{u}}(\mathbf{x}) + \oint_S \mathbf{T}_e(\mathbf{y},\mathbf{x})\tilde{\mathbf{u}}(\mathbf{y})\ ds_y - \int_{B_i} \mathbf{U}(\mathbf{y},\mathbf{x})\mathbf{b}(\mathbf{y})\ dv_y \quad (4.87)$$

and the problem $[T_i,\tilde{\mathbf{t}}]$ with $\mathbf{Lu} = -\mathbf{b}$ in B_i leads to the integral equation

$$\dot{\mathbf{C}}_i(\mathbf{x})\tilde{\mathbf{u}}(\mathbf{x}) + \oint_S \mathbf{T}_e(\mathbf{y},\mathbf{x})\tilde{\mathbf{u}}(\mathbf{y})\ ds_y = \oint_S \mathbf{U}(\mathbf{y},\mathbf{x})\tilde{\mathbf{t}}(\mathbf{y})\ ds_y + \int_{\mathbf{B}_i} \mathbf{U}(\mathbf{y},\mathbf{x})\mathbf{b}(\mathbf{y})\ dv_y \quad (4.88)$$

The integral equation for the exterior problems are obtained by changing the subscript i to e and vice versa in any symbol, that appears above.

By using Miklihn's results on the integral operators of the indirect method, see Section 4.3, which appear again in the integral equation of the direct method, Equations 4.87 and 4.88, we derive the following theorem.

Theorem 4.4

Suppose S is a smooth regular surface
(1) If $\mathbf{u}\in C^{1,\gamma}(S)$, $\mathbf{b}\in C^2(B)$, then there exists in $C^{0,\gamma}(S)$ a unique solution $\hat{\mathbf{t}}$ of Equation 4.87.
(2) If $\hat{\mathbf{t}}\in C^{0,\gamma}(S)$, $\mathbf{b}\in C^2(B)$ and the two functions satisfy the equilibrium condition

$$\oint_S \hat{\mathbf{t}}^T\cdot\mathbf{s}\ ds+\int_B \mathbf{b}^T\cdot\mathbf{s}\ dv=0,\quad \text{for all } \mathbf{s} \tag{4.89}$$

where $\mathbf{s}=\mathbf{a}+\mathbf{d}\wedge\mathbf{x}$ is a rigid body movement ($\mathbf{a}$ and $\mathbf{d}$ being two arbitrary constant vectors), then there exists in $C^{0,\gamma}(S)$ a solution $\hat{\mathbf{u}}$ of Equation 4.88 which is unique up to rigid body movements.

Exterior problems are subject to the same theory, but the equilibrium condition, Equation 4.89, is no necessary condition for exterior problems. For a proof we refer to Hartmann[5], Theorems 8.3–8.6.

The interior mixed problem leads to the integral equation

$$\oint_S \mathbf{U}(\mathbf{y},\mathbf{x})\hat{\mathbf{t}}(\mathbf{y})\ ds_y=\dot{\mathbf{C}}_i(\mathbf{x})\mathbf{u}(\mathbf{x})+\oint_S \mathbf{T}_e(\mathbf{y},\mathbf{x})\mathbf{u}(\mathbf{y})\ ds_y-\int_{B_i}\mathbf{U}(\mathbf{y},\mathbf{x})\mathbf{b}(\mathbf{y})\ dv_y\quad \mathbf{x}\in S_1 \tag{4.90}$$

$$\dot{\mathbf{C}}_i(\mathbf{x})\hat{\mathbf{u}}(\mathbf{x})+\oint_S \mathbf{T}_e(\mathbf{y},\mathbf{x})\hat{\mathbf{u}}(\mathbf{y})\ ds_y=\oint_S \mathbf{U}(\mathbf{y},\mathbf{x})\hat{\mathbf{t}}(\mathbf{y})\ ds_y+\int_{B_i}\mathbf{U}(\mathbf{y},\mathbf{x})\mathbf{b}(\mathbf{y})\ dv_y\ \ \mathbf{x}\in S_2 \tag{4.91}$$

$$S=S_1\cup S_2$$

This integral equation can again, as far as we know, be only studied in the framework of general results on mixed problems in elasticity and the results on the behaviour of solutions near points where the boundary condition changes, see Section 4.3 and the references there.

So as not to be misunderstood we want to stress one point. Theorems 4.3 and 4.4 might seem superfluous to someone who argues as follows: Given (say) boundary data $\hat{\mathbf{u}}$ and the problem $[D_i,\hat{\mathbf{u}}]$, then there exists a regular solution of this problem, necessarily the integral equation which determines the unknown traction $\mathbf{t}=\tilde{\tau}(\mathbf{u})(\mathbf{y})$ has a solution and the Somigliana identity holds, i.e. the function (Equation 4.5) assumes the boundary values $\hat{\mathbf{u}}$ and $\hat{\mathbf{t}}$.

This is correct — if we know that there exists a regular solution corresponding to the boundary data. Theorems 4.3 and 4.4 were originally intended to avoid the difficult question of the existence of a regular solution and intended to serve to establish that the solution procedure of the direct method leads to a correct solution, if the layers satisfy some smoothness conditions and naturally the integral relation, Equation 4.83 on S.

Using the latest results of Kupradze *et al.*[9] we could tackle this question now directly: Kupradze *et al.* established under which conditions the elastic potentials on smooth surfaces are regular solutions (see Kupradze[9], Theorems 7.1 and 7.2 on p. 317 and Theorem 10.9 on p. 342). Furthermore, employing the results concerning the integral equations of the indirect method, see Section 4.3.2, we could now derive existence theorems for regular solutions on smooth surfaces for given boundary values and then use the above stated argument.

Before closing this section, we should compare the integral equation 4.88 of the problem $[T_i,\tilde{\mathbf{t}}]$ with the integral equation of the same problem when we treat it with the indirect method, Equation 4.58. Equation 4.88 offers two advantages:

(1) Even if the traction $\mathbf{t}(\mathbf{x})$ on S is only piecewise continuous, as might happen in practical problems, its **U**-transform becomes a continuous function on S, cf. Theorem 3.2., i.e. the right-hand side of the integral Equation 4.88 is a continuous function, contrary to the right-hand side of Equation 4.58 and

(2) the left-hand side of the integral equation of the direct method exists also at non-smooth points.

4.4.3 Internal stresses

Once the integral equations are solved, the function $\mathbf{u}(\mathbf{x})$, Equation 4.83, can be constructed and values of $\mathbf{u}(\mathbf{x})$ at interior points can be calculated by simply evaluating the expression 4.83.

To derive formulae for internal stresses, we have to apply the expressions 4.3 and 4.4 to the kernels of the potentials differentiating with respect to the field point $\mathbf{x}$ under the integral. Keeping in mind that the stress tensor $\mathbf{S}=[\sigma_{ij}]$ is symmetric, we obtain for any of the three potentials six kernels. Because the potential of the first kind and the volume potential have the same kernel $\mathbf{U}(\mathbf{x},\mathbf{y})$, twelve different kernels suffice to calculate the six stresses at any interior point:

$$\sigma_{ij}(\mathbf{x})=\oint_S[\mathbf{D}_{ij}^T(\mathbf{y},\mathbf{x})\tilde{\mathbf{t}}(\mathbf{y})-\mathbf{S}_{ij}^T(\mathbf{y},\mathbf{x})\tilde{\mathbf{u}}(\mathbf{y})]\,\mathrm{ds}_{\mathbf{y}}+\int_B\mathbf{D}_{ij}^T(\mathbf{y},\mathbf{x})\mathbf{b}(\mathbf{y})\,\mathrm{dv}_{\mathbf{y}} \tag{4.92}$$

where $\mathbf{D}_{ij}=[D_{1ij},D_{2ij},D_{3ij}]^T$ is the vector

$$\begin{aligned}D_{kij}=\frac{1}{r^\alpha}[(1-2\nu)\{\delta_{ki}\cdot r_{,j}+\delta_{kj}\cdot r_{,i}-\delta_{ij}\cdot r_{,k}\}+\\+\beta\cdot r_{,i}\cdot r_{,j}\cdot r_{,k}]\frac{1}{4\alpha\pi(1-\nu)}\qquad k=1,2,3\end{aligned} \tag{4.93}$$

and $\mathbf{S}_{ij} = [S_{1ij}, S_{2ij}, S_{3ij}]^T$ is the vector

$$S_{kij} = \frac{2\mu}{r^\beta}\Bigg[\beta \cdot \frac{\partial r}{\partial n_y}\{(1-2v)\delta_{ij}\cdot r_{,k} + v(\delta_{ik}\cdot r_{,j} + \delta_{jk}\cdot r_{,i}) - \gamma r_{,i}\cdot r_{,j}\cdot r_{,k}\} + \beta v(n_i\cdot r_{,j}\cdot r_{,k} + n_j\cdot r_{,i}\cdot r_{,k}) + (1-2v)(\beta n_k\cdot r_{,i}\cdot r_{,j}\cdot + n_j\delta_{ik} + n_i\cdot\delta_{jk}) - (1-4v)n_k\cdot\delta_{ij}\Bigg]\frac{1}{4\alpha\pi(1-v)}, \qquad k=1,2,3 \tag{4.94}$$

with $r_{,i} = \dfrac{\partial r}{\partial y_i}$ and $\mathbf{n}(\mathbf{y}) = [n_1, n_2, n_3]^T$.

These equations apply for two and three dimensions. In two dimensions it is $\alpha=1$, $\beta=2$, $\gamma=4$. In three dimensions it is $\alpha=2$, $\beta=3$, $\gamma=5$. Naturally the vectors $\mathbf{D}_{ij}$ and $\mathbf{S}_{ij}$ have in two dimensions only two components. The equations above can be found in Brebbia[32] or in Cruse[28].

4.4.4 Stresses on the boundary

If $\mathbf{x}$ lies on S the formulae above are no longer applicable in computations. We proceed as follows: At any smooth point $\mathbf{x}$ on S, 15 quantities are defined: Six stress componenets σ_{ij} and nine derivatives u_{i,x_j} $i,j=1,2,3$.

On the other hand there exist 15 equations, namely:

$$\begin{aligned} \sigma_{ij} &= 2\mu\varepsilon_{ij} + \frac{2\mu v}{1-2v}\delta_{ij}\varepsilon_{kk} \\ t_i &= \sigma_{ij}n_j \\ \frac{\mathrm{d}u_i}{\mathrm{d}s_l} &= u_{i,x_j}\frac{\mathrm{d}x_j}{\mathrm{d}s_l} \end{aligned} \tag{4.95}$$

where s_l $l=1,2$ are the parameters of the surface S, and t_i, $i=1,2,3$, and $\dfrac{\mathrm{d}u_i}{\mathrm{d}s_l}$ are known quantities on S.

The 15 equations lead to a regular linear system for the unknown quantities. The values $\dfrac{\mathrm{d}u_i}{\mathrm{d}s_l}$ therein can be calculated on S by finite differences or by directly differentiating the functions u_i, if they are polynomials of higher degree or by using an integral representation of the derivatives $\dfrac{\mathrm{d}u_i}{\mathrm{d}s_l}$ (see Warlo[33], p. 19, Equation 29).

In plane problems the number 15 reduces to 7 and the same approach leads to the following linear system for the unknown quantities σ_{ij} and $u_{i,j}$:

$$\begin{bmatrix} 1 & 0 & 0 & a & 0 & 0 & -\lambda \\ n_1 & 0 & n_2 & 0 & 0 & 0 & 0 \\ 0 & n_2 & n_1 & 0 & 0 & 0 & 0 \\ 0 & 1 & 0 & -\lambda & 0 & 0 & a \\ 0 & 0 & 0 & -n_2 & 0 & n_1 & 0 \\ 0 & 0 & 1 & 0 & -\mu & -\mu & 0 \\ 0 & 0 & 0 & 0 & -n_2 & 0 & n_1 \end{bmatrix} \begin{bmatrix} \sigma_{11} \\ \sigma_{22} \\ \sigma_{12} \\ u_{1,1} \\ u_{2,1} \\ u_{1,2} \\ u_{2,2} \end{bmatrix} = \begin{bmatrix} 0 \\ t_1 \\ t_2 \\ 0 \\ u_{1,s} \\ 0 \\ u_{2,s} \end{bmatrix} \quad (4.96)$$

It is $a = -\lambda - 2\mu$, and $\lambda = \dfrac{2\mu v}{1-2v}$ and $\mathbf{n} = [n_1, n_2]^T$ is the unit normal on the curve S whose parameter s is the arc length. For n_1 and n_2 not identically zero, this matrix is regular (see Warlo[33], p. 18).

4.4.5 Numerical treatment

If we solve integral equations numerically we encounter new mathematical problems which call for new theories and proofs. Some of the preceding results can be extended to cover the numerical solution of integral equations but others cannot be used. Let us give an example: while the converse of the Somigliana formula has been established, cf. Theorem 4.3, for smooth surfaces and smooth functions, it is not at all clear, in which sense this theorem, which is essential for the boundary element method, holds for the solutions of a given numerical procedure. To get an idea what the tools and means are to tackle such questions and how mathematicians cope with these problems we refer to the works of Atkinson[34], Ben Noble[35], Ciarlet[36], Hsiao and Wendland[37], Nedelec[38], Prössdorf[39], Wendland[12,40] and Chapter 9.

In this section we give a detailed description of the steps necessary to discretize the integral equations, if they are solved with a collocation method, which is up to now the standard procedure, though the Galerkin method might become a serious alternative, due to the latest theoretical and practical results, see Wendland[40] and Chapter 9.

Throughout this section we assume, that the body force $\mathbf{b}$ is zero, so the volume integral in the Somigliana identity vanishes. It is, we believe, widely understood, that the boundary element method draws its main advantage from the reduction of the dimension. This advantage gets lost (in parts) if body forces must be considered. Body forces, which are the gradient of a scalar valued function ψ which satisfies a Poisson equation in B, as e.g. the mass forces of a body, allow the replacement of the volume integral by a surface integral, see Theorem 3.1.

Prior to discretizing the Somigliana identity for $\mathbf{x}$ on S, we modify it slightly. We have, see Equation 4.80†, for any fixed $\mathbf{x} \in S$

† Equation 4.97 is easily derived from Equation 4.80 by identifying the vector $\mathbf{u}(\mathbf{x})$ with a rigid body movement, whose traction is zero on S.

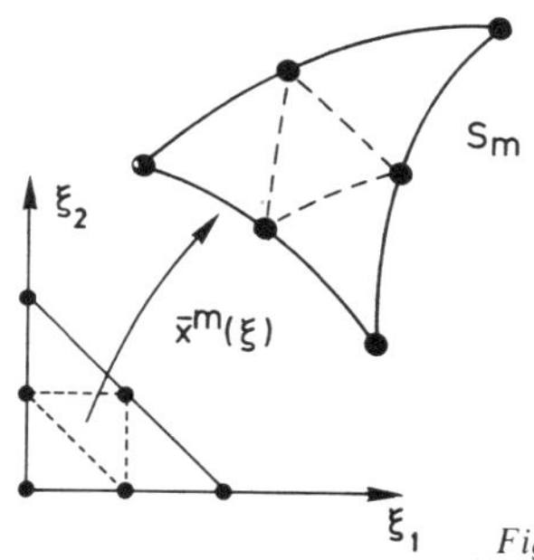

Figure 4.14 A reference triangle S_0 with six nodes

$$\dot{\mathbf{C}}_i(\mathbf{x})\mathbf{u}(\mathbf{x}) = -\int_S \mathbf{T}_e(\mathbf{y},\mathbf{x})\mathbf{u}(\mathbf{x})\, ds_y \tag{4.97}$$

and so the Somigliana identity, Equation 4.82 becomes

$$\int_S \mathbf{U}(\mathbf{y},\mathbf{x})\tilde{\mathbf{t}}(\mathbf{y})\, ds_y = \int_S \mathbf{T}_e(\mathbf{y},\mathbf{x})[\tilde{\mathbf{u}}(\mathbf{y}) - \tilde{\mathbf{u}}(\mathbf{x})]\, ds_y \tag{4.98}$$

This equation has the advantage, that $\dot{\mathbf{C}}(\mathbf{x})$ has formally disappeared (it hides in the integral on the right-hand side) and the integral on the right-hand side, as the one on the left, is for $\tilde{\mathbf{u}}(\mathbf{y}) \in C^{0,\gamma}(S)$ weakly singular in any point $\mathbf{x}$ on S.

For the integrals in Equation 4.98 to exist on regular surfaces, $\mathbf{t}(\mathbf{y})$ must be at least piecewise continuous and $[\tilde{\mathbf{u}}(\mathbf{y}) - \tilde{\mathbf{u}}(\mathbf{x})] \in C^{0,\gamma}(S)$ for any fixed $\mathbf{x}$, see Theorem 3.2. The two conditions are satisfied if we use isoparametric boundary elements with polynomial shape functions and guarantee that the function which interpolates $\tilde{\mathbf{u}}$ is continuous on adjacent sides of the boundary elements. Concerning the function $\tilde{\mathbf{t}}$ we will deliberately exploit the chance to use piecewise continuous tractions.

The isoparatmetric approach with triangular elements starts then this way: the surface S is partitioned in triangular elements R_m. To the triangulation corresponds a reference triangle S_0 in the plane ξ_1,ξ_2 (Figure 4.14) and on S_0 exist L nodes $\boldsymbol{\xi}^l = (\xi_1^l,\xi_2^l)$ $l = 1,2,\ldots L$ and L scalar valued functions $s_l(\boldsymbol{\xi})$ with the property:

$$s_l(\boldsymbol{\xi}^m) = \delta_{lm} \quad 1 \leqslant l,m \leqslant L \tag{4.99}$$

Suppose the function $\mathbf{x}^m(\xi) = [x_1^m, x_2^m, x_3^m]_T$ maps the element S_0 on $R_m \subset S$. The function $\bar{\mathbf{x}}^m(\boldsymbol{\xi}) = [\bar{x}_1^m, \bar{x}_2^m, \bar{x}_3^m]^T$ with

$$\bar{x}_i^m(\xi) = \sum_{l=1}^{L} s_l(\boldsymbol{\xi}) \cdot x_i^m(\boldsymbol{\xi}^l) \tag{4.100}$$

interpolates $\mathbf{x}^m(\boldsymbol{\xi})$ on S_0 and the surface element S_m represented by $\bar{\mathbf{x}}^m(\boldsymbol{\xi})$, $\boldsymbol{\xi} \in S_0$, is called together with the functions $p_n^m(\mathbf{x})$, see below, a boundary element. The set $s_l(\boldsymbol{\xi})$, $l=1,2,\ldots L$ and triangulation are supposed to satisfy the usual regularity condition, cf. Ciarlet[36] Chapter 2, so the surface $S'=\bigcup\limits_m S_m$ is a regular surface, In general S' does not coincide with the original surface S at any point $\mathbf{x}\in S'$.

Let us assume in the following that S' consists of M elements S_m. By $J^m(\boldsymbol{\xi})$ we denote the Jacobian of the transformation $\bar{\mathbf{x}}^m(\xi)$

$$J^m(\xi)=(\det\ [g_{ik}^m])^{1/2} \tag{4.101}$$

with $g_{ik}^m=\bar{\mathbf{x}}_{\xi_i}^{m^T},\ \bar{\mathbf{x}}_{\xi_k}^m$. It is

$$\int_{S_m} \mathrm{d}s=\int_{S_0} J^m(\boldsymbol{\xi})\ \mathrm{d}\boldsymbol{\xi} \tag{4.102}$$

with $\mathrm{d}\boldsymbol{\xi}=\mathrm{d}\xi_1\mathrm{d}\xi_2$. If in Equation 4.100, l runs through $1,2,\ldots L$ and m through $1,2,\ldots M$ the values of the vector valued function $\bar{\mathbf{x}}^m(\xi)$ run through the set of the coordinate vectors of all nodes $\mathbf{x}^k=\bar{\mathbf{x}}^m(\xi^l)$ on S'_q where $k=k(m,l)$ is the global number of the node with the coordinate vector $\bar{\mathbf{x}}^m(\xi^l)$. We assume that there are K different nodes $\mathbf{x}^k$, $k=1,2,\ldots K$ on S'. Besides the function $k=k(m,l)$ there exists a function $l=l(k,m)1\leqslant k\leqslant K$, $1\leqslant m\leqslant M$ which names the number l of the node ξ^l on S_0 for which $\mathbf{x}^k=\bar{\mathbf{x}}^m(\xi^l)$ holds.

Next we construct a set P of K scalar valued nodal functions $p_k(\mathbf{x})$, $\mathbf{x}\in S'$, with the property

$$p_k(\mathbf{x}^l)=\delta_{kl},\qquad 1\leqslant k,\ l\leqslant K \tag{4.103}$$

Let us designate by $f^m(\mathbf{x}):=f(\mathbf{x})|_{S_m}$ the restriction of a function $f(\mathbf{x})$, $\mathbf{x}\in S'$, to $S_m\subset S'$. We define $p_k(\mathbf{x})$ by its restriction on any element S_m, $m=1,2,\ldots M$.

$$p_k^m(\mathbf{x})=p_k(\mathbf{x})|_{S_m}=\begin{cases}0, \text{ if } \mathbf{x}^k\notin S_m\\ s_l(\boldsymbol{\xi}), \text{ if } \mathbf{x}^k\in S_m\end{cases} \tag{4.104}$$

where $l=l(k,m)$ and $\mathbf{x}=\bar{\mathbf{x}}^m(\xi)$. Usually the function $s_l(\xi)$ is on S_0 a polynomial with degree r, so $p_k^m\in C^\infty(S_m)$. The global properties of the functions $p_k(\mathbf{x})$, $k=1\ldots K$, i.e. $p_k(\mathbf{x})\in C^n(S')$ with $n=0,1$ or else, depend on the functions $s_l(\boldsymbol{\xi})$ and their continuation over adjacent sides. (We refer to Ciarlet[36], Chapter 2, for a detailed discussion). We need at least $p_k(\mathbf{x})\in C^{0,\gamma}(S')$, which can, e.g. be achieved with the standard reference triangle S_0 with three nodes and three corresponding linear functions $s_l(\boldsymbol{\xi})$, $l=1,2,3$.

The components of the displacement vector $\tilde{\mathbf{u}}(\mathbf{x})=[\tilde{u}_1,\tilde{u}_2,\tilde{u}_3]^T$ are interpolated on S' by the functions

$$u_i(\mathbf{x})=\sum_{n=1}^{K} p_n(\mathbf{x})u_{in},\qquad i=1,2,3 \tag{4.105}$$

where u_{in} denotes the nodal values $u_{in}=\tilde{u}_i(\mathbf{x}^n)$ and the tractions $t_i(\mathbf{x})$ $i=1,2,3$ are interpolated by the functions

$$t_i^m(\mathbf{x})=t_i(\mathbf{x})|_{S_m}=\sum_{n=1}^{K} p_n^m(\mathbf{x})t_{in}^m \qquad \begin{matrix} i=1,2,3 \\ m=1,2,\ldots M \end{matrix} \tag{4.106}$$

where t_{in}^m denotes the nodal values $t_{in}^m=\tilde{t}_i^m(\mathbf{x}^n)$. Note that $\tilde{t}_i^m(\mathbf{x})$ is the restriction of $\tilde{t}_i$ to S_m. The function $t_i(\mathbf{x})$ as the function $\tilde{t}_i(\mathbf{x})$ which it interpolates is solely defined by its restrictions to S_m, $m=1\ldots M$ and therefore in general discontinuous over adjacent sides of triangles S_m and multivalued at a node $\mathbf{x}^k$.

Inserting the expressions, Equations 4.105 and 4.106 in Equation 4.98, this equation becomes, if we evaluate it at a node $\mathbf{x}^k$, and for an index i;

$$\sum_{n=1}^{K}\sum_{m=1}^{M}\left\{\int_{S_m} U_{ij}(\mathbf{y},\mathbf{x}^k)p_n^m(\mathbf{y})\;\mathrm{d}s_y\right\}t_{jn}^m=\sum_{n=1}^{K}\left\{\oint_{S'} T_{ij}(\mathbf{y},\mathbf{x}^k)[p_n(\mathbf{y})-\delta_{nk}]\;\mathrm{d}s_y\right\}u_{jn} \tag{4.107}$$

(we have used $p_n(\mathbf{x}^k)=\delta_{nk}$ and summation over j is implied) or

$$\sum_{n=1}^{K}\sum_{m=1}^{K} G_{ij}^{nm}(\mathbf{x}^k)t_{jn}^m=\sum_{n=1}^{K} H_{ij}^n(\mathbf{x}^k)u_{jn} \tag{4.108}$$

with

$$G_{ij}^{nm}(\mathbf{x}^k)=\int_{S_m} U_{ij}(\mathbf{y},\mathbf{x}^k)p_n^m(\mathbf{y})\;\mathrm{d}s_y=\int_{S_0} U_{ij}(\dot{\mathbf{y}}^m(\boldsymbol{\eta}),\mathbf{x}^k)p_n^m(\boldsymbol{\eta})J^m(\boldsymbol{\eta})\;\mathrm{d}\boldsymbol{\eta} \tag{4.109}$$

and

$$H_{ij}^n(\mathbf{x}^k)=\oint_{S'} T_{ij}(\mathbf{y},\mathbf{x}^k)[p_n(\mathbf{y})-\delta_{nk}]\;\mathrm{d}s_y=\sum_{m=1}^{M}\int_{S_0} T_{ij}(\dot{\mathbf{y}}^m(\boldsymbol{\eta}),\mathbf{x}^k)[p_n^m(\boldsymbol{\eta})-\delta_{nk}]J^m(\boldsymbol{\eta})\;d\boldsymbol{\eta} \tag{4.110}$$

If i runs in Equation 4.108 through 1,2,3 and k through $1,2,\ldots K$ we get a set of $3\times K$ linear equations between the $3\times K$ values u_{jn}, $j=1,2,3$; $n=1\ldots K$ on the right-hand side and $3\times K'$ values t_{jn}^m on the left-hand side. K is the number of nodes on S' and we have $K=K_1+K_2+\ldots K_c$ where K_i is the total number of nodes, that lie on i elements S_m, and it follows

$$K'=\sum_{i=1}^{c} i\cdot K_i \tag{4.111}$$

Clearly it is $K'>K$ and Equation 4.108 is an under-determined system which must be modified prior to its solution if the values u_{jn} are known and the values

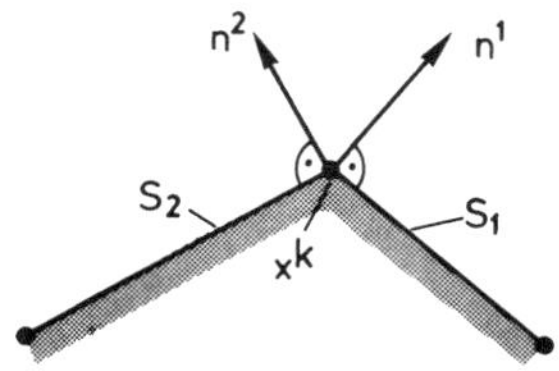

Figure 4.15 Case 2(ii). The traction vector is in general discontinuous in non-smooth boundary points due to $n^1 \neq n^2$; $t_i^1(x^k) = t_i^2(x^k)$, $i = 1,2$ (plane problem)

t_{jn}^m are unknown. The under-determination can be removed by reducing in any node the number of nodal unknowns to 3. To learn how this is done it suffices to study the situation in one arbitrary node. So let k be a fixed number. We designate for clarity by $t_i^m(\mathbf{x}^k)$ the value of the restriction $t_i^m(\mathbf{x})$ in the node $\mathbf{x}^k \in S_m$ and assume for simplicity that $\mathbf{x}^k$ lies on the first c' elements S_m, $m = 1,2,\ldots c'$. (In the previous notation it is $t_i^m(\mathbf{x}^k) = t_{ik}^m$).

We distinguish three cases:

(1) $t_i^m(\mathbf{x}^k)$ is prescribed for $m = 1,2,\ldots c'$, $i = 1,2,3$

(2) $u_{ik} = u_i(\mathbf{x}^k)$, $i = 1,2,3$ is prescribed and $t_i^m(\mathbf{x}^k)$, $m = 1,2,\ldots c'$; $i = 1,2,3$ is unknown.

(3) $u_{ik} = u_i(\mathbf{x}^k)$, $i = 1,2,3$ is prescribed and $t_i^m(\mathbf{x}^k)$ is prescribed for $m = 1,2,\ldots b$; $1 \leqslant b \leqslant c'$; $i = 1,2,3$ and $t_i^m(\mathbf{x}^k) m = b+1$; $b+2,\ldots c'$; $i = 1,2,3$ is unknown.

Case (1) corresponds to the 2nd bvp. The three traction components are prescribed functions on all elements S_m, $m = 1,2,\ldots c'$. The nodal unknowns at $\mathbf{x}^k$ are the three components $u_{ik} = u_i(\mathbf{x}^k)$, $i = 1,2,3$. The case (2) may be encountered in two different situations:

(i) at the node $\mathbf{x}^k$ the displacement components $u_i(\mathbf{x}^k)$, $i = 1,2,3$ are prescribed and the traction is unknown on S_m, $m = 1,2,\ldots c'$, and the traction is assumed to be continuous at $\mathbf{x}^k$. In this case it is $t_i^m(\mathbf{x}^k) = t_i(\mathbf{x}^k)$ for $1 \leqslant m \leqslant c'$, $i = 1,2,3$ and the three unknowns at $\mathbf{x}^k$ are $t_i(\mathbf{x}^k)$, $i = 1,2,3$.

(ii) The node $\mathbf{x}^k$ is a non-smooth point on S', i.e. the normal $\mathbf{n}$ is discontinuous. At such points a smooth displacement field possesses a discontinuous traction vector $\boldsymbol{\tau}(\mathbf{u})(\mathbf{x})$, which must be modelled by choosing different values $t_i^m(\mathbf{x}^k)$ for the function $t_i(\mathbf{x})$ at the same node on different elements S_m, see Figure 4.15.

The case (3) is encountered if the boundary condition changes from displacement to traction at the node, see Figure 4.16.

We could discuss a fourth case corresponding to the 3rd or 4th bvp when the nodal unknown are components of both $\mathbf{u}$ and $\mathbf{t}$ but the treatment would be essentially as in case (3).

In case (3) proceed as follows: pick three values, e.g. $t_i^b(\mathbf{x}^k)$, $i = 1,2,3$ as nodal unknowns, establish a linear relation that expresses the terms $t_i^{b+1}(\mathbf{x}^k)$, $t_i^{b+2}(\mathbf{x}^k)\ldots$ as functions of $t_i^b(\mathbf{x}^k)$, $i = 1,2,3$ and eliminate the terms $t_i^{b+1}(\mathbf{x}^k)$, $t_i^{b+2}(\mathbf{x}^k)$, ...etc. according to the established relations.

These relations are based on the properties of continuous stress and strain fields and we refer to Chaudonneret[41] and Wardle and Crotty[42] for details.

The same elimination procedure must be applied in the case 2(ii) when the traction is discontinuous and the number of nodal unknowns exceeds the number of equations in the integral equation which corresponds to the node $\mathbf{x}^k$.

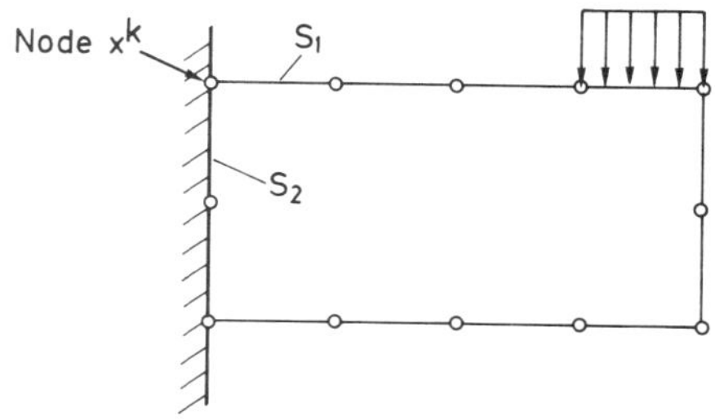

Figure 4.16 Case 3: The boundary condition changes at x^k; $u_i(x^k)=0$, $t_i^1(x^k)=0$, $t_i^2(x^k)$ = unknown, $i=1,2$ (plane problem)

If we follow the lines above, Equation 4.108 becomes

$$\mathbf{G}\{\mathbf{t}\}=\mathbf{H}\{\mathbf{u}\} \tag{4.112}$$

where **G** and **H** are square matrices $3K \times 3K$, $\{\mathbf{u}\}$ is the vector of the nodal values of $\mathbf{u}(\mathbf{x})$ at $\mathbf{x}^k$, $k=1,2,\ldots K$, and $\{\mathbf{t}\}$ the vector of the nodal values of $\mathbf{t}(\mathbf{x})$ in $\mathbf{x}^k$, $k=1,2,\ldots K$, or if $\mathbf{t}(\mathbf{x})$ is discontinuous and unknown at $\mathbf{x}^k$, three selected values $t_i^m(\mathbf{x}^k)$, $i=1,2,3$ on one element S_m, which belongs to the neighbourhood of $\mathbf{x}^k$.

Equation 4.112 becomes after a rearrangement, which depends on the kind of bvp we have to solve, a linear system for $3 \times K$ nodal values. The coefficient matrix of this linear system is fully populated and non-symmetric. After solution the functions 4.105 and 4.106 are inserted in Equation 4.83 (or its counterpart for exterior regions), which then represents an approximate solution of the bvp.

Up to here we have proposed to use one set of shape functions to model the boundary S and interpolate the functions $\tilde{\mathbf{t}}$ and $\tilde{\mathbf{u}}$ on S. We have seen that the strong singularity of the matrix **T** on S called for a Hölder-continuous layer $\tilde{\mathbf{u}}$, but with respect to the components of **t** we are free to choose singular functions like $t_i^m(\mathbf{y})=0(|\mathbf{y}-\mathbf{x}^k|^\alpha)$, $0<\alpha<1$, $\mathbf{y}$, $\mathbf{x}^k \in S_m$ because the **U**-transform of such a function still exists at the point $\mathbf{x}^k$:

$$\left|\int_{S_m} \mathbf{U}(\mathbf{y},\mathbf{x}^k)\tilde{\mathbf{t}}^m(\mathbf{y})\, ds_\mathbf{y}\right| < \infty \tag{4.113}$$

The introduction of singular shape functions is necessary because the traction $\tilde{\mathbf{t}}$, contrary to the trace $\tilde{\mathbf{u}}$, can become singular at critical points of the boundary, (cf. the example at the end of Section 4.3), and it seems therefore reasonable to approximate $\tilde{\mathbf{t}}$ near such points by (mildly) singular functions. (Studying the Equation 4.113 one might conjecture that the existence of the **U**-transform on S of the traction $\tilde{\mathbf{t}}(\mathbf{y})$ is a necessary condition for the traction of any well posed problem. This would set a limit, as indicated above, to the order of possible singularities on S).

The foremost practical problem in the numerical process is the calculation of the coefficients

$$G_{ij}^{nm}(\mathbf{x}^k), \qquad H_{ij}^{nm}(\mathbf{x}^k)$$

and it is easy to understand that the quality of the solution depends mainly on how exactly these can be calculated. For $\mathbf{x}^k \notin S_m$ the integrals in Equation 4.109 and 4.110 are regular and allow in principle the application of numerical quadrature without modification, though a modification seems reasonable, which takes into account the distance between the source point $\mathbf{x}^k$ and the integration point $\mathbf{y} \in S_m$.

For $\mathbf{x}^k \in S_m$ the integrands are weakly singular and prior to using a quadrature rule they must be modified. For a detailed discussion of the methods, which have become public up to now, we refer to Lachat and Watson[43] and Rizzo and Shippy[44] and also to Cristescu and Loubignac[45] and Wendland[12].

4.5 AXISYMMETRIC ELASTIC STATES

In this section we derive by a formal transformation the Somigliana identity for axisymmetric problems. We will see that an axisymmetric elastic state is uncoupled: Radial (r) and axial (z) quantities do not depend on tangential (φ) quantities (and vice versa). Axisymmetric problems therefore reduce to integral equations for at most two quantities on the curve C that generates the axisymmetric body B on rotating around the vertical axis z. This is quite a simplification in contrast to the general three-dimensional case.

The explicit formulae in Section 4.5.3 and 4.5.4 stem from the works of Mayr[46] and Mayr and Neureiter[47]. They were not derived by a step by step transformation of 'cartesian kernels' into 'cylindrical kernels', but by using from the start the equations of fundamental solutions in cylindrical coordinates and basis vectors and repeating the process of the formulation of the Somigliana identity.

Mayr and Neureiter actually work with these formulae in a computer program which is used successfully, so we may trust that their formulae are correct. Dr Mayr has been so kind to check their reproduction in this book. Nevertheless, we would strongly recommend to anyone who wants to use them in a computer program to contact Dr Kuhn[72] for a final check.

4.5.1 **Representations in cylindrical coordinates**

Cartesian coordinates with respect to the unit vectors $\mathbf{e}_1$, $\mathbf{e}_2$, $\mathbf{e}_3$ are designated by x_1, x_2, x_3 and cylindrical coordinates by r,φ,z, where

$$x_1 = r\cos\varphi,\ x_2 = r\sin\varphi,\ x_3 = z \tag{4.114}$$

At any point $\mathbf{P} = (r,\varphi,z)$ there exists three unit vectors $\mathbf{e}_r, \mathbf{e}_\phi, \mathbf{e}_z$

$$\left\{\begin{array}{c}\mathbf{e}_r\\ \mathbf{e}_\phi\\ \mathbf{e}_z\end{array}\right\}=\left[\begin{array}{ccc}\cos\varphi & r\sin\varphi & 0\\ -r\sin\varphi & \cos\varphi & 0\\ 0 & 0 & 1\end{array}\right]\left\{\begin{array}{c}\mathbf{e}_1\\ \mathbf{e}_2\\ \mathbf{e}_3\end{array}\right\}=\mathbf{A}^T\left\{\begin{array}{c}\mathbf{e}_1\\ \mathbf{e}_2\\ \mathbf{e}_3\end{array}\right\} \tag{4.115}$$

If $\hat{\mathbf{v}}=[v_r,v_\phi,v_z]^T$ are the components of a vector $\vec{\mathbf{v}}$ at the point (r,φ,z) with respect to $\mathbf{e}_r,\mathbf{e}_\phi,\mathbf{e}_z$, then its cartesian components $\mathbf{v}=[v_1,v_2,v_3]^T$ at the point (r,φ,z) are

$$\mathbf{v}=\mathbf{A}\cdot\hat{\mathbf{v}} \tag{4.116}$$

Note that $\mathbf{A}^T=\mathbf{A}^{-1}$ wherever $\mathbf{A}^{-1}$ exists.

Suppose there is a vector valued function $\mathbf{u(x)}$ in a domain B such that the Somigliana identity, Equations 4.80 and 4.81, holds and let $\hat{\mathbf{u}}(\mathbf{x})$, $\hat{\boldsymbol{\tau}}(\mathbf{u})(\mathbf{x})$ designate the coordinate vectors of $\mathbf{u(x)}$ and $\mathbf{t(x)}$ with respect to the cylindrical coordinates. Then the following equation is equivalent with Equations 4.80 and 4.81.

$$\mathbf{C(P)A(P)\hat{u}(P)}=\oint_S[\mathbf{U(P,Q)A(Q)\hat{\tau}(u)(Q)}-\mathbf{T(P,Q)A(Q)\hat{u}(Q)}]\,ds_\mathbf{Q}+\int_B\mathbf{U(P,Q)A(Q)\hat{b}(Q)}\,dv_\mathbf{Q} \tag{4.117}$$

with $\mathbf{U(P,Q)}$ and $\mathbf{T(P,Q)}$ as in Equations 4.10 and 4.13 and $\mathbf{Q}=(\rho,\psi,\zeta)$, $\mathbf{P}=(r,\varphi,z)$.

Multiplying the equation on both sides from the left with the matrices $\mathbf{C(P)}^{-1}$, $\mathbf{A(P)}^T$ and $\mathbf{C(P)}$ consecutively, we get

$$\mathbf{C(P)\hat{u}(P)}=\mathbf{C(P)\hat{A}(P)}^T\mathbf{C(P)}^{-1}\left[\oint_S[\mathbf{UA\hat{\tau}(u)}-\mathbf{TA\hat{u}}]\,ds_\mathbf{Q}+\int_B\mathbf{UA\hat{b}}\,dv_\mathbf{Q}\right] \tag{4.118}$$

If $\mathbf{P}$ is an interior point or a smooth boundary point, then $\mathbf{C(P)}=\mathbf{I}$ or $\frac{1}{2}\mathbf{I}$ and we obtain

$$\mathbf{C(P)\hat{u}(P)}=\oint_S[\mathbf{\hat{U}(P,Q)\hat{\tau}(u)(Q)}-\mathbf{\hat{T}(P,Q)\hat{u}(Q)}]\,ds_\mathbf{Q}+\int_B\mathbf{\hat{U}(P,Q)\hat{b}(Q)}\,dv_\mathbf{Q} \tag{4.119}$$

with

$$\mathbf{\hat{U}(P,Q)}=\mathbf{A(P)}^T\mathbf{U(P,Q)A(Q)}$$

$$\mathbf{\hat{T}(P,Q)}=\mathbf{A(P)}^T\mathbf{T(P,Q)A(Q)}$$

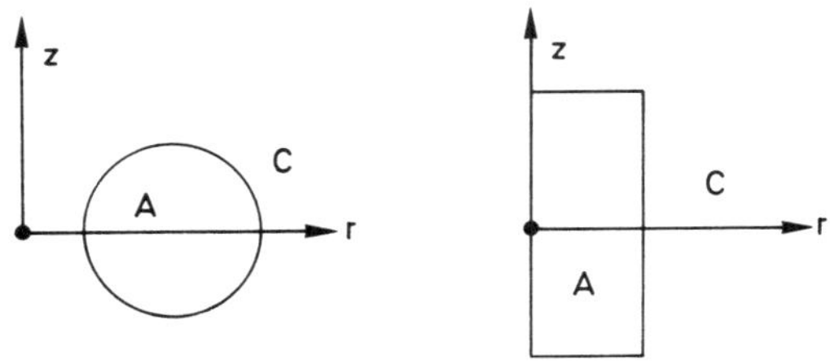

Figure 4.1[[The first curve generates a torus, the second a cylinder.

and

$$\mathbf{C(P)} = \begin{cases} \mathbf{I}, & \mathbf{x} \in B \\ 1/2\ \mathbf{I}, & \mathbf{x} \in S \text{ (smooth point)} \\ \mathbf{0}, & \mathbf{x} \in \bar{B} \end{cases} \tag{4.120}$$

Note that the three columns of $\hat{\mathbf{U}}(\mathbf{P},\mathbf{Q})$ represent the components of the displacement at the point $\mathbf{P}=(r,\varphi,z)$ with respect to the unit vectors $\mathbf{e}_r,\mathbf{e}_\phi,\mathbf{e}_z$ due to a single load $\mathbf{e}_\rho,\mathbf{e}_\psi$ or $\mathbf{e}_\zeta$ at the point $\mathbf{Q}=(\rho,\psi,\zeta)$.

4.5.2 Axisymmetric problems

Suppose the body B is axisymmetric, i.e. there exists a closed curve C in the plane r, z such that C on rotating around the axis z generates the surface S of B, see Figure 4.17. Let A designate the area bounded by C in the plane r,z. It is (let dS with a capital S denote here the surface element)

$$\mathrm{d}S=\rho\ \mathrm{d}\psi\ \mathrm{d}s \qquad \mathrm{d}v=\rho\ \mathrm{d}\psi\ \mathrm{d}\rho\ \mathrm{d}\zeta=\rho\ \mathrm{d}\psi\ \mathrm{d}A \tag{4.121}$$

where s is the arc length on C. Suppose the displacement field $\hat{\mathbf{u}}$ in B is axisymmetric, i.e. $\hat{\mathbf{u}}=\hat{\mathbf{u}}(r,z)$. Then $\hat{\boldsymbol{\tau}}(\mathbf{u})$ on S is also axisymmetric due to S and Equation 4.119 becomes, after integrating with respect to ψ from $\psi=0$ to $\psi=2\pi$,

$$\begin{aligned}\mathbf{C(P')}\hat{\mathbf{u}}(\mathbf{P'}) = 2\pi\oint_C[\dot{\mathbf{U}}(\mathbf{P'},\mathbf{Q'})\hat{\boldsymbol{\tau}}(\mathbf{u})(\mathbf{Q'}) - \dot{\mathbf{T}}(\mathbf{P'},\mathbf{Q'})\hat{\mathbf{u}}(\mathbf{Q'})]\rho\ \mathrm{d}s_{\mathbf{Q}} + \\ 2\pi\int_A \dot{\mathbf{U}}(\mathbf{P'},\mathbf{Q'})\hat{\mathbf{b}}(\mathbf{Q'})\rho\ \mathrm{d}A_{\mathbf{Q'}}\end{aligned} \tag{4.122}$$

where $\mathbf{Q'}=(\rho,\zeta)$, $\mathbf{P'}=(r,z)$, and

$$2\pi\cdot\dot{\mathbf{U}}(\mathbf{P'},\mathbf{Q'}) = \int_0^{2\pi}\hat{\mathbf{U}}(\mathbf{P},\mathbf{Q})\ \mathrm{d}\psi \tag{4.123}$$

$$2\pi \cdot \dot{\mathbf{T}}(\mathbf{P}',\mathbf{Q}') = \int_0^{2\pi} \hat{\mathbf{T}}(\mathbf{P},\mathbf{Q})\, \mathrm{d}\psi \tag{4.124}$$

The right-hand side in Equations 4.123 and 4.124 does not depend on φ anymore† after integration over ψ, so we may write $\mathbf{P}=\mathbf{P}'$ on the left-hand side. It is

$$\dot{\mathbf{U}}(\mathbf{P}',\mathbf{Q}') = [\dot{\mathbf{g}}^r,\dot{\mathbf{g}}^\phi,\dot{\mathbf{g}}^z] = \begin{bmatrix} u_r^r & u_r^\phi & u_r^z \\ u_\phi^r & u_\phi^\phi & u_\phi^z \\ u_z^r & u_z^\phi & u_z^z \end{bmatrix} = \begin{bmatrix} * & 0 & * \\ 0 & * & 0 \\ * & 0 & * \end{bmatrix} \tag{4.125}$$

$$\dot{\mathbf{T}}(\mathbf{P}',\mathbf{Q}') = [\dot{\mathbf{t}}^r,\dot{\mathbf{t}}^\phi,\dot{\mathbf{t}}^z] = \begin{bmatrix} t_r^r & t_r^\phi & t_r^z \\ t_\phi^r & t_\phi^\phi & t_\phi^z \\ t_z^r & t_z^\phi & t_z^z \end{bmatrix} = \begin{bmatrix} * & 0 & * \\ 0 & * & 0 \\ * & 0 & * \end{bmatrix} \tag{4.126}$$

with

$$t_k^j = \tau_{kl}^j \hat{n}_l, \qquad j,k,l = r,\varphi,z \tag{4.127}$$

$\hat{\mathbf{n}} = [n_r,n_\phi,n_z]^T = [n_r,0,n_z]^T$ the outward normal, and integration on C in Equation 4.122 is done with A lying to the left. It is evident, after what has been said at the end of Section 4.5.1, that the vectors $\dot{\mathbf{g}}^r$, $\dot{\mathbf{g}}^\phi$ and $\dot{\mathbf{g}}^z$ times $2\pi\rho$ denote the displacement of an arbitrary point $\mathbf{P}$ on the circle $c_{\mathbf{P}'}(r,z)$ with center $\mathbf{x} = (0,0,z)$ and radius r due to a constant distribution of loads (force/unit length) of magnitude $q=1$ in radial, axial and tangential direction on a circle $c_{\mathbf{Q}'}(\rho,\zeta)$ with center $\mathbf{x}=(0,0,\zeta)$ and radius ρ. The vectors $\dot{\mathbf{t}}^r$, $\dot{\mathbf{t}}^\phi$ and $\dot{\mathbf{t}}^z$ times $2\pi\rho$ denote the corresponding tractions at $\mathbf{P}$ with respect to a normal $\mathbf{n}$ at $\mathbf{P}$. The distribution of the zero elements in the matrices $\dot{\mathbf{U}}(\mathbf{P}',\mathbf{Q}')$ and $\dot{\mathbf{T}}(\mathbf{P}',\mathbf{Q}')$ cf. Equations 4.125 and 4.126, is a result of the next Section, 4.5.3. Instead of source points where a single load is located we now encounter rings, where loads are distributed in different directions. These ring sources are called 'radial source', 'axial source' and 'tangential source' (Mayr).

Equation 4.122 is the Somigliana identity for axisymmetric regular solutions in axisymmetric bodies in terms of vector components which refer to a basis corresponding to cylindrical coordinates.

Studying Equations 4.125 and 4.126 we see indeed that the displacement components u_r, u_z on one side and u_ϕ on the other side (the same holds for the tractions t_r, t_z and t_ϕ) of any axisymmetric elastic state in an axisymmetric body are uncoupled.

4.5.3 The kernels

Next we list the elements of the matrix $\dot{\mathbf{U}}$ of Equation 4.125 and the elements of the stress tensors $\mathbf{S}^k = [\tau_{ij}^k]$ which are needed to calculate (see Equation

† This is easily understood if we realize the physical meaning of Equations 4.123 and 4.124. See also the remarks following Equation 4.127.

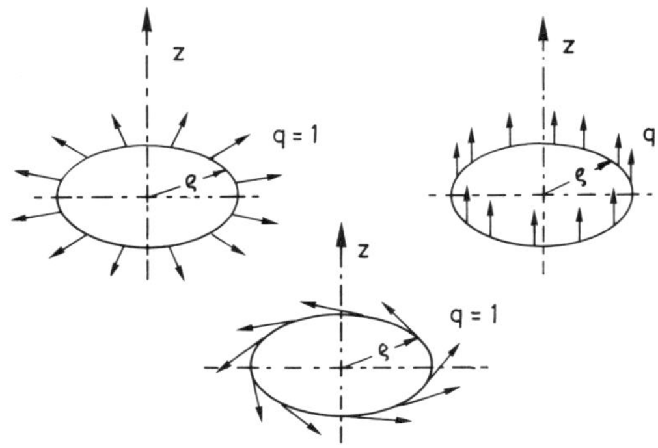

Figure 4.18 Ring sources

4.127) the elements of the matrix $\dot{\mathbf{T}}$ in Equation 4.126. Mayr denotes the general term of a stress tensor **S** by τ_{ij} and the terms on the diagonal of S by $\sigma_r, \sigma_\phi, \sigma_z$. We will follow this convention here. The letter e, which is to designate r in his formulae has been replaced by the more natural choice r. The number of the equations coincide with the numbers in Mayr[46]; Equation 2.2.1/12 is incomplete but printed correctly here. Finally the abbreviation

$$\tilde{\mathbf{z}} = z - \zeta \tag{4.128}$$

is used throughout the formulae.

$$u_r^r = \frac{1}{16\pi^2\mu(1-\nu)} \frac{1}{r\rho\sqrt{(r+\rho)^2+\tilde{z}^2}} \left\{ \left\{ -4(1-\nu)[(r+\rho)^2+\tilde{z}^2] + \frac{(r^2-\rho^2)+\tilde{z}^2(r^2+\rho^2)}{(r-\rho)^2+\tilde{z}^2} \right\} E(\kappa) + \{4(1-\nu)(r^2+\rho^2+\tilde{z}^2)-(r^2+\rho^2)\} K(\kappa) \right\} \tag{2.2.1/12}$$

$$u_\phi^r = 0 \tag{2.2.1/13}$$

$$u_z^r = \frac{1}{16\pi^2\mu(1-\nu)} \frac{\tilde{z}}{\rho\sqrt{(r+\rho)^2+\tilde{z}^2}} \left\{ \frac{r^2-\rho^2+\tilde{z}^2}{(r-\rho)^2+\tilde{z}^2} E(\kappa) - K(\kappa) \right\} \tag{2.2.1/14}$$

$$u_\phi^\phi = \frac{1}{4\pi^2\mu} \frac{1}{r\rho\sqrt{(r+\rho)^2+\tilde{z}^2}} \{(r^2+\rho^2+\tilde{z}^2)K(\kappa) - [(r+\rho)^2+\tilde{z}^2]E(\kappa)\}$$

((2) in Mayr[47])

$$u_r^\phi = u_z^\phi = 0$$

$$u_r^z = \frac{1}{16\pi^2\mu(1-v)} \frac{\tilde{z}}{r\sqrt{(r+\rho)^2+\tilde{z}^2}} \left\{ \frac{r^2-\rho^2-\tilde{z}^2}{(r-\rho)^2+\tilde{z}^2} E(\kappa) + K(\kappa) \right\} \quad (2.2.2/11)$$

$$u_\phi^z = 0 \quad (2.2.2/12)$$

$$u_z^z = \frac{1}{8\pi^2\mu(1-v)} \frac{1}{\sqrt{(r+\rho)^2+\tilde{z}^2}} \left\{ \frac{\tilde{z}^2}{(r-\rho)^2+\tilde{z}^2} E(\kappa) + (3-4v)K(\kappa) \right\} \quad (2.2.2/13)$$

$$\sigma_r^r = \frac{1}{32\pi^2(1-v)} \frac{1}{\rho r^2\sqrt{(r+\rho)^2+\tilde{z}^2}} \left\{ \left\{ (7-8v)[(r+\rho)^2+\tilde{z}^2] \right.\right.$$

$$+ \frac{4(1-v)[\rho^2(\rho^2-r^2)+\tilde{z}^2(2\rho^2+r^2+\tilde{z}^2)] - 3(r^2-\rho^2)^2 - 6\tilde{z}^2(r^2-\rho^2) + 9\tilde{z}^4}{(r-\rho)^2+\tilde{z}^2}$$

$$\left. + \frac{4[(\rho+r)^3(\rho-r)^3 - \tilde{z}^6 + \tilde{z}^2(r^2-\rho^2)(\tilde{z}^2+r^2-\rho^2)](r^2+\rho^2+\tilde{z}^2)}{[(r+\rho)^2+\tilde{z}^2][(r-\rho)^2+\tilde{z}^2]^2} \right\} E(\kappa)$$

$$+ \left\{ 3(r^2-\rho^2) - 9\tilde{z}^2 - 4(1-2v)(r^2+2\rho^2+2\tilde{z}^2) \right.$$

$$\left.\left. - \frac{(\rho+r)^3(\rho-r)^3 - \tilde{z}^6 + \tilde{z}^2(r^2-\rho^2)(\tilde{z}^2+r^2-\rho^2)}{[(r+\rho)^2+\tilde{z}^2][(r-\rho)^2+\tilde{z}^2]} \right\} K(\kappa) \right\} \quad (2.2.1/15)$$

$$\sigma_\phi^r = \frac{1}{32\pi^2(1-v)} \frac{1}{\rho r^2\sqrt{(r+\rho)^2+\tilde{z}^2}} \left\{ \left\{ -(7-8v)[(r+\rho)^2+\tilde{z}^2] \right.\right.$$

$$\frac{+4(1-v)[\rho^2(r^2-\rho^2) - \tilde{z}^2(r^2+2\rho^2+\tilde{z}^2)] + 6r^2\rho^2 - 9r^4 + 3\rho^2 - 3\tilde{z}^2(6r^2+2\rho^2+3\tilde{z}^2)}{(r-\rho)^2+\tilde{z}^2}$$

$$\left. \frac{+4[(r-\rho)^3(r+\rho)^3 + 3r^2\tilde{z}^2(r^2+\tilde{z}^2) + \rho^2\tilde{z}^2(\tilde{z}^2-\rho^2-2r^2) - \tilde{z}^6](r^2+\rho^2+\tilde{z}^2)]}{(r^2+\rho^2+\tilde{z}^2)][(r+\rho)^2+\tilde{z}^2][(r-\rho)^2+\tilde{z}^2]^2} \right\} E(\kappa)$$

$$+ \left\{ 4(1-2v)[r^2+2\rho^2+2\tilde{z}^2] + 3(3r^2+\rho^2+3\tilde{z}^2) \right.$$

$$\left.\left. - \frac{(r-\rho)^3(r+\rho)^3 + 3r^2\tilde{z}^2(r^2+\tilde{z}^2) + \rho^2\tilde{z}^2(\tilde{z}^2-\rho^2-2r^2) - \tilde{z}^6}{[(r+\rho)^2+\tilde{z}^2][(r-\rho)^2+\tilde{z}^2]} \right\} K(\kappa) \right\} \quad (2.2.1/16)$$

$$\sigma_z^r = \frac{1}{8\pi^2(1-\nu)} \frac{1}{\rho\sqrt{(r+\rho)^2+\tilde{z}^2}} \left\{ \left\{ \frac{(1-2\nu)(r^2-\rho^2+\tilde{z}^2)+3\tilde{z}^2}{(r-\rho)^2+\tilde{z}^2} \right. \right.$$

$$\left. + \frac{4\tilde{z}^2(\rho^2+r^2+\tilde{z}^2)(\rho^2-r^2-\tilde{z}^2)}{[(r+\rho)^2+\tilde{z}^2][(r-\rho)^2+\tilde{z}^2]^2} \right\} E(\kappa) \tag{2.2.1/17}$$

$$\left. - \left\{ 1-2\nu + \frac{\tilde{z}^2(\rho^2-r^2-\tilde{z}^2)}{[(r+\rho)^2+\tilde{z}^2][(r-\rho)^2+\tilde{z}^2]} \right\} K(\kappa) \right\}$$

$$\tau_{rz}^r = \frac{1}{16\pi^2(1-\nu)} \frac{\tilde{z}}{\rho r\sqrt{(r+\rho)^2+\tilde{z}^2}} \left\{ \left\{ -\frac{6\tilde{z}^2+2(1-2\nu)(r^2+\rho^2+\tilde{z}^2)}{(r-\rho)^2+\tilde{z}^2} \right. \right.$$

$$\left. + \frac{4[\tilde{z}^4-(r^2-\rho^2)^2](r^2+\rho^2+\tilde{z}^2)}{[(r+\rho)^2+\tilde{z}^2][(r-\rho)^2+\tilde{z}^2]^2} \right\} E(\kappa) \tag{2.2.1/18}$$

$$\left. + \left\{ 3+2(1-2\nu) - \frac{\tilde{z}^4-(r^2-\rho^2)^2}{[(r+\rho)^2+\tilde{z}^2][(r-\rho)^2+\tilde{z}^2]} \right\} K(\kappa) \right\}$$

$$\tau_{r\phi}^r = \tau_{z\phi}^r = 0$$

$$\sigma_r^\phi = \sigma_\phi^\phi = \sigma_z^\phi = 0$$

$$\tau_{r\phi}^\phi = \frac{1}{8\pi^2} \frac{1}{r^2\rho\sqrt{(r+\rho)^2+\tilde{z}^2}} \left\{ -2(2\rho^2+r^2+2\tilde{z}^2)K(\kappa) + \right.$$

((3) in Mayr[47])

$$\left. + \left[3((r+\rho)^2+\tilde{z}^2) + \frac{(r^2+\rho^2+\tilde{z}^2)(\rho^2-r^2+\tilde{z}^2)}{(r-\rho)^2+\tilde{z}^2} \right] E(\kappa) \right\}$$

$$\tau_{z\phi}^\phi = \frac{1}{4\pi^2} \frac{\tilde{z}}{r\rho\sqrt{(r+\rho)^2+\tilde{z}^2}} \left\{ K(\kappa) - \frac{r^2+\rho^2+\tilde{z}^2}{(r-\rho)^2+\tilde{z}^2} E(\kappa) \right\}$$

((4) in Mayr[47])

$$\tau_{rz}^\phi = 0$$

$$\sigma_r^z = \frac{1}{16\pi^2(1-\nu)} \frac{\tilde{z}}{r^2\sqrt{(r+\rho)^2+\tilde{z}^2}} \left\{ \left\{ \frac{4(1-2\nu)r^2+6(\rho^2-r^2+\tilde{z}^2)}{(r-\rho)^2+\tilde{z}^2} \right. \right.$$

$$\left. - \frac{4(r^2+\rho^2+\tilde{z}^2)(\rho^2-r^2+\tilde{z}^2)^2}{[(r+\rho)^2+\tilde{z}^2][(r-\rho)^2+\tilde{z}^2]^2} \right\} E(\kappa) + \left\{ -3 \right. \tag{2.2.2/14}$$

$$\left. \left. + \frac{(\rho^2-r^2+\tilde{z}^2)^2}{[(r+\rho)^2+\tilde{z}^2][(r-\rho)^2+\tilde{z}^2]} \right\} K(\kappa) \right\}$$

$$\sigma_\phi^z = \frac{1}{16\pi^2(1-\nu)} \frac{\tilde{z}}{r^2\sqrt{(r+\rho)^2+\tilde{z}^2}} \left\{ \left\{ \frac{4(1-2\nu)r^2 - 6(r^2+\rho^2+\tilde{z}^2)}{(r-\rho)^2+\tilde{z}^2} \right.\right.$$

$$+ \left. \frac{4(r^2+\rho^2+\tilde{z}^2)[(r^2-\rho^2)^2+\tilde{z}^2(2r^2+2\rho^2+\tilde{z}^2)]}{[(r+\rho)^2+\tilde{z}^2][(r-\rho)^2+\tilde{z}^2]^2} \right\} E(\kappa) \tag{2.2.2/15}$$

$$+ \left. \left\{ 3 - \frac{(r^2-\rho^2)^2+\tilde{z}^2(2r^2+2\rho^2+\tilde{z}^2)}{[(r+\rho)^2+\tilde{z}^2][(r-\rho)^2+\tilde{z}^2]} \right\} K(\kappa) \right\}$$

$$\sigma_z^z = \frac{1}{4\pi^2(1-\nu)} \frac{\tilde{z}}{\sqrt{(r+\rho)^2+\tilde{z}^2}} \left\{ \left\{ -\frac{1-2\nu}{(r-\rho)^2+\tilde{z}^2} \right.\right.$$

$$- \left. \frac{4\tilde{z}^2(r^2+\rho^2+\tilde{z}^2)}{[(r+\rho)^2+\tilde{z}^2][(r-\rho)^2+\tilde{z}^2]^2} \right\} E(\kappa) \tag{2.2.2/16}$$

$$+ \left. \frac{\tilde{z}^2}{[(r+\rho)^2+\tilde{z}^2][(r-\rho)^2+\tilde{z}^2]} K(\kappa) \right\}$$

$$\tau_{rz}^z = \frac{-1}{8\pi^2(1-\nu)} \frac{1}{r\sqrt{(r+\rho)^2+\tilde{z}^2}} \left\{ \left\{ \frac{3\tilde{z}^2+(1-2\nu)(r^2-\rho^2-\tilde{z}^2)}{(r-\rho)^2+\tilde{z}^2} \right.\right.$$

$$+ \left. \frac{4\tilde{z}^2(r^2+\rho^2+\tilde{z}^2)(r^2-\rho^2-\tilde{z}^2)}{[(r+\rho)^2+\tilde{z}^2][(r-\rho)^2+\tilde{z}^2]^2} \right\} E(\kappa) + \left\{ 1-2\nu \right. \tag{2.2.2/17}$$

$$+ \left. \left. \frac{\tilde{z}^2(\rho^2-r^2+\tilde{z}^2)}{[(r+\rho)^2+\tilde{z}^2][(r-\rho)^2+\tilde{z}^2]} \right\} K(\kappa) \right\}$$

$$\tau_{r\phi}^z = \tau_{\phi z}^z = 0 \tag{2.2.2/18}$$

The expressions

$$K(\kappa) = \int_0^{\pi/2} [1-\kappa^2 \sin^2\alpha]^{-1/2} \, \mathrm{d}\alpha \tag{4.129}$$

$$E(\kappa) = \int_0^{\pi/2} [1-\kappa^2 \sin^2\alpha]^{1/2} \, \mathrm{d}\alpha \tag{4.130}$$

denote the complete elliptic integral of the first and second kind. It is

$$\kappa^2 = \frac{4r\rho}{(r+\rho)^2 + \tilde{z}^2} \tag{4.131}$$

For $r=0$ the functions u_j^i, τ_{jk}^k (naturally) exist too. Their value at $r=0$ is the limit when r approaches zero which is calculated by expanding $E(\kappa)$ and $K(\kappa)$ into series and letting r tend to zero. This way we get for the point $\mathbf{P}'=(0,z)$,

$$u_r^r = u_\phi^r = 0$$

$$u_z^r = \frac{-1}{16\pi\mu(1-\nu)} \frac{\tilde{z}\rho}{\sqrt{\rho^2+\tilde{z}^2}^3}$$

$$u_r^\phi = u_\phi^\phi = u_z^\phi = 0 \qquad \text{(see Mayer}^{47}\text{)}$$

$$u_r^z = u_\phi^z = 0$$

$$u_z^z = \frac{1}{16\pi\mu(1-\nu)} \frac{1}{\sqrt{\rho^2+\tilde{z}^2}} \left\{ \frac{\tilde{z}^2}{\rho^2+\tilde{z}^2} + 3 - 4\nu \right\}$$

$$\sigma_r^r = \frac{3}{16\pi(1-\nu)} \frac{1}{\rho\sqrt{\rho^2+\tilde{z}^2}} \left\{ 1 - \frac{\tilde{z}^2}{\rho^2+\tilde{z}^2} - \frac{\tilde{z}^2\rho^2}{(\rho^2+\tilde{z}^2)^2} \right\}$$

$$\sigma_\phi^r = \sigma_r^r$$

$$\sigma_z^r = \frac{1}{16\pi(1-\nu)} \frac{1}{\rho\sqrt{\rho^2+\tilde{z}^2}} \left\{ -1 + 2\nu + \frac{(1-2\nu)(\tilde{z}^2-\rho^2)+6\tilde{z}^2}{\rho^2+\tilde{z}^2} - \frac{6\tilde{z}^4}{(\rho^2+\tilde{z}^2)^2} \right\}$$

$$\tau_{rz}^r = 0$$

$$\tau_{r\phi}^r = \tau_{\phi z}^r = 0$$

$$\sigma_r^\phi = \sigma_\phi^\phi = \sigma_z^\phi = \tau_{r\phi}^\phi = \tau_{rz}^\phi = \tau_{\phi z}^\phi = 0 \qquad \text{(see Mayer}^{47}\text{)}$$

$$\sigma_r^z = \frac{1}{8\pi(1-\nu)} \frac{\tilde{z}}{\sqrt{\rho^2+\tilde{z}^2}} \left\{ \frac{1-2\nu}{\rho^2+\tilde{z}^2} - \frac{3\rho^2}{2(\rho^2+\tilde{z}^2)^2} \right\}$$

$$\sigma_\phi^z = \sigma_r^z$$

$$\sigma_z^z = \frac{-1}{8\pi(1-\nu)} \frac{\tilde{z}}{\sqrt{\rho^2+\tilde{z}^2}} \left\{ \frac{1-2\nu}{\rho^2+\tilde{z}^2} + \frac{3\tilde{z}^2}{(\rho^2+\tilde{z}^2)^2} \right\}$$

$$\tau_{rz}^z = 0$$

$$\tau_{r\phi}^z = \tau_{\phi z}^z = 0$$

4.5.4 Stresses on the surface

To calculate stresses τ_{ij} at internal points, we could make use of Equation 4.92:

$$\tau_{ij}(\mathbf{x}) = \oint_B [\mathbf{D}_{ij}^T(\mathbf{y},\mathbf{x})\tilde{\mathbf{t}}(\mathbf{y}) - \mathbf{S}_{ij}^T(\mathbf{y},\mathbf{x})\tilde{\mathbf{u}}(\mathbf{y})]\ \mathrm{d}s_y + \int_B \mathbf{D}_{ij}^T(\mathbf{y},\mathbf{x})\mathbf{b}(\mathbf{y})\ \mathrm{d}v_y \qquad (4.132)$$

and by applying the transformation matrices $\mathbf{A}(\mathbf{P})$ and $\mathbf{A}(\mathbf{Q})$ derive an integral representation for the components τ_{ij} of the stress tensor with respect to the cylindrical coordinates and then use the axisymmetric properties of $\mathbf{t}$ and $\mathbf{u}$ to reduce the integrals to a curve resp. area integral in the sense above. But we would like to leave the actual calculation as an 'exercise' to the reader and rather state formulae for the calculation of stresses on the surface, because this requires no further determination of special kernels for axisymmetric stresses.

First we want to calculate the elements of the stress tensor at the point $\mathbf{P}$ on S from the data $\tilde{t}_r,\tilde{t}_z,\tilde{u}_r,\tilde{u}_z,\mathbf{n},\lambda,\mu$ $\left(\text{it is } \lambda = \dfrac{2\mu\nu}{(1-2\nu)}\right)$, where the first four quantities are the non-tangential components of the traction vector and displacement vector on S. Employing the relations

$$\tau_{ij} = \lambda\delta_{ij}\varepsilon_{kk} + 2\mu\varepsilon_{ij} \qquad i,j,k = r,\varphi,z \qquad (4.133)$$

(summation over repeated indices is implied in Equation 4.133) and

$$\varepsilon_{rr} = u_{r,r} \quad \varepsilon_{\phi\phi} = \frac{u_r}{r} \quad \varepsilon_{zz} = u_{z,z} \qquad (4.134)$$

$$\varepsilon_{rz} = u_{r,z} + u_{z,r} \quad \varepsilon_{r\phi} = \varepsilon_{z\phi} = 0$$

$$\begin{aligned} t_r &= \sigma_r \cos\beta + \tau_{rz} \sin\beta \\ t_z &= \tau_{rz} \cos\beta + \sigma_z \sin\beta \end{aligned} \qquad (4.135)$$

$$\begin{aligned} u_{r,s} &= -u_{r,r} \sin\beta + u_{r,z} \cos\beta \\ u_{z,s} &= -u_{z,r} \sin\beta + u_{z,z} \cos\beta \end{aligned} \qquad (4.136)$$

we get as in Section 4.4.4 the system

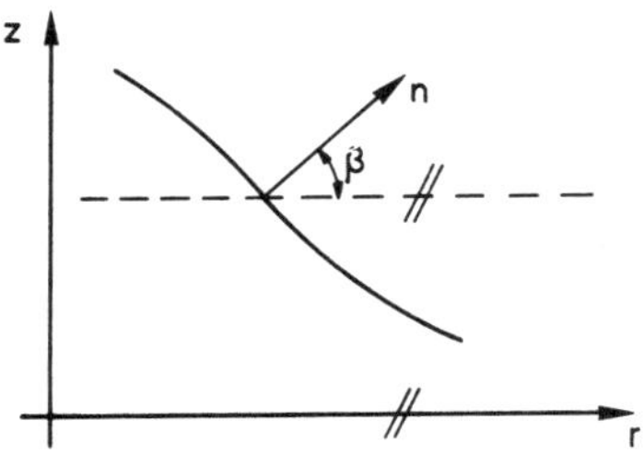

Figure 4.19

$$\begin{bmatrix} 1 & 0 & 0 & 0 & -\omega & 0 & 0 & -\lambda \\ \xi & 0 & 0 & \eta & 0 & 0 & 0 & 0 \\ 0 & 0 & \eta & \xi & 0 & 0 & 0 & 0 \\ 0 & 1 & 0 & 0 & -\lambda & 0 & 0 & -\lambda \\ 0 & 0 & 1 & 0 & -\lambda & 0 & 0 & -\omega \\ 0 & 0 & 0 & 1 & 0 & -\mu & -\mu & 0 \\ 0 & 0 & 0 & 0 & -\eta & 0 & \xi & 0 \\ 0 & 0 & 0 & 0 & 0 & -\eta & 0 & 0 \end{bmatrix} \cdot \begin{bmatrix} \sigma_r \\ \sigma_\phi \\ \sigma_z \\ \tau_{rz} \\ u_{r,r} \\ u_{z,r} \\ u_{r,z} \\ u_{z,z} \end{bmatrix} = \begin{bmatrix} \frac{\tilde{u}_r}{r}\lambda \\ \tilde{t}_r \\ \tilde{t}_z \\ \frac{\tilde{u}_r}{r}\omega \\ \frac{\tilde{u}_r}{r}\lambda \\ 0 \\ \tilde{u}_{r,s} \\ \tilde{u}_{z,s} \end{bmatrix} \qquad (4.137)$$

where

$$\xi = \cos\beta \qquad \eta = \sin\beta$$

$$\omega = 2\mu + \lambda$$

and β is the angle of the normal **n**, see Figure 4.19.

The derivatives $u_{i,s}$, $i=r,z$ with respect to the arc length are calculated by finite differences or by directly differentiating $u_i(s)$ if it is a polynomial of higher degree in s. Solving the system of Equations 4.137 we obtain the desired quantities τ_{ij}.

Secondly we calculate the stresses on S due to the tangential components $\tilde{u}_\phi$ and $\tilde{t}_\phi$ of the displacement vector and traction vector and obtain:

$$\tau_{r\phi} = \mu\left[\left(\frac{\tilde{t}_\phi}{\mu} + \frac{\tilde{u}_\phi}{r}\cos\beta\right)\cos\beta - \tilde{u}_{\phi,s}\sin\beta - \frac{\tilde{u}_\phi}{r}\right] \qquad (4.138)$$

$$\tau_{z\phi} = \mu\left[\left(\frac{\tilde{t}_\phi}{\mu} + \frac{\tilde{u}_\phi}{r}\cos\beta\right)\sin\beta + \tilde{u}_{\phi,s}\cos\beta\right] \qquad (4.139)$$

where the derivatives $\tilde{u}_{\phi,s}$ are again to be calculated by (e.g.) finite differences.

Figure 4.20 presents the problem of a cylindrical bar under tension, which was solved by Mayr[46]. Stresses at internal points were calculated with the help

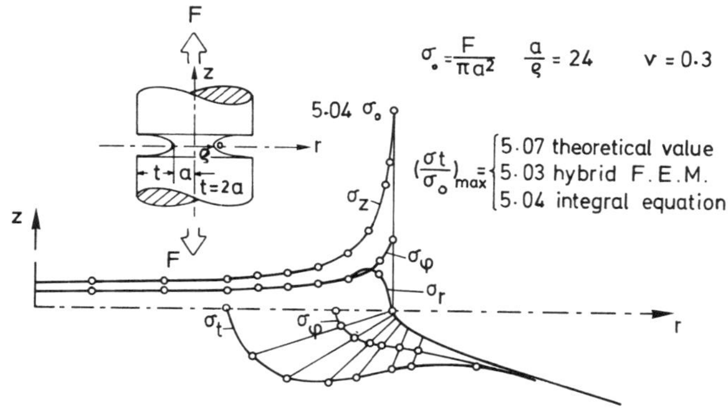

Figure 4.20

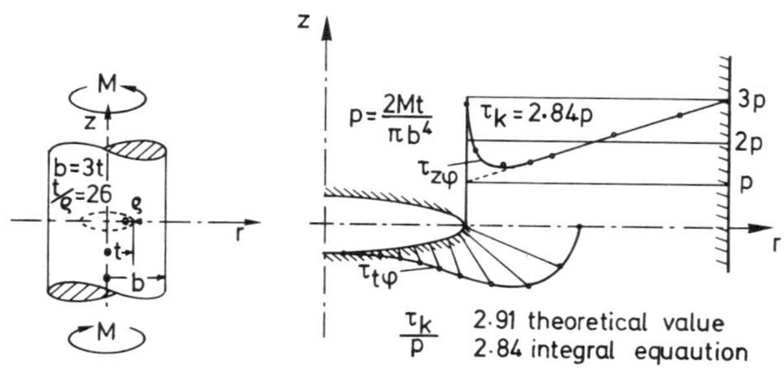

Figure 4.21

of finite differences. The curved line under the straight dotted line (axis r) is the curve C that generates the notch and the values on the curves σ_t and σ_ϕ refer to the points on C. It is σ_t the tension in the direction of the tangent $\mathbf{t}$ on C.

Figure 4.21 presents the problem of a cylindrical bar with an ellipsoidal cavity in the center under tension, Mayr and Neureiter[47].

4.5.5 An extension to 'arbitrary' bvp in axisymmetric bodies

Axisymmetric displacement fields are invariant with respect to rotation, i.e. φ. A displacement field that exhibits a symmetry under rotation has the components

$$u_{rn}(\mathbf{P}) = u'_{rn}(\mathbf{P}')\cos(n\varphi)$$
$$u_{\phi n}(\mathbf{P}) = u'_{\phi n}(\mathbf{P}')\sin(n\varphi) \tag{4.140}$$
$$u_{zn}(\mathbf{P}) = u'_{zn}(\mathbf{P}')\cos(n\varphi)$$

where n is any natural number and $\mathbf{P} = (r,\varphi,z)$, $\mathbf{P}' = (r,z)$. For the three Equations 4.140 we use the abbreviations

$$u_{in}(\mathbf{P}) = u'_{in}(\mathbf{P}')\ a_{in}(\varphi) \tag{4.141}$$

with $i = r,\varphi,z$ and the terms $a_{rn}(\varphi) = a_{zn}(\varphi) = \cos(n\varphi)$, $a_{\phi n}(\varphi) = \sin(n\varphi)$.

An anti-symmetric displacement field has the components

$$u_{in}(\mathbf{P}) = u'_{in}(\mathbf{P}')\ b_{in}(\varphi) \tag{4.142}$$

with $i = r,\varphi,z$ and $b_{rn}(\varphi) = b_{zn}(\varphi) = \sin(n\varphi)$, $b_{\phi n}(\varphi) = \cos(n\varphi)$. Suppose: (1) there is a symmetric displacement field

$$\begin{aligned} &\hat{\mathbf{u}}(\mathbf{P}) = \sum_{n=1}^{N} \hat{\mathbf{u}}_n(\mathbf{P}), \text{ with } \hat{\mathbf{u}}_n = [u_{rn'} u_{\varphi n'} u_{zn}]^T \text{ and} \\ &u_{in}(\mathbf{P}) = u'_{in}(\mathbf{P}')\ a_{in}(\varphi), \qquad i = r,\varphi,z \end{aligned} \tag{4.143}$$

and any $\hat{\mathbf{u}}_n$ satisfies the equation $\mathbf{L}\mathbf{u}_n = \mathbf{0}$ and is a regular solution, (2) B is axisymmetric then $\hat{\boldsymbol{\tau}}(\mathbf{u}_n)(\mathbf{P})$ must have the form

$$\boldsymbol{\tau}(\hat{\mathbf{u}}_n)(\mathbf{P}) = [t_{rn}, t_{\phi n}, t_{zn}]^T \tag{4.144}$$

with

$$t_{in}(\mathbf{P}) = t'_{in}(\mathbf{P}') a_{in}(\varphi) \quad i = r,\varphi,z \tag{4.145}$$

i.e. it must preserve the symmetry. If (1) and (2) hold, the Somigliana identity must hold for any $\hat{\mathbf{u}}_n$ and have consequently the form

$$\begin{aligned} \dot{C}_{ij}(\mathbf{P}) u'_{jn}(\mathbf{P}') a_{jn}(\varphi) = \oint_S [&\hat{U}_{ij}(\mathbf{P},\mathbf{Q}) t'_{jn}(\mathbf{Q}') a_{jn}(\psi) - \\ &\hat{T}_{ij}(\mathbf{P},\mathbf{Q}) u'_{jn}(\mathbf{Q}') a_{jn}(\psi)]\ \mathrm{d}S_{\mathbf{Q}} \qquad i = r,\varphi,z \end{aligned} \tag{4.146}$$

(dS denotes the surface element, ds the differential of the arc length on C (see below) and summation over j is above and in the following equations always implied.)

The approach in this section rests essentially on the remarkable fact that there exist matrices $\mathring{\mathbf{U}}^n = [\mathring{U}^n_{ij}]$ and $\mathring{\mathbf{T}}^n = [\mathring{T}^n_{ij}]$ (see Mayr[46], p. 85) such that the equations

$$\int_0^{2\pi} \hat{U}_{ij}(\mathbf{P}.\mathbf{Q}) a_{jn}(\psi)\ \mathrm{d}\psi = \mathring{U}^n_{ij}(\mathbf{P}',\mathbf{Q}') a_{jn}(\varphi) \tag{4.147}$$

$$\int_0^{2\pi} \hat{T}_{ij}(\mathbf{P},\mathbf{Q}) a_{jn}(\psi)\ \mathrm{d}\psi = \mathring{T}^n_{ij}(\mathbf{P}',\mathbf{Q}') a_{jn}(\varphi) \tag{4.148}$$

hold. These matrices have in all points $\mathbf{P}',\mathbf{Q}'$ for any n the same pattern.

$$\text{Pattern}\ (\mathring{\mathbf{U}}^n)=\text{Pattern}\ (\mathring{\mathbf{T}}^n)=\begin{bmatrix} * & 0 & * \\ 0 & * & 0 \\ * & 0 & * \end{bmatrix} \tag{4.149}$$

(The zeros denote null elements). Inserting Equations 4.147 and 4.148 in Equations 4.146 we obtain

$$\begin{aligned}\dot{C}_{ij}(\mathbf{P})u'_{jn}(\mathbf{P}')a_{jn}(\varphi)=\Bigg[\oint_C [\mathring{U}^n_{ij}(\mathbf{P}',\mathbf{Q}')t'_{jn}(\mathbf{Q}')-\\ \mathring{T}^n_{ij}(\mathbf{P}',\mathbf{Q}')u'_{jn}(\mathbf{Q}')]\rho\ ds_{\mathbf{Q}'}\Bigg]\cdot a_{jn}(\varphi) \qquad i=r,\varphi,z\end{aligned} \tag{4.150}$$

Now it is (1) $\dot{C}_{ij}=\frac{1}{2}\delta_{ij}$ in smooth points and (2) $\mathring{\mathbf{U}}^n$ and $\mathring{\mathbf{T}}^n$ have the pattern depicted in Equation 4.149, so it follows that in the equations where $i=r$ or $i=z$ the coefficients of $a_{\phi n}=\sin(n\varphi)$ disappear on both sides. If $i=\varphi$ disappear the coefficients of $a_{rn}=a_{zn}=\cos(n\varphi)$.

The three Equations 4.150 are therefore equivalent with the following three equations:

$$\begin{aligned}\dot{C}_{ij}(\mathbf{P}')u'_{jn}(\mathbf{P}')=\oint_C[\mathring{U}^n_{ij}(\mathbf{P}',\mathbf{Q}')t'_{jn}(\mathbf{Q}')-\mathring{T}^n_{ij}(\mathbf{P}',\mathbf{Q}')u'_{jn}(\mathbf{Q}')]\rho\ ds_{\mathbf{Q}'}\\ i=r,\varphi,z\end{aligned} \tag{4.151}$$

Note that this relation is due to Equation 4.149 in parts uncoupled: The terms $u'_{rn},u'_{zn},t'_{rn},t'_{zn}$ do not depend on $u'_{\phi n},t'_{\phi n}$ and vice versa. It is now obvious how to proceed to solve a bvp where the boundary values on the surface of the axisymmetric body can be expanded in a Fourier series with respect to φ, see Equation 4.152 below. Suppose there is (e.g.) posed the 1st bvp, $[D_i,\tilde{\mathbf{u}}]$ with $\tilde{\mathbf{u}}=[\tilde{u}_r,\tilde{u}_\phi,\tilde{u}_z]^T$

$$\tilde{u}_i(\mathbf{P})=\sum_{n=1}^{N} u'_{in}(\mathbf{P}')a_{in}(\varphi)\,,\ \mathbf{P}\in S,\ i=r,\varphi,z \tag{4.152}$$

then we assume, that there exists a layer $\tilde{\mathbf{t}}=[\tilde{t}_r,\tilde{t}_\phi,\tilde{t}_z]^T$

$$\tilde{t}_i(\mathbf{P})=\sum_{n=1}^{N} t'_{in}(\mathbf{P}')a_{in}(\varphi)\,,\ \mathbf{P}\in S,\ i=r,\varphi,z \tag{4.153}$$

such that the integral relation, Equation 4.150, holds and the function

$$\hat{\mathbf{u}} = [u_r, u_\phi, u_z]^T = \left[\sum_n u_{rn}, \sum_n u_{\phi n}, \sum_n u_{zn}\right]^T$$

with the components

$$u_{in}(\mathbf{P}) = a_{jn}(\varphi) \cdot \oint_C [(\mathring{U}^n_{ij}(\mathbf{P}',\mathbf{Q}')t'_{jn}(\mathbf{Q}') - \mathring{T}^n_{ij}(\mathbf{P}',\mathbf{Q}')u'_{jn}(\mathbf{Q}')]\rho \; ds_{\mathbf{Q}'} \tag{4.154}$$

$$i = r, \varphi, z$$

solves the bvp.

The integral relation, Equation 4.150, or to be more precise its equivalent form Equation 4.151 becomes then an integral equation for any set $(t_{rn}, t_{\phi n}, t_{zn}]^T$, $n = 1,2,\ldots N$, i.e. we have to solve N separate systems of integral equations on C:

$$\oint_C \mathring{U}^n_{ij}(\mathbf{P}',\mathbf{Q}')t'_{jn}(\mathbf{Q}')\rho \; ds_{\mathbf{Q}'} = \tag{4.155}$$

$$\oint_C \mathring{T}^n_{ij}(\mathbf{P}',\mathbf{Q}')u'_{jn}(\mathbf{Q}')\rho \; ds_{\mathbf{Q}'} + \dot{C}_{ij}(\mathbf{P}')u'_{jn}(\mathbf{P}'), \; \mathbf{P}' \in C$$

$$i = r, \varphi, z; \; n = 1,2,\ldots N$$

This is no easy task and so the extension of the method is restricted in effect to bvp where a few Fourier terms suffice to model the boundary data to the required precision.

The handling of anti-symmetric problems follows naturally the same pattern.

4.6 PROBLEMS IN HALFSPACES

There is an ample literature on problems in elastic halfspaces, such as rigid punch problems, crack problems, etc. These can be treated with a wide spectrum of mathematical tools, e.g. Fourier series and integral transforms, which otherwise encounter serious difficulties, cf. Sneddon[48], so there seems to be no real need to study the approach of potential theory to such problems. Nevertheless, we will in Section 4.6.1 briefly discuss Kupradze's[9] approach because this approach (where it works) is simple and extends also to anisotropic halfspace. In Section 4.6.2 we return to the notion of a complementary problem to solve some special bvp by using continuation properties of 'exterior' and 'interior' solutions, i.e. solutions which have equal displacement or equal traction on the infinite plane S, that separates the lower and upper elastic halfspaces. This way we can avoid the task of having to solve an integral equation and write down the solution of some bvp immediately.

4.6.1 Special potentials

The technique to solve the bvp

$$\mathbf{Lu} = \mathbf{0}, \qquad \lim_i \mathbf{b}(\mathbf{u}) = \tilde{\mathbf{u}} \tag{4.156}$$

in the bounded domain B with boundary S, where $\mathbf{L}$ is the differential operator and $\mathbf{b}(\cdot)$ the boundary operator (= condition), can be summed up in this way: Find a kernel $\mathbf{K}(\mathbf{x},\mathbf{y})$ such, that (1) the transform

$$\mathbf{u}(\mathbf{x}) = \oint_S \mathbf{K}(\mathbf{x},\mathbf{y})\bar{\mathbf{s}}(\mathbf{y})\ ds_y \tag{4.157}$$

satisfies the equation $\mathbf{Lu} = \mathbf{0}$ in B for any function $\bar{\mathbf{s}}(\mathbf{y})$, $\mathbf{y} \in S$, and (2) the limit of $\mathbf{b}(\mathbf{u})$ has the form

$$\lim_i \mathbf{b}(\mathbf{u}) = \tfrac{1}{2}\bar{\mathbf{s}}(\mathbf{x}) + \oint_S \mathbf{K}(\mathbf{x},\mathbf{y})\bar{\mathbf{s}}(\mathbf{y})\ ds_y \tag{4.158}$$

This leads to an integral equation of the second kind for the layer $\bar{\mathbf{s}}(\mathbf{y})$. If one is lucky, one finds a kernel $\mathbf{K}(\mathbf{x},\mathbf{y})$ with the property (1), but the integral in Equation 4.158 vanishes for arbitrary functions $\bar{\mathbf{s}}(\mathbf{y})$. Then the solution of the problem is simply

$$\mathbf{u}(\mathbf{x}) = 2 \cdot \oint_S \mathbf{K}(\mathbf{x},\mathbf{y})\tilde{\mathbf{u}}(\mathbf{y})\ ds_y \tag{4.159}$$

Naturally such kernels $\mathbf{K}(\mathbf{x},\mathbf{y})$ depend on the domain B and are in general unknown. One domain for which such kernels are known are halfspaces. There the first four bvp, see Section 4.2.3, can be solved this way. Concerning the formulae of these kernels and the conditions which are imposed on the layers (= boundary values) for the potentials to exist we refer to Kupradze[9], Chapter XIV §2. The approach extends also to anisotropic halfspaces.

4.6.2 Some selected boundary value problems

In this section we will discuss some bvp in elastic halfspaces whose common feature is that their solutions in terms of elastic potentials can be written down without solving an integral equation. While the boundary value problems discussed are themselves of limited practical interest the method showing how the solutions are derived has certainly a value of its own and sheds some light on the nature of elastic potentials.

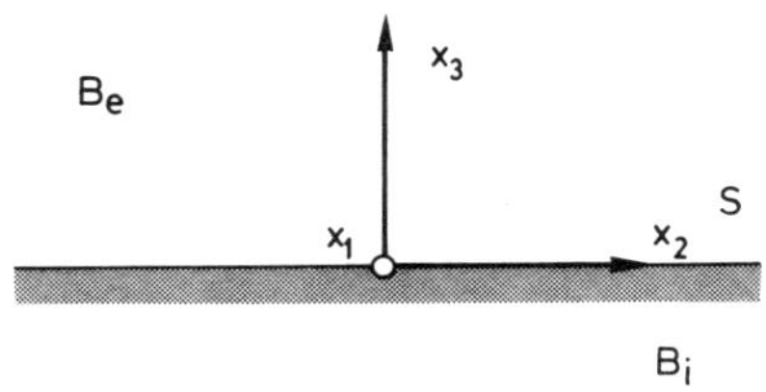

Figure 4.22

The motivation for this section has come through the study of a paper by Maiti, Bela Das and Palit[49] who used the potential of case (a), see below, to solve some bvp. We want to stress the theoretical background of their approach and extent it furthermore to case (b), as will be seen below.

Let B_i and B_e denote now the lower and upper halfspaces and S the infinite plane, that separates these (Figure 4.22). Suppose there exist two regular solutions $\mathbf{u}^i(\mathbf{x})$, $\mathbf{u}^e(\mathbf{x})$, $\mathbf{Lu}^i=\mathbf{0}$, $\mathbf{x}\in B_i$, $\mathbf{Lu}^e=\mathbf{0}$, $\mathbf{x}\in B_e$ that satisfy the radiation condition (A), see Section 4.2.3. Then it is easy to see that the Somigliana identity holds for both solutions:

$$\mathbf{C}_i(\mathbf{x})\mathbf{u}^i(\mathbf{x})=\oint_S[\mathbf{U}(\mathbf{y},\mathbf{x})\boldsymbol{\tau}_e(\mathbf{u}^i)(\mathbf{y})-\mathbf{T}_e(\mathbf{y},\mathbf{x})\mathbf{u}^i(\mathbf{y})]\ ds_{\mathbf{y}} \tag{4.160}$$

$$\mathbf{C}_e(\mathbf{x})\mathbf{u}^e(\mathbf{x})=\oint_S[\mathbf{U}(\mathbf{y},\mathbf{x})\boldsymbol{\tau}_i(\mathbf{u}^e)(\mathbf{y})-\mathbf{T}_i(\mathbf{y},\mathbf{x})\mathbf{u}^e(\mathbf{y})]\ ds_{\mathbf{y}} \tag{4.161}$$

For $\mathbf{x}\in B_i$ it is $\mathbf{C}_i(\mathbf{x})=\mathbf{I}$, $\mathbf{C}_e(\mathbf{x})=\mathbf{0}$ and adding Equation 4.161 to Equation 4.160 we obtain

$$\mathbf{u}^i(\mathbf{x})=\oint_S\{\mathbf{U}(\mathbf{y},\mathbf{x})[\boldsymbol{\tau}_e(\mathbf{u}^i)+\boldsymbol{\tau}_i(\mathbf{u}^e)](\mathbf{y})-\mathbf{T}_e(\mathbf{y},\mathbf{x})[\mathbf{u}^i-\mathbf{u}^e](\mathbf{y})\}\ ds_{\mathbf{y}} \qquad \mathbf{x}\in B_i \tag{4.162}$$

and in the same way for $\mathbf{x}\in B_e$

$$\mathbf{u}^e(\mathbf{x})=\oint_S\{\mathbf{U}(\mathbf{y},\mathbf{x})[\boldsymbol{\tau}_e(\mathbf{u}^i)+\boldsymbol{\tau}_i(\mathbf{u}^e)](\mathbf{y})-\mathbf{T}_e(\mathbf{y},\mathbf{x})[\mathbf{u}^i-\mathbf{u}^e](\mathbf{y})\}\ ds_{\mathbf{y}} \qquad \mathbf{x}\in B_e \tag{4.163}$$

(We have used the identity $\mathbf{T}_i(\mathbf{y},\mathbf{x})=-\mathbf{T}_e(\mathbf{y},\mathbf{x})$, $\mathbf{x}\in R^3$, $\mathbf{y}\in S$).

We study two possible cases:

Case (a) $$\boldsymbol{\tau}_i(\mathbf{u}^e)+\boldsymbol{\tau}_e(\mathbf{u}^i)=\mathbf{0},\ \text{on } S \tag{4.164}$$

The tractions on S are in equilibrium.

Case (b) $$\mathbf{u}^i(\mathbf{x})-\mathbf{u}^e(\mathbf{x})=\mathbf{0} \tag{4.165}$$

The displacements on S are equal.

We begin with case (a). Equation 4.162 becomes

$$\mathbf{u}^i(\mathbf{x})=-\oint_S \mathbf{T}_e(\mathbf{y},\mathbf{x})[\mathbf{u}^i-\mathbf{u}^e](\mathbf{y})\ ds_y \tag{4.166}$$

Passing to S in Equation 4.166 from the inside we obtain, if we consider the jump that occurs on S, see Equation 4.49

$$\lim_i \mathbf{u}^i(\mathbf{x})=\frac{1}{2}[\mathbf{u}^i-\mathbf{u}^e](\mathbf{x})-\oint_S \mathbf{T}_e(\mathbf{y},\mathbf{x})[\mathbf{u}^i-\mathbf{u}^e](\mathbf{y})\ ds_y=\mathbf{u}^i(\mathbf{x}) \tag{4.167}$$

Passing to S from the outside in Equation 4.163 and using Equation 4.50 we obtain

$$\lim_e \mathbf{u}^e(\mathbf{x})=-\frac{1}{2}[\mathbf{u}^i-\mathbf{u}^e](\mathbf{x})-\oint_S \mathbf{T}_e(\mathbf{y},\mathbf{x})[\mathbf{u}^i-\mathbf{u}^e](\mathbf{y})\ ds_y=\mathbf{u}^e(\mathbf{x}) \tag{4.168}$$

Adding Equation 4.167 to Equation 4.168 we obtain

$$-2\oint_S \mathbf{T}_e(\mathbf{y},\mathbf{x})[\mathbf{u}^i-\mathbf{u}^e](\mathbf{y})\ ds_y=[\mathbf{u}^i+\mathbf{u}^e](\mathbf{x}) \tag{4.169}$$

(Equation 4.167–Equation 4.168 is the identity $\mathbf{u}^i-\mathbf{u}^e=\mathbf{u}^i-\mathbf{u}^e$). Equation 4.169 is a necessary condition for two functions $\mathbf{u}^i$ and $\mathbf{u}^e$ on S to be the traces ($=$ value of $\mathbf{u}(\mathbf{x})$ on S) of two displacement fields, whose tractions are in equilibrium on S, i.e. Equation 4.164 holds.

Let M denote the class of layers $\mathbf{w}(\mathbf{y})\in C^{1,\gamma}(S)$ with the property that

$$\mathbf{W}(\mathbf{x})=\oint_S \mathbf{T}_e(\mathbf{y},\mathbf{x})\mathbf{w}(\mathbf{y})\ ds_y$$

exists and satisfies the radiation condition (A) (see Section 4.2.3) in B_i and B_e.

Suppose it is

$$\mathbf{w} \in M \text{ and } \mathbf{u}^i = \lim_i \mathbf{W} \text{ and } \mathbf{u}^e = \lim_e \mathbf{W},$$

then Equation 4.169 is satisfied, as can be easily verified by inserting the above expressions in Equation 4.169 and using Equations 4.49 and 4.50.

Due to $\mathbf{w} \in C^{1,\gamma}(S)$ we have

$$\lim_i \boldsymbol{\tau}_e(\mathbf{W}) = \lim_e \boldsymbol{\tau}_e(\mathbf{W}),$$

cf. Equation 4.51, and so we see that any $\mathbf{w} \in M$ generates two elastic states whose boundary values (traction and displacement) satisfy Equations 4.164 and 4.169. A closer look at the necessary conditions in Equation 4.169 will lead us next to two sets of layers.

Equation 4.169 assumes, due to the special pattern of the matrix $\mathbf{T}(\mathbf{y},\mathbf{x})$ for $\mathbf{x} \in S$, the form

$$-2\oint_S \begin{bmatrix} 0 & 0 & * \\ 0 & 0 & * \\ * & * & 0 \end{bmatrix} [\mathbf{u}^i - \mathbf{u}^e](\mathbf{y})\, ds_{\mathbf{y}} = [\mathbf{u}^i + \mathbf{u}^e] \tag{4.170}$$

or

$$-2\oint_S \begin{Bmatrix} T_{13}[u_3^i - u_3^e] \\ T_{23}[u_3^i - u_3^e] \\ T_{31}[u_1^i - u_1^e] + T_{32}[u_2^i - u_2^e] \end{Bmatrix} ds_{\mathbf{y}} = \begin{bmatrix} u_1^i + u_1^e \\ u_2^i + u_2^e \\ u_3^i + u_3^e \end{bmatrix} \tag{4.171}$$

Sufficient for these equations to hold is that one of the following two sets of equations is satisfied:

(1) $$u_1^i - u_1^e = 0, \quad u_2^i - u_2^e = 0, \quad u_3^i = -u_3^e \tag{4.172}$$

(2) $$u_1^i = -u_1^e, \quad u_2^i = -u_2^e, \quad u_3^i - u_3^e = 0 \tag{4.173}$$

These equations exhibit a certain symmetry in the displacements in $\mathbf{u}$ on S and we can state: Any potential

$$\mathbf{W}(\mathbf{x}) = \oint_S \mathbf{T}_e(\mathbf{y},\mathbf{x})\mathbf{w}(\mathbf{y})\, ds_{\mathbf{y}}$$

with the layer

$$\mathbf{w}(\mathbf{x}) = [0,0 - 2u_3]^T \in M \tag{4.174}$$

and

$$\tilde{u}_3 = u_3^i = -u_3^e \tag{4.175}$$

satisfies the equations of set (1), and any potential

$$\mathbf{W}(\mathbf{x}) = \oint_{S} \mathbf{T}_e(\mathbf{y},\mathbf{x})\mathbf{w}(\mathbf{y})\, ds_y$$

with the layer

$$\mathbf{w}(\mathbf{x}) = [-2\tilde{u}_1, -2\tilde{u}_2, 0]^T \in M \tag{4.176}$$

and

$$\tilde{u}_1 = u_1^i = -u_1^e; \quad \tilde{u}_2 = u_2^i = -u_2^e \tag{4.177}$$

satisfies the equations of set (2). These statements can be verified by simple calculation. Due to the fact that the functions $\tilde{u}_3$ resp. $\tilde{u}_1$ and $\tilde{u}_2$ are arbitrary in Equation 4.174 resp. Equation 4.176 we understand that these equations constitute two sets of layers.

By a formal application ($\mathbf{x}$ lies on S) of Equation 4.92 to the potential $\mathbf{W}(\mathbf{x})$ with the layer of Equation 4.174 we furthermore obtain

$$\sigma_{ij} = \oint_{S} S_{3ij} 2\tilde{u}_3\, ds_y = 0, \quad \mathbf{x} \in S \tag{4.178}$$

if $\sigma_{ij} = \sigma_{13}$ or $\sigma_{ij} = \sigma_{23}$, and it readily follows that with $\mathbf{n} = [0,0,1]^T$ the first two components of the traction $\mathbf{t}$ are zero, i.e. $t_1 = t_2 = 0$ on S.

Turning to boundary value problems we can state the results in this way: The solution of the bvp

$$\mathbf{L}\mathbf{u} = \mathbf{0}, \quad \mathbf{x} \in B_i:$$

$$\lim_{i} u_3(\mathbf{x}) = \tilde{u}_3(\mathbf{x}) \tag{4.179}$$

$$\lim_{i} [\boldsymbol{\tau}_e(\mathbf{u})]_j = 0, \quad j = 1,2$$

is the potential

$$\mathbf{W}(\mathbf{x}) = -2\oint_{S} \mathbf{T}_e(\mathbf{y},\mathbf{x}) \begin{bmatrix} 0 \\ 0 \\ \tilde{u}_3(\mathbf{y}) \end{bmatrix} ds_y \tag{4.180}$$

By arguments as above it follows that the solution of the bvp

$$\mathbf{Lu}=\mathbf{0}, \quad \mathbf{x}\in B_i$$

$$\lim_i \begin{cases} u_1(\mathbf{x})=\tilde{u}_1(\mathbf{x}) \\ u_2(\mathbf{x})=\tilde{u}_2(\mathbf{x}) \end{cases} \tag{4.181}$$

$$\lim_i\ [\boldsymbol{\tau}_e(\mathbf{u})]_3=0$$

is the potential

$$\mathbf{W}(\mathbf{x})=-2\oint_S \mathbf{T}_e(\mathbf{y},\mathbf{x}) \begin{bmatrix} \tilde{u}_1(\mathbf{y}) \\ \tilde{u}_2(\mathbf{y}) \\ 0 \end{bmatrix} ds_{\mathbf{y}} \tag{4.182}$$

(The layers in Equations 4.180 and 4.182 are naturally supposed to belong to the class M, see above.)

Next we will discuss case (b) whose treatment is very similar to case (a). We assume the layers to be elements of $C^{0,\gamma}(S)$ and decreasing rapidly enough for the integrals over the infinite plane S to exist.

The equations 4.162 and 4.163 become, with $\mathbf{u}^i-\mathbf{u}^e=\mathbf{0}$,

$$\mathbf{u}^i(\mathbf{x})=\oint_S \mathbf{U}(\mathbf{y},\mathbf{x})[\boldsymbol{\tau}_e(\mathbf{u}^i)+\boldsymbol{\tau}_i(\mathbf{u}^e)](\mathbf{y})\ ds_{\mathbf{y}}, \quad \mathbf{x}\in B_i \tag{4.183}$$

$$\mathbf{u}^e(\mathbf{x})=\oint_S \mathbf{U}(\mathbf{y},\mathbf{x})[\boldsymbol{\tau}_e(\mathbf{u}^i)+\boldsymbol{\tau}_i(\mathbf{u}^e)](\mathbf{y})\ ds_{\mathbf{y}}, \quad \mathbf{x}\in B_e \tag{4.184}$$

and we obtain, if we apply the traction operator $\boldsymbol{\tau}(\cdot)$ to both sides, pass to S and consider Equations 4.52 and 4.53,

$$\lim_i\ \boldsymbol{\tau}_e(\mathbf{u}^i(\mathbf{x}))=\frac{1}{2}[\mathbf{t}^i+\mathbf{t}^e]+\oint_S \mathbf{T}_e^*(\mathbf{x},\mathbf{y})[\mathbf{t}^i+\mathbf{t}^e]\ ds_{\mathbf{y}}=\mathbf{t}^i \tag{4.185}$$

$$\lim_e\ \boldsymbol{\tau}_e(\mathbf{u}^e(\mathbf{x}))=-\frac{1}{2}[\mathbf{t}^i+\mathbf{t}^e]+\oint_S \mathbf{T}_e^*(\mathbf{x},\mathbf{y})[\mathbf{t}^i+\mathbf{t}^e]\ ds_{\mathbf{y}}=\mathbf{t}^e \tag{4.186}$$

(We use the abbreviations $\mathbf{t}^i=\boldsymbol{\tau}_e(\mathbf{u}^i),\mathbf{t}^e=\boldsymbol{\tau}_i(\mathbf{u}^e)$). Note that the negative sign on the rigth side of Equation 4.186 is due to

$$\lim_e\ \boldsymbol{\tau}_e(\mathbf{u}^e(\mathbf{x}))=-\lim_e\ \boldsymbol{\tau}_i(\mathbf{u}^e(\mathbf{x}))=-\mathbf{t}^e.$$

Adding Equation 4.185 to Equation 4.186 we obtain

$$2\oint_S \mathbf{T}_e^*(\mathbf{x},\mathbf{y})[\mathbf{t}^i+\mathbf{t}^e](\mathbf{y})\ ds_y=[\mathbf{t}^i-\mathbf{t}^e](\mathbf{x}) \tag{4.187}$$

Equation 4.187 states again a necessary condition: If two problems in B_i and B_e have the same trace on S, i.e. $[\mathbf{u}^i-\mathbf{u}^e]=\mathbf{0}$ on S, then their tractions must satisfy Equation 4.187.

Due to the pattern of the matrix $\mathbf{T}^*$ on S, Equation 4.187 assumes the form

$$2\oint_S \begin{bmatrix} 0 & 0 & * \\ 0 & 0 & * \\ * & * & 0 \end{bmatrix} [\mathbf{t}^i+\mathbf{t}^e](\mathbf{y})\ ds_y=[\mathbf{t}^i-\mathbf{t}^e](\mathbf{x}) \tag{4.188}$$

or

$$2\oint_S \left\{ \begin{matrix} T_{13}^*[t_3^i+t_3^e] \\ T_{23}^*[t_3^i+t_3^e] \\ T_{31}^*[t_1^i+t_1^e]+T_{32}^*[t_2^i+t_2^e] \end{matrix} \right\} ds_y=\begin{bmatrix} t_1^i-t_1^e \\ t_2^i-t_2^e \\ t_3^i-t_3^e \end{bmatrix} \tag{4.189}$$

Sufficient for these equations to hold is that one of the two sets of equations is satisfied:

(1) $$t_1^i+t_1^e=0, \quad t_2^i+t_2^e=0, \quad t_3^i=t_3^e \tag{4.190}$$

(2) $$t_1^i=t_1^e, \quad t_2^i=t_2^e, \quad t_3^i+t_3^e=0 \tag{4.191}$$

The equations in the first set are satisfied by the traction of the potential

$$\mathbf{V}^{(1)}(\mathbf{x})=\oint_S \mathbf{U}(\mathbf{y},\mathbf{x})\mathbf{v}(\mathbf{y})\ ds_y \tag{4.192}$$

with the layer

$$\mathbf{v}(\mathbf{y})=[0,0,2\tilde{t}_3]^T \tag{4.193}$$

and $\tilde{t}_3=t_3^i=t_3^e$, and the equations in the second set by the potential

$$\mathbf{V}^{(2)}(\mathbf{x})=\oint_S \mathbf{U}(\mathbf{y},\mathbf{x})\mathbf{v}(\mathbf{y})\ ds_y$$

with the layer

$$\mathbf{v}(\mathbf{y})=[2\tilde{t}_1,2\tilde{t}_2,0]^T \tag{4.194}$$

For **y**, **x**∈S the matrix **U(y,x)** has the pattern (cf. Equation 4.10)

$$\mathbf{U}=\begin{bmatrix} * & * & 0 \\ * & * & 0 \\ 0 & 0 & * \end{bmatrix} \tag{4.195}$$

and so it is $V_i^{(1)}=0$, $i=1,2$, and $V_3^{(2)}=0$ on S. We conclude that the solution of the bvp

$$\mathbf{Lu}=\mathbf{0}, \quad \mathbf{x}\in B_i:$$

$$\lim_i \begin{cases} u_1=0 \\ u_2=0 \\ [\boldsymbol{\tau}_e(\mathbf{u})]_3=\tilde{t}_3 \end{cases} \tag{4.196}$$

is the potential

$$\mathbf{V}(\mathbf{x})=2\oint_S \mathbf{U}(\mathbf{y},\mathbf{x}) \begin{bmatrix} 0 \\ 0 \\ \tilde{t}_3(\mathbf{y}) \end{bmatrix} ds_{\mathbf{y}} \tag{4.197}$$

and the solution of the bvp

$$\mathbf{Lu}=0, \quad \mathbf{x}\in B_i$$

$$\lim_i \begin{cases} [\boldsymbol{\tau}_e(\mathbf{u})]_1=\tilde{t}_i \\ [\boldsymbol{\tau}_e(\mathbf{u})]_2=\tilde{t}_2 \\ u_3=0 \end{cases} \tag{4.198}$$

is the potential

$$\mathbf{V}(\mathbf{x})=2\oint_S \mathbf{U}(\mathbf{y},\mathbf{x})\cdot \begin{bmatrix} \tilde{t}_1(\mathbf{y}) \\ \tilde{t}_2(\mathbf{y}) \\ 0 \end{bmatrix} ds_{\mathbf{y}} \tag{4.199}$$

Maiti, Bela Das and Palit[49] used the results of case (a) in 2-D to solve problems in halfplanes and expanded the method to problems which can be reduced to one of the two bvp of case (a) and so were able to avoid the solution of integral equations.

4.7 PLATE BENDING PROBLEMS

The biharmonic operator $\Delta\Delta u$, which governs Airy's stress function and plate bending problems usually has been for engineers the first encounter with integral equation methods in elasticity. (Contrary to other fields of physics, there is no standard problem in elasticity — besides Saint Vernant's torsion — which is governed by the Laplace equation $\Delta u = f$.) The well known approach of Muskhelishvili to solve boundary value problems of Airy's stress function rests essentially on Goursat's theorem, that any biharmonic function can be represented as the real part of the sum of two analytic complex functions. This theorem is also applied by Jaswon, Maiti and Symm[50,51], who reduced plate problems to bvp for Laplace potentials, solving them with two single layer potentials. The approach of Maiti and Chakrabarty[52] is also essentially a layer approach, though they start with the Rayleigh-Green identity to derive a 'bipotential', which turns out to be the sum of a biharmonic and harmonic potential well suited for the kind of bvp they want to solve. A similar approach is the paper of Holzemer[53], who constructs special biharmonic potentials by modifying the integral identity of the biharmonic operator (Rayleigh-Green identity). Pahnke[21] is concerned with the problem of corners and he constructs a method to overcome the difficulties, which arise in the layer approach in such points.

For a further discussion of the ways and means to handle plate problems in the general framework of elliptic boundary value problems we refer to Wendland[54].

This section is primarily concerned with the direct method in plate problems. The method hinges here on the Rayleigh-Green identity, which is the analogue of Green's second identity and Betti's theorem. We know of four papers where the direct method has been discussed: Forbes and Robinson[55] (1969), Bezine and Gamby[56] (1978), Stern[57] (1979) and Tottenham[58] (1979). It seems that all four papers were written independently from another. The first paper by Forbes and Robinson already covered all the details of the method for plates with smooth boundaries. Bezine and Gamby repeated the approach and extended the Rayleigh-Green identity to domains with piecewise smooth boundaries but did not place collocation points at non-smooth points. This was finally done by Stern who extended the equations to cover the most common case of plates with a piecewise smooth (straight lines) boundary. The final paper, Tottenham's paper, is more of a synopsis though containing some valuable hints.

4.7.1 **Basic equations**

The vertical displacement $w(x,y)$ of a plate B with thickness t and elastic constants E (Young's modulus) and $\bar{v}$ (Poisson's ratio) is a solution of the differential equation

$$\Delta\Delta w(x,y) = \frac{1}{D} \cdot p(x,y) \tag{4.200}$$

where

$D=\frac{E}{12}\frac{t^3}{(1-\bar{\nu})^2}$ is the bending stiffness and $p(x,y)$ the vertical load. At a point **P** $=(x,y)$ of the boundary S with the outward unit normal $\mathbf{n}=[n_1,n_2]^T$ and the tangent $\mathbf{t}=[t_1,t_2]^T$ four mechanical quantities are defined (besides $(w(x,y))$. The (negative) slope $\theta_n=-\partial w/\partial n$ the bending moment M_n, the twisting moment M_{nt} and the shear force Q_n. From a mathematical point of view only four quantities are admissible, if it comes to solving a bvp, and so the quantities Q_n and M_{nt} are replaced by the Kirchhoff shear V_n

$$V_n=Q_n+\frac{\partial}{\partial s}M_{nt} \tag{4.201}$$

(s denotes the arc length on S) which also appears in the Rayleigh-Green identity. The equations of the single quantities in terms of w and the unit normal $\mathbf{n}=[n_1,n_2]^T$ are:

$$\theta_n(w)=-\frac{\partial w}{\partial n}=-[w,_x\cdot n_1+w,_y\cdot n_2] \tag{4.202}$$

$$M_n(w)=-D\{(1-\bar{\nu})[w,_{xx}n_1^2+2w,_{xy}\cdot n_1\cdot n_2+w,_{yy}\cdot n_2^2]+\bar{\nu}\cdot\Delta w\} \tag{4.203}$$

$$M_m(w)=-D(1-\bar{\nu})\{(w,_{xx}-w,_{yy})n_2\cdot n_1-w,_{xy}(n_1^2-n_2^2)\} \tag{4.204}$$

$$Q_n(w)=-D\{\Delta w,_x\cdot n_1+\Delta w,_y\cdot n_2\} \tag{4.205}$$

$$\begin{aligned}V_n(w)=D\Big\{&(1-\bar{\nu})\frac{\partial}{\partial s}[(w,_{xx}-w,_{yy})n_1\cdot n_2-\\&w,_{xy}(n_1^2-n_2^2)]-(w,_{xxx}+w,_{xyy})n_1-\\&(w,_{yyy}+w,_{xxy})n_2\Big\}\end{aligned} \tag{4.206}$$

The boundary conditions for these quantities are:

$V_n=M_n=0,\quad$ on a free boundary

$w=M_n=0,\quad$ on a simply supported boundary

$w=\theta_n=0,\quad$ on a clamped boundary

The function

$$v_1(x,y;\ \xi,\eta)=r^2\ \ln\ r \tag{4.207}$$

with

$$r=[(x-\xi)^2+(y-\eta)^2]^{1/2}$$

represents the deflection at the point $\mathbf{Q}=(\xi,\eta)$ if an infinite plate with stiffness D due to a single load of magnitude $8\pi D$ at the point $\mathbf{P}=(x,y)$.

Next consider two single loads of magnitude $8\pi D/\varepsilon$ and opposite signs. One at the point $\mathbf{P}=(x,y)$ and the other at the point $\mathbf{P}'=(x',y')$, which are distance ε apart. Let $\mathbf{n}=[n_1,n_2]^T$ denote the unit vector of the direction $\mathbf{PP}'$. The limit of the deflection caused by the couple $M=(8\pi D/\varepsilon)\varepsilon$ at the point $\mathbf{Q}=(\xi,\eta)$ if ε tends to zero, is the function

$$v_2(x,y;\ \xi,\eta)=-r(1+2\ \ln\ r)r_n \tag{4.208}$$

Here, and in what follows, a subscript on r denotes the directional derivative with respect to the vector named by the subscript and the variables, which run with the vector

$$r_n=r_{,x}\cdot n_1+r_{,y}\cdot n_2 \tag{4.209}$$

($\mathbf{n}$ runs with (x,y)). The two functions v_1 and v_2 are homogeneous solutions of the equation $\Delta\Delta w=0$ at any point $\mathbf{P}\neq\mathbf{Q}$ and singular solutions in the following sense: For an arbitrary smooth function $f(\mathbf{Q})$ holds

$$\frac{1}{8\pi D}\cdot\int_B \Delta\Delta v_1(\mathbf{P},\mathbf{Q})f(\mathbf{Q})\ \mathrm{d}a_{\mathbf{Q}}=\delta(|\mathbf{P}-\mathbf{Q}|)f(\mathbf{P}) \tag{4.210}$$

$$\frac{1}{8\pi D}\cdot\int_B \Delta\Delta v_2(\mathbf{P},\mathbf{Q})f(\mathbf{Q})\ \mathrm{d}a_{\mathbf{Q}}=-\delta(|\mathbf{P}-\mathbf{Q}|)\frac{\partial f}{\partial n}(\mathbf{P}) \tag{4.211}$$

Suppose the point $\mathbf{Q}=(\xi,\eta)$ lies on the curve S, which has the curvature κ at $\mathbf{Q}$ and the unit normal $\boldsymbol{\nu}$ and the tangent $\boldsymbol{\tau}$. Let $\mathbf{n}$ and $\mathbf{t}$ denote a system of orthogonal axis at the source point $\mathbf{P}=(x,y)$ (Figure 4.23). Then we obtain†

$$U_{11}=v_1(\mathbf{P},\mathbf{Q}) \tag{4.212}$$

$$U_{12}=\theta_\nu(v_1)=-r\cdot r_\nu(1+2\ \ln\ r) \tag{4.213}$$

$$T_{11}=V_\nu(v_1)=-\frac{2D}{r}[2r_\nu+(1-\bar{\nu})(r_\nu-\kappa r)(r_\nu^2-r_\tau^2)] \tag{4.214}$$

$$T_{12}=M_\nu(v_1)=-D[2(1+\bar{\nu})\ \ln\ r+(3+\bar{\nu})r_\nu^2+(1+3\bar{\nu})r_\tau^2] \tag{4.215}$$

† For convenience and later use we identify the functions with matrix elements U_{ij}, T_{ij}, $i,j=1,2$.

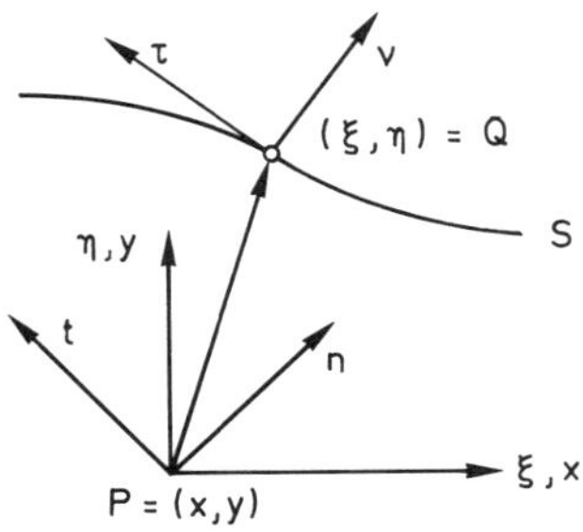

Figure 4.23

and

$$U_{21}=v_2(\mathbf{P},\mathbf{Q}) \tag{4.216}$$

$$U_{22}=\theta_\nu(v_2)=2(r_\nu r_n+r_\tau r_t)\ \ln\ r+3r_\nu r_n+r_t r_\tau \tag{4.217}$$

$$T_{21}=V_\nu(v_2)=\frac{2D}{r^2}\{[3-\bar{\nu}-2(1-\bar{\nu})r_\tau^2](r_\tau r_t-r_\nu r_n)+4(1-\bar{\nu})(r_\nu-\kappa r)r_\nu r_\tau r_t\} \tag{4.218}$$

$$T_{22}=M_\nu(v_2)=\frac{2D}{r}[(1+\bar{\nu})r_n+2(1-\bar{\nu})r_\nu r_\tau r_t] \tag{4.219}$$

The Equations 4.212 to 4.219 stem from Forbes and Robinson[55], p. 197. They use the symbols K_{ij} and J_{ij} where $U_{ij}=K_{i+2,j+2}$, $T_{ij}=J_{i+2,j+2}$ in their notation in pp. 197 and 198.

4.7.2 Biharmonic potentials

Suppose the boundary S is smooth and $m(\mathbf{Q})$, $k(\mathbf{Q})$ and $p(\mathbf{Q})$ are sufficiently smooth functions, then the three functions

$$p_1(\mathbf{P})=\oint_S v_1(\mathbf{P},\mathbf{Q})m(\mathbf{Q})\ \mathrm{d}s_\mathbf{Q}=\oint_S r^2\ \ln\ r\ m(\mathbf{Q})\ \mathrm{d}s_\mathbf{Q} \tag{4.220}$$

$$p_2(\mathbf{P})=\oint_S v_2(\mathbf{P},\mathbf{Q})k(\mathbf{Q})\ \mathrm{d}s_\mathbf{Q}=\oint_S[-r(1+2\ \ln\ r)r_\nu]k(\mathbf{Q})\ \mathrm{d}s_\mathbf{Q} \tag{4.221}$$

$$p_3(\mathbf{P})=\int_B v_1(\mathbf{P},\mathbf{Q})p(\mathbf{Q})\ \mathrm{d}a_\mathbf{Q}=\int_B r^2\ \ln\ r\ p(\mathbf{Q})\ \mathrm{d}a_\mathbf{Q} \tag{4.222}$$

satisfy the equations

$$\Delta\Delta p_1 = 0, \quad \Delta\Delta p_2 = 0, \quad \mathbf{P} \in B/S \tag{4.223}$$

$$\Delta\Delta p_3 = 8\pi D \cdot p(\mathbf{P}), \quad \mathbf{P} \in B \tag{4.224}$$

Note that $\boldsymbol{\nu}$ and $\boldsymbol{\tau}$ are the outward normal and tangent on the curve S at the integration point $\mathbf{Q}$. While the same quantities will be referred to as $\mathbf{n}$ and $\mathbf{t}$ at the free point if $\mathbf{P}$ lies on a curve inside $\bar{B}$.

The solution of plate problems with these biharmonic potentials proceeds as the indirect method in Section 4.3. The reproducing kernel of the potential p_3 allows again the reduction of problems with non-zero load $p(\mathbf{Q})$ to homogeneous problems $\Delta\Delta w = 0$. Because two boundary values are specified at any point on S, we would choose a sum of two potentials to represent the solution $w(\mathbf{P})$:

$$w(\mathbf{P}) = \oint_S [v_1(\mathbf{P},\mathbf{Q}) m(\mathbf{Q}) + v_2(\mathbf{P},\mathbf{Q}) k(\mathbf{Q})] \, ds_{\mathbf{Q}} \tag{4.225}$$

The function $w(\mathbf{P})$, Equation 4.225, satisfies the equation $\Delta\Delta w = 0$ with arbitrary layers $m(\mathbf{Q})$ and $k(\mathbf{Q})$. Applying the boundary condition operator to $w(\mathbf{P})$ and passing to the boundary, we would derive two coupled integral equations for the layers $m(\mathbf{Q})$ and $k(\mathbf{Q})$. In passing to S due respect must be paid to jumps that might occur on S. We do not list these jumps here, but it should not be too difficult to derive their equations, because the kernels have the form $r \ln r$, $\ln r$, r^{-1}, r^{-2} times bounded functions.

Instead of generating the potentials on the boundary S of the plate, Altiero and Sikarskie[59] and Freitag[60] use an auxiliary curve S' outside the plate B. This offers at least the advantage that the kernels are regular on the edge of the plate where the boundary conditions are posed and therefore lead to regular integral equations.

4.7.3 The direct method

Following Bergman and Shiffers[61], p. 235, Bezine and Gamby[56] derive an identity for two functions v, $w \in C^3(\bar{B}) \cap C^4(B)$:

$$\int_B [v\Delta\Delta w - \Delta v \Delta w + \bar{p}(v,w)] \, da =$$

$$= \frac{1}{D} \int_S \left[\frac{\partial v}{\partial \nu} M_\nu(w) + \frac{\partial v}{\partial s} M_{\nu\tau}(w) + D \cdot v \frac{\partial \Delta w}{\partial \nu} \right] ds \tag{4.226}$$

where $\bar{p}(v,w)$ denotes the term

$$\bar{p}(v,w)=(1-\bar{\nu})[v_{,xx}\cdot w_{,yy}-2v_{,xy}\cdot w_{,xy}+v_{,yy}\cdot w_{,xx}] \tag{4.227}$$

It is, cf. Equation 4.201,

$$V_\nu(w)=Q_\nu(w)+\frac{\partial}{\partial s}M_{\nu\tau}(w)=-D\frac{\partial}{\partial\nu}\Delta w+\frac{\partial}{\partial s}M_{\nu\tau}(w) \tag{4.228}$$

so we obtain

$$\begin{aligned}\frac{\partial v}{\partial s}M_{\nu\tau}(w)&=\frac{\partial}{\partial s}[vM_{\nu\tau}(w)]-v\frac{\partial M_{\nu\tau}(w)}{\partial s}=\\&=\frac{\partial}{\partial s}[vM_{\nu\tau}(w)]-v\left[V_\nu(w)+D\frac{\partial}{\partial\nu}\Delta w\right]\end{aligned} \tag{4.229}$$

Inserting this expression in Equation 4.226 we obtain

$$\begin{aligned}&\int_B[v\Delta\Delta w-\Delta v\Delta w+\bar{p}(v,w)]\,\mathrm{da}=\\&=\frac{1}{D}\oint_S\left\{\frac{\partial v}{\partial\nu}M_\nu(w)+\frac{\partial}{\partial s}[vM_{\nu\tau}(w)]-\right.\\&\left.v\left[V_\nu(w)+D\frac{\partial}{\partial\nu}\Delta w\right]+Dv\frac{\partial}{\partial\nu}\Delta w\right\}\mathrm{ds}\\&=\frac{1}{D}\oint_S\left[\frac{\partial v}{\partial\nu}M_\nu(w)-vV_\nu(w)\right]\mathrm{ds}+\sum_{k=1}^{N}[\![vM_{\nu\tau}(w)]\!](c_k)\end{aligned} \tag{4.230}$$

Where the term $[\![vM_{\nu\tau}(w)]\!](c_k)$ denotes the difference between

$$[vM_{\nu\tau}(w)](c_k^-)-[vM_{\nu\tau}(w)](c_k^+) \tag{4.231}$$

the leftsided and rightsided value of the function $[vM_{\nu\tau}(w)](s)$ at a point on S whose arc length is c_k and where this function is discontinuous. Naturally c_k, $k=1,2,\ldots N$ are the parameters of all N corner points on S. Repeating the

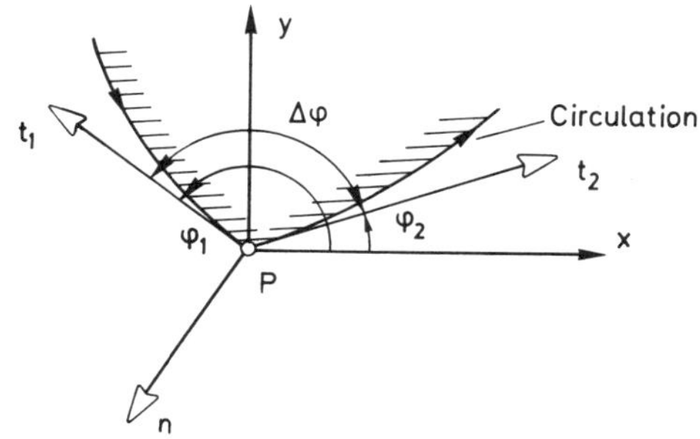

Figure 4.24 A non-smooth point

equation with v and w interchanging and subtracting the second equation from the first, we obtain

$$\int_B [v\Delta\Delta w - w\Delta\Delta v]\ \mathrm{da} =$$

$$= \frac{1}{D}\oint_S \left[\frac{\partial v}{\partial \nu} M_\nu(w) - vV_\nu(w) - \frac{\partial w}{\partial \nu} M_\nu(v) + V_\nu(w)\right] \mathrm{ds} \tag{4.232}$$

$$+ \frac{1}{D}\sum_{k=1}^{N} [\![vM_{\nu\tau}(w) - wM_{\nu\tau}(v)]\!](c_k)$$

This is the Rayleigh-Green identity for plane domains with piecewise smooth boundaries, as given in Bezine and Gamby[56].

Choosing for v the solution v_1 and then the solution v_2 and setting $\Delta\Delta w = p/D$, we derive from the Rayleigh-Green identity the following two equations for a function $w \in C^3(\bar{B}) \cap C^4(B)$:

$$c(\mathbf{P})w(\mathbf{P}) = \frac{1}{8\pi D}\oint_S [v_1(\mathbf{P},\mathbf{Q})V_\nu(w)(\mathbf{Q}) + \theta_\nu(v_1)(\mathbf{P},\mathbf{Q})M_\nu(w)(\mathbf{Q}) -$$

$$V_\nu(v_1)(\mathbf{P},\mathbf{Q})w(\mathbf{Q}) - M_\nu(v_1)(\mathbf{P},\mathbf{Q})\theta_\nu(w)(\mathbf{Q})]\ \mathrm{ds}_{\mathbf{Q}} + \tag{4.233}$$

$$\frac{1}{8\pi D}\int_B v_1(\mathbf{P},\mathbf{Q})\frac{p}{D}(\mathbf{Q})\ \mathrm{da}_{\mathbf{Q}} - \frac{1}{8\pi D}\sum_{k=1}^{N} [\![v_1(\mathbf{P},\mathbf{Q}_k)M_{\nu\tau}(w)(\mathbf{Q}_k) -$$

$$w(\mathbf{Q}_k)M_{\nu\tau}(v_1)(\mathbf{P},\mathbf{Q}_k)]\!]$$

$$c_x(\mathbf{P})w,_x(\mathbf{P})+c_y(\mathbf{P})w,_y(\mathbf{P})=\frac{1}{8\pi D}\oint_S[v_2(\mathbf{P},\mathbf{Q})V_v(w)(\mathbf{Q})+$$

$$\theta_v(v_2)(\mathbf{P},\mathbf{Q})M_v(w)(\mathbf{Q})-V_v(v_2)(\mathbf{P},\mathbf{Q})[w(\mathbf{Q})-w(\mathbf{P})]- \tag{4.234}$$

$$M_v(v_2)(\mathbf{P},\mathbf{Q})\theta_v(w)(\mathbf{Q})]\ ds_\mathbf{Q}+\frac{1}{8\pi D}\int_B v_2(\mathbf{P},\mathbf{Q})\frac{p(\mathbf{Q})}{D}\ da_\mathbf{Q}-$$

$$\frac{1}{8\pi D}\sum_{k=1}^{N}[\![v_2(\mathbf{P},\mathbf{Q}_k)M_{v\tau}(w)(\mathbf{Q}_k)-w(\mathbf{Q}_k)M_{v\tau}(v_2)(\mathbf{P},\mathbf{Q}_k)]\!]$$

with

$$c(\mathbf{P})=\begin{cases}1, & \mathbf{P}\in B\\ \dfrac{\Delta\varphi}{2\pi}, & \mathbf{P}\in S\\ 0, & \mathbf{P}\notin \bar{B}\end{cases} \tag{4.235}$$

$$c_x(\mathbf{P})=\begin{cases}1, & \mathbf{P}\in B\\ \dot{c}_x, & \mathbf{P}\in S\\ 0, & \mathbf{P}\notin \bar{B}\end{cases} \tag{4.236}$$

$$c_y(\mathbf{P})=\begin{cases}1, & \mathbf{P}\in B\\ \dot{c}_y, & \mathbf{P}\in S\\ 0, & \mathbf{P}\notin \bar{B}\end{cases} \tag{4.237}$$

$$\dot{c}_x(\mathbf{P})=-\frac{\Delta\varphi}{2\pi}n_1-\frac{\bar{v}}{2\pi}\left[\frac{1}{2}\sin 2\varphi n_1+\sin^2\varphi n_2\right]\Bigg|_{\varphi_2}^{\varphi_1} \tag{4.238†}$$

$$\dot{c}_y(\mathbf{P})=-\frac{\Delta\varphi}{2\pi}n_2-\frac{\bar{v}}{2\pi}\left[\sin 2\varphi n_1-\frac{1}{2}\sin^2\varphi n_2\right]\Bigg|_{\varphi_2}^{\varphi_1} \tag{4.239†}$$

The following rule applies: if Equations 4.233 and 4.234 are used at a non-smooth point **P** the contribution of the point **P** itself to the sum $\Sigma[\![\cdots]\!]$ is to be deleted.

The vector components n_1 and n_2 in Equations 4.238 and 4.239 belong to the vector $\mathbf{n}=[n_1,n_2]^T$ which appears in the kernel v_2 and it is in Equations 4.233 and 4.234

$$M_{v\tau}(v_1)=D(1-\bar{v})\sin 2\beta \tag{4.240}$$

† Cf. Figure 4.24 for a definition of φ_1, φ_2 and $\Delta\varphi=|\varphi_1-\varphi_2|$. Compare the two equations with Equation 4.29!

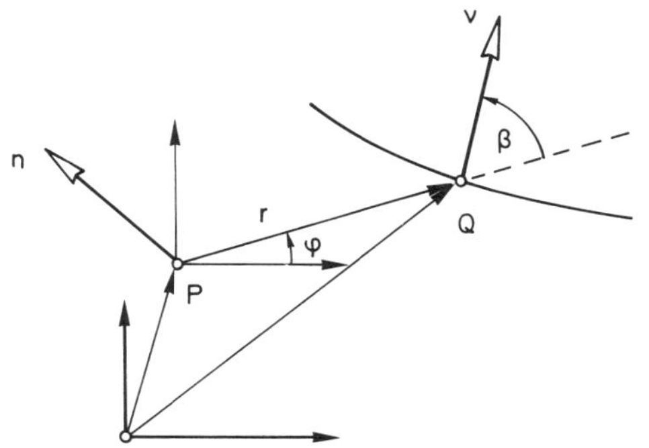

Figure 4.25 Geometrical definitions

$$M_{\nu\tau}(v_2) = -\frac{D(1-\bar{\nu})}{r} \cdot 2[-\sin\varphi n_1 + \cos\varphi n_2]\cos 2\beta \qquad (4.241)$$

where β denotes the angle between the outward normal $\boldsymbol{\nu}$ and the radius vector $\mathbf{r} = \mathbf{Q} - \mathbf{P}$ of the point $\mathbf{Q}$, see Figure 4.25.

The first term in the sum $\Sigma[\![\cdots]\!]$ represents the influence of concentrated vertical forces of magnitude F_k

$$M_{\nu\tau}(w)(Q_k^-) - M_{\nu\tau}(w)(Q_k^+) = F_k \qquad (4.242)$$

which can appear at non-smooth points of a plate (see, e.g. Timoshenko and Woinowsky-Krieger[62], p. 85). Any engineer knows that a rectangular simply supported plate needs four additional concentrated forces at the corners to keep its ends down. These single forces are just the terms F_k, $k = 1,2,3,4$. If we do not provide these forces the surface of the plate, i.e. the function w, takes such a shape that no discontinuity in $M_{\nu\tau}$ occurs in the corners.

The derivation of the Equations 4.233 and 4.234 follows the usual pattern and is not too complicated if we suppose the boundary S to consist of straight lines near the point $\mathbf{P} \in S$. This is no real restriction, because it follows easily that the equations also apply for the case of a boundary S consisting of piecewise smooth curves in the sense of Section 4.2.2. We refer to Stern[57] for a proof where you might find the derivation of the first Equation 4.233 and the derivation of a second equation which is similar to the second Equation 4.234. Stern uses in his second equation the kernel

$$v_2 = \frac{1}{2}\left[\frac{\partial}{\partial n}(v_1) - r\frac{\partial r}{\partial n}\right] \cdot \frac{1}{2\pi D} = [r \ln r \cos\varphi]\frac{1}{2\pi D}$$

with $\partial/\partial n = -\partial/\partial x$ instead of

$$v_2 = -\frac{\partial}{\partial n}(v_1) = -r(1 + 2\ln r)r_n$$

but the general rule how to proceed with the proof remains valid.

If the boundary S is smooth at **P**, i.e. $\Delta\varphi = |\varphi_2 - \varphi_1| = \pi$ then the terms in the brackets in Equation 4.238 and 4.239 vanish, which means the singular solution $v_2 = -\partial/\partial n(v_1)$ yields the normal derivative times (-1) or more generally speaking the directional derivatives times (-1) with respect to an arbitrary direction, see Figure 4.24.

If **P** is a non-smooth point the additional terms do not vanish and the right-hand side $R(\mathbf{n})$ in Equation 4.234, which as the coefficients c_x and c_y depends on the direction $\mathbf{n} = \mathbf{n}'$, assumes the value

$$R(\mathbf{n}') = w,_x c_x(\mathbf{n}') + w,_y c_y(\mathbf{n}') \neq \frac{\partial w}{\partial n}(\mathbf{P}) \tag{4.243}$$

Therefore we have to repeat the second Equation 4.234 with a kernel $v_2 = -\partial/\partial n''(v_1) = -r(1 + 2\ln r)r_{,n''}$, i.e. replace $\mathbf{n} = \mathbf{n}'$ by a second direction $\mathbf{n} = \mathbf{n}''$ to obtain

$$R(\mathbf{n}'') = w,_x c_x(\mathbf{n}'') + w,_y c_y(\mathbf{n}'') \tag{4.244}$$

If $\mathbf{n}' \neq \mathbf{n}''$ and $\Delta\varphi \neq 0$ the two equations should constitute a regular linear system for the two values $w,_x$ and $w,_y$ and after its solution we can calculate any directional derivative at the non-smooth point **P**.

A further remark concerns the calculation of the integrals with the kernels

$$V_\nu(v_2) = 0(r^{-2} r_\tau r_t) = 0(r^{-2}) \text{ and } M_\nu(v_2) = 0(r^{-1} r_n) = 0(r^{-1+\alpha})$$

with $\alpha > 0$ if $\mathbf{n}$ is the normal on S (otherwise it is $M_\nu(v_2) = 0(r^{-1})$). If the collocation point **P** is a smooth point we need $w \in C^{1,\gamma}$ in a neighbourhood of **P** while $\theta_\nu(w)$ could be a piecewise continuous function which would be sufficient for the single integrals to exist. At a non-smooth point the situation is somewhat different: in deriving Equation 4.234 we have to expand the layer

$$w(\mathbf{Q}) - w(\mathbf{P}) = w,_x(\mathbf{P}) \cdot x + w,_y(\mathbf{P}) \cdot y + 0(r^2)$$

and the layer

$$\theta_\nu(w) = -w,_x(\mathbf{P}) \cdot v_1 - w,_y(\mathbf{P}) \cdot v_2 + 0(r)$$

in Taylor series near the point to prove that the sum of the two integrals exists as a Cauchy principal value. (The single integrals do not exist as a Cauchy principle value†.) The sum exists because the constant and linear terms in the expansion of the two layers balance each other and as such bridge the singularity while the error terms yield regular integrands after multiplication with the kernels.

† This should not come as a surprise and it seems to happen whenever an influence function is represented by a boundary integral. Note, e.g. that the normal derivative of a Laplace potential of the 1st kind does not exist at such points and that the traction of an elastic potential of the 1st kind which involves the normal derivative does not exist at non-smooth points, see Theorem 3.2.

If we use indicial notations and agree to read in the second equation $(i=2)$ on the right side $u_1(\mathbf{Q})=w(\mathbf{Q})-w(\mathbf{P})$, the Equations 4.233 and 4.234 become

$$\left.\begin{array}{l} c(\mathbf{P})w(\mathbf{P}) \\ c_x(\mathbf{P})w_{,x}(\mathbf{P})+c_y(\mathbf{P})w_{,y}(\mathbf{P}) \end{array}\right\} = \frac{1}{8\pi D}\oint_S [U_{ij}(\mathbf{P},\mathbf{Q})t_j(\mathbf{Q})-$$

$$T_{ij}(\mathbf{P},\mathbf{Q})u_j(\mathbf{Q})]\ ds_{\mathbf{Q}}+\frac{1}{8\pi D}\int_B U_{i1}(\mathbf{P},\mathbf{Q})\frac{p(\mathbf{Q})}{D}\ da_{\mathbf{Q}}- \tag{4.245a}$$

$$\frac{1}{8\pi D}\sum_{k=1}^{N}[\![v_i(\mathbf{P},\mathbf{Q}_k)M_{\nu\tau}(w)(\mathbf{Q}_k)-w(\mathbf{Q}_k)M_{\nu\tau}(v_i)(\mathbf{P},\mathbf{Q}_k)]\!] \tag{4.245b}$$

with $u_1=w$, $u_2=\theta_\nu(w)$, $t_1=V_\nu(w)$ and $t_2=M_\nu(w)$ and the terms $U_{ij},T_{ij};\ i,j=1,2$ as defined in Equation 4.212 through 4.219. If the point $\mathbf{P}$ lies on the boundary S and if a load $p(\mathbf{Q})$ is prescribed the Equations 4.245(a) and 4.245(b) become a relation between three objects: the vector valued function $\mathbf{u}=[w,\theta_\nu(w)]$, $\mathbf{t}=[V_\nu(w),\ M_\nu(w)]$ and the N numbers $F_k=M_{\nu\tau}(w)(\mathbf{Q}_k^-)-M_{\nu\tau}(w)(\mathbf{Q}_k^+)$, $k=1,2,\ldots N$, (we do not list the N numbers $w(\mathbf{Q}_k)=w(\mathbf{Q}_k^-)=w(\mathbf{Q}_k^+)$ separately from the function $w(\mathbf{Q})$) and we presume again: if two vector valued functions $\tilde{\mathbf{u}}=[\tilde{u}_1,\tilde{u}_2]^T$ and $\tilde{\mathbf{t}}=[\tilde{t}_1,\tilde{t}_2]^T$ and N numbers $\tilde{F}_k$ satisfy this relation on S then the function

$$w(\mathbf{P})=\frac{1}{8\pi D}\oint_S [U_{1j}(\mathbf{P},\mathbf{Q})\tilde{t}_j(\mathbf{Q})-T_{1j}(\mathbf{P},\mathbf{Q})\tilde{u}_j(\mathbf{Q})]\ ds_{\mathbf{Q}}+$$

$$\frac{1}{8\pi D}\int_B U_{11}(\mathbf{P},\mathbf{Q})\frac{p(\mathbf{Q})}{D}\ da_{\mathbf{Q}}-\frac{1}{8\pi D}\sum_{k=1}^{N}\{v_1(\mathbf{P},\mathbf{Q}_k)\tilde{F}_k- \tag{4.246}$$

$$[\![\tilde{u}_1(\mathbf{Q}_k)M_{\nu\tau}(v_1)(\mathbf{P},\mathbf{Q}_k)]\!]\}$$

assumes the boundary values

$$\lim_i w=\tilde{u}_1;\quad \lim_i \theta_n(w)=\tilde{u}_2$$

$$\lim_i V_n(w)=\tilde{t}_1;\quad \lim_i M_n(w)=\tilde{t}_2$$

and then

$$M_{\nu\tau}(w)(\mathbf{Q}_k^-)-M_{\nu\tau}(w)(\mathbf{Q}_k^+)=\tilde{F}_k,\quad k=1,2,\ldots N$$

$$\Delta\Delta w(\mathbf{P})=\frac{p(\mathbf{P})}{D},\quad \mathbf{P}\in B$$

Because two of the four components $\tilde{u}_1, \tilde{u}_2, \tilde{t}_1, \tilde{t}_2$ are in general prescribed at any point of S the integral relation becomes an integral equation for the two unknown components and the corner forces $\tilde{F}_k$ and solved approximately by employing the usual numerical means.

The results achieved by Forbes and Robinson[55], Bezine[56] and Stern[57] with these equations surely justify this direct approach to the solution of plate bending problems and we refer to their papers for illustrative examples.

Forbes and Robinson[55] provide a complete coverage of the numerical details if the layers and the boundary are interpolated by polynomials of the second degree. However, as their equations include only smooth boundaries it might be helpful if we discuss some details of the numerical procedure for a plate with a piecewise smooth boundary.

We mentioned above that we need near a smooth collocation point $w \in C^{1,\gamma}$ and that $\theta_v(w)$ must be at least piecewise continuous for the integrals to exist. We achieve this if we interpolate w and $\theta_v(w)$ by functions w and θ whose restrictions on any boundary element S_i are polynomials and which satisfy $w \in C^1(L)$ and $\theta \in C(L)$ on any arbitrary smooth connected part L of S.

For the restriction of w on a boundary element S_i we choose a cubic polynomial which allows a continuous continuation of the first derivative in the nodes of neighbouring elements and for the restrictions of θ, M and V we choose linear functions. Let us for simplicity assume that the edge of the plate has a polygonal shape. We partition the edge in K straight boundary elements S_i, $i=1,2,\ldots K$ whose endpoints are the nodes $\mathbf{P}^i$, $i=1,2,\ldots K$. To the partition corresponds a reference element S_0, i.e. the interval $[0,1]$ on the axis ξ, with the nodes $\xi_1=0$ and $\xi_2=1$.

The following functions are defined on S_0:

$$\begin{aligned} g_1(\xi) &= \xi & g_2(\xi) &= 1-\xi \\ s_1(\xi) &= \xi^2(3-2\xi) & s_2(\xi) &= (\xi-1)^2(1+2\xi) \\ t_1(\xi) &= \xi^2(\xi-1) & t_2(\xi) &= \xi(\xi-1)^2 \end{aligned} \tag{4.247}$$

with the properties

$$\begin{gathered} g_i(\xi_j) = 1-\delta_{ij} \\ s_i(\xi_j) = 1-\delta_{ij}, \quad \frac{\mathrm{d}s_i}{\mathrm{d}\xi}(\xi_j) = 0 \quad i,j=1,2 \\ t_i(\xi_j) = 0, \quad \frac{\mathrm{d}t_i}{\mathrm{d}\xi}(\xi_j) = 1-\delta_{ij} \end{gathered} \tag{4.248}$$

The boundary element S_i extends from the node $\mathbf{P}^{i-1}=(x_{i-1},y_{i-1})$ to the node $\mathbf{P}^i=(x_i,y_i)$ and the function

$$\bar{\mathbf{P}}^i(\xi) = \begin{bmatrix} x^i(\xi) \\ y^i(\xi) \end{bmatrix} = \begin{bmatrix} x_{i-1}g_2 + x_i g_1 \\ y_{i-1}g_2 + y_i g_1 \end{bmatrix} \tag{4.249}$$

maps S_0 on the element S_i. The Jacobian of this transform is just $[(x_i - x_{i-1})^2 + (y_i - y_{i-1})^2]^{1/2} = \Delta s_i$, the length of the element S_i and we have for any function $f(\mathbf{P})$

$$\int_{S_i} f(\mathbf{P})\, \mathrm{d}s = \int_0^1 f(\bar{\mathbf{P}}^i(\xi))\Delta s_i \, \mathrm{d}\xi \tag{4.250}$$

While the deflection w is continuous on S the functions $\partial w/\partial s(s)$ and $\partial w/\partial n(s)$ are discontinuous at non-smooth points. The same can happen to the quantities $M_n(w)$ and $V_n(w)$ at these points and also at smooth points where the boundary conditions changes†. Thus, a bounded function $f(\mathbf{P})$ interpolated on S has in general, at a node $\mathbf{P}^i$ (arc length $s = c_i$), two nodal values $f_1(\mathbf{P}^i)$ and $f_2(\mathbf{P}^i)$:

$$f_1(\mathbf{P}^i) = \lim_{\substack{s \to c_i \\ s < c_i}} f(\mathbf{P}(s)) \quad f_2(\mathbf{P}^i) = \lim_{\substack{s \to c_i \\ s > c_i}} f(\mathbf{P}(s))$$

and its nodal value is therefore a vector with two components. Likewise the notation of the nodal shape functions must follow this vector notation. We introduce the functions:

$$\mathbf{l}_i(\mathbf{P}) = \begin{bmatrix} l_{i1}(\mathbf{P}) \\ l_{i2}(\mathbf{P}) \end{bmatrix}, \quad i = 1,2,\ldots K \tag{4.251}$$

with

$$l_{i1}(\mathbf{P}) = g_1(\xi), \quad \text{if } \mathbf{P} \in S_i, \text{ and 0 otherwise}$$

$$l_{i2}(\mathbf{P}) = g_2(\xi), \quad \text{if } \mathbf{P} \in S_{i+1}, \text{ and 0 otherwise}$$

$$p_i(\mathbf{P}) = \begin{cases} s_1(\xi), & \mathbf{P} \in S_i \\ s_2(\xi), & \mathbf{P} \in S_{i+1} \\ 0, & \text{otherwise} \end{cases} \qquad i = 1,2,\ldots K \tag{4.252}$$

$$\mathbf{q}_i(\mathbf{P}) = \begin{bmatrix} q_{i1}(\mathbf{P}) \\ q_{i2}(\mathbf{P}) \end{bmatrix}, \qquad i = 1,2,\ldots K \tag{4.253}$$

with

$$q_{i1}(\mathbf{P}) = t_1(\xi), \quad \mathbf{P} \in S_i, \text{ and 0 otherwise}$$

$$q_{i2}(\mathbf{P}) = t_2(\xi), \quad \mathbf{P} \in S_{i+1}, \text{ and 0 otherwise}$$

† Naturally the engineer also approximates solutions with the direct method which contain singular points on the boundary, e.g. points where the boundary conditions change, see Rannacher and Melzer[64].

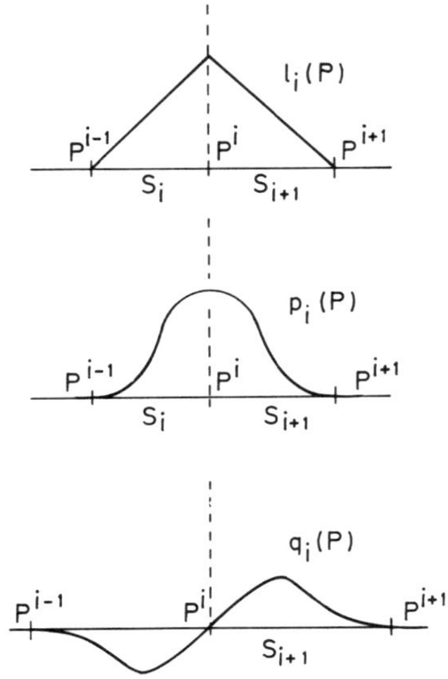

Figure 4.26 The functions $l_i(P)$,$p_i(P)$ and $q_i(P)$

whose finite supports extent only over the boundary elements S_i and S_{i+1}, i.e. they vanish outside of $S_i \cup S_{i+1}$ and the two components are the restrictions of the function on S_i and S_{i+1} respectively, see Figure 4.26.

Introducing for the nodal values the notations:

$$
\begin{aligned}
w^i &= w(\mathbf{P}^i) \\
\mathbf{w}_i' &= [w_1'^i, w_2'^i]^T = \left[\left(\frac{\mathrm{d}w}{\mathrm{d}s}\right)_1, \left(\frac{\mathrm{d}w}{\mathrm{d}s}\right)_2\right]^T(\mathbf{P}^i) \\
\boldsymbol{\theta}_i &= [\theta_1^i \theta_2^i]^T = [\theta_v(w)_1, \theta_v(w)_2]^T(\mathbf{P}^i) \\
\mathbf{M}_i &= [M_1^i, M_2^i]^T = [M_v(w)_1, M_v(w)_2]^T(\mathbf{P}^i) \\
\mathbf{V}_i &= [V_1^i, V_2^i]^T = [V_v(w)_1, V_v(w)_2]^T(\mathbf{P}^i)
\end{aligned} \tag{4.254}
$$

we can write for the interpolates to the layers

$$
\begin{aligned}
w(\mathbf{P}) &= \sum_{i=1}^{K} w^i p_i(\mathbf{P}) + w_i'^T \mathbf{q}_i(\mathbf{P}) \\
\theta(\mathbf{P}) &= \sum_{i=1}^{K} \boldsymbol{\theta}_i^T \cdot \mathbf{l}_i(\mathbf{P}) \\
M_v(\mathbf{P}) &= \sum_{i=1}^{K} \mathbf{M}_i^T \cdot \mathbf{l}_i(\mathbf{P})
\end{aligned} \tag{4.255}
$$

$$V_v(\mathbf{P}) = \sum_{i=1}^{K} \mathbf{V}_i^{T} \cdot \mathbf{l}_i(\mathbf{P})$$

Solving the integral equation with a collocation procedure which uses the nodes $\mathbf{P}^i$ we must evaluate at any node the equivalent of four boundary integrals (we assume that always two of the four quantities w, θ_n, M_n, V_n are zero on any part of S) and two integrals that extend over the plate B if the vertical load $p(\mathbf{Q})$ is not zero. The kernels in these latter integrals are $v_1(\mathbf{P}^i, \mathbf{Q})$ and $v_2(\mathbf{P}^i, \mathbf{Q})$ which are regular on B and they allow therefore the application of a numerical quadratur, see e.g. Davis and Rabinowitz[63], Chapter 5.

With numerical quadrature we evaluate also any boundary integral on an element S_m which does not contain the point $\mathbf{P}^i$, i.e. $S_m \neq S_i$ or S_{i+1}. Integration over the two elements $T_i = S_i \cup S_{i+1}$ depends on the kernel in question and whether $\mathbf{P}^i$ is a smooth point or not. The kernels $v_1(\mathbf{P}^i, \mathbf{Q})$, $v_2(\mathbf{P}^i, \mathbf{Q})$, $\theta_v(v_1)(\mathbf{P}^i, \mathbf{Q})$ and $\theta_v(v_2)(\mathbf{P}^i, \mathbf{Q})$ are sufficiently regular to allow numerical quadrature at any point $\mathbf{P}^i$.

The calculation of the integrals over T_i with the kernels

$$\begin{aligned} &M_v(v_1) = 0(\ln r), \quad M_v(v_2) = 0(r^{-1} r_n) \\ &V_v(v_1) = 0(r^{-1} r_v), \quad V_v(v_2 = 0(r^{-2}) \end{aligned} \tag{4.256}$$

differs only slightly. The lower terms in the expansion of the layers $w(\mathbf{Q}) - w(\mathbf{P}^i)$, $\theta_v(\mathbf{Q})$ and $w(\mathbf{Q})$ yield, after multiplication with the kernels, singular integrands whose corresponding Cauchy principal value is calculated separately (it is always zero) while the higher terms yield regular integrands whose integral can be evaluated as before with numerical quadrature.

It might be interesting if we discuss this procedure in more details. Let us first assume that $\mathbf{P}^i$ is a smooth point and $\mathbf{n}$ the normal at $\mathbf{P}^i$ then holds on T_i:$r_v = r_n = 0$ and $r_t = \pm 1$, $r_\tau = \mp 1$ and we obtain

$$\begin{aligned} &M_v(v_1) = -D[2(1+\bar{v}) \ln r + (1+3\bar{v})] \\ &M_v(v_2) = 0, \quad V_v(v_1) = 0 \\ &V_v(v_2) = -\frac{2D}{r^2} \end{aligned} \tag{4.257}$$

Assuming that both elements have the same length Δs we get

$$\int_{T_i} V_v(v_2)(\mathbf{P}^i, \mathbf{Q})[w(\mathbf{Q}) - w(\mathbf{P}^i)] \, ds_{\mathbf{Q}} =$$

$$= -2D\left\{\int_0^1 \frac{1}{(1-\xi)^2}[(w^{i-1} - w^i)s_2 + w_1'^i t_1 + w_2'^{i-1} t_2] \, d\xi + \right.$$

$$+\int_0^1 \frac{1}{\xi^2}[(w^{i+1}-w^i)s_1+w_1'^{i+1}t_1+w_2'^i t_2]\ d\xi\Big\}\Delta s \tag{4.258}$$

Due to $w \in \mathcal{Z}^1(L)$ it holds that $w_1'^i = w_2'^i$ and therefore the Cauchy principal value

$$\begin{aligned}&\left\{\int_0^1 \frac{1}{(1-\xi)^2}\cdot w_1'^i t_1(\xi)\ d\xi + \int_0^1 \frac{1}{\xi^2} w_2'^i t_2(\xi)\ d\xi\right\} = \\ &\lim_{\varepsilon\to 0}\left\{\int_0^{1-\varepsilon} \frac{1}{(1-\xi)^2}\xi^2(\xi-1)\ d\xi + \int_\varepsilon^1 \frac{1}{\xi^2}\xi(\xi-1)^2\ d\xi\right\} w_1'^i\end{aligned} \tag{4.259}$$

exists and is zero so that we end with

$$\begin{aligned}&\int_{T_i} V_\nu(v_2)(\mathbf{P}^i,\mathbf{Q})[w(\mathbf{Q})-w(\mathbf{P}^i)]\ ds_{\mathbf{Q}} = \\ &-2D\left\{\int_0^1 \frac{1}{(1-\xi)^2}[(w^{i-1}-w^i)s_2+w_2'^{i-1}t_2]\ d\xi + \right. \\ &\left.\int_0^1 \frac{1}{\xi^2}[(w^{i+1}-w^i)s_1+w_1'^{i+1}t_1]\ d\xi\right\}\Delta s = \\ &= -2D\left\{\int_0^1 [(3-2\xi)\{w^{i-1}-2w^i+w^{i+1}\} + \right. \\ &\left.(\xi-1)(w_1'^{i+1}-w_2'^{i-1})]\ d\xi\right\}\Delta s\end{aligned} \tag{4.260}$$

where we find only regular integrands which can be evaluated by quadrature.

Analogously we derive

$$\begin{aligned}&\int_{T_i} M_\nu(v_1)(\mathbf{P}^i,\mathbf{Q})\theta(\mathbf{Q})\ ds_{\mathbf{Q}} = D(1-\bar{\nu})2\theta^i\Delta s - \\ &D(\theta_2^{i-1}-2\theta^i+\theta_1^{i+1})\int_0^1 [2(1+\bar{\nu})\xi\ \ln\ \xi+(1+3\bar{\nu})\xi]\ d\xi\ \Delta s\end{aligned} \tag{4.261}$$

(At a smooth point $\mathbf{P}^i$ it is naturally $\theta_1^i = \theta_2^i = \theta^i$). Since $M_\nu(v_2) = V_\nu(v_1) = 0$ on T_i, it is furthermore

$$\int_{T_i} M_\nu(v_2)(\mathbf{P}^i,\mathbf{Q})\theta(\mathbf{Q})\,\mathrm{d}s_\mathbf{Q} = \int_{T_i} V_\nu(v_1)(\mathbf{P}^i,\mathbf{Q})w(\mathbf{Q})\,\mathrm{d}s_\mathbf{Q} = 0 \qquad (4.262)$$

Turning now to the integration if $\mathbf{P}^i$ is a non-smooth point we should remember that the integrals with the kernels $M_\nu(v_2)$ and $V_\nu(v_2)$ must be evaluated simultaneously as mentioned above.

Since $r_\nu = 0$ on S_i and S_{i+1}, we have

$$V_\nu(v_2) = \frac{2D}{r^2}(1+\bar{\nu})r_\tau r_t$$
$$M_\nu(v_2) = \frac{2D}{r}(1+\bar{\nu})r_n \qquad (4.263)$$

Now, $r_\tau = -1$, $r_t = r_t^i$ and $r_n = r_n^i$ on S_i, and $r_\tau = +1$, $r_t = r_t^{i+1}$ and $r_n = r_n^{i+1}$ on S_{i+1}, so that we obtain:

$$A(\mathbf{P}^i) = \int_{T_i} V_\nu(v_2)(\mathbf{P}^i,\mathbf{Q})[w(\mathbf{Q}) - w(\mathbf{P}^i)]\,\mathrm{d}s_\mathbf{Q} +$$

$$\int_{T_i} M_\nu(v_2)(\mathbf{P}^i,\mathbf{Q})\theta(\mathbf{Q})\,\mathrm{d}s_\mathbf{Q} =$$

$$2D(1+\bar{\nu})\left\{\int_0^1 \frac{-1}{(1-\xi)^2} r_t^i \left[(w^{i-1} - w^i)s_2 + \underline{w_1'^i t_1} + w_2'^{i-1} t_2\right]\mathrm{d}\xi + \right. \qquad (4.264)$$

$$\int_0^1 \frac{1}{\xi^2} r_t^{i+1} \left[(w^{i+1} - w^i)s_1 + w_1'^{i+1} t_1 + \underline{w_2'^i t_2}\right]\mathrm{d}\xi +$$

$$\left.\int_0^1 \frac{1}{(1-\xi)} r_n^i \left[\underline{\theta_1^i g_1} + \theta_2^{i-1} g_2\right]\mathrm{d}\xi + \int_0^1 \frac{1}{\xi} r_n^{i+1}\left[\theta_1^{i+1} g_1 + \underline{\theta_2^i g_2}\right]\mathrm{d}\xi\right\}\Delta s$$

The functions underlined in Equation 4.264 do not match the denominator in front of the bracket and contribute singular terms. The singular terms alone yield the expression

$$J = 2D(1+\bar{\nu})\int_0^1 \frac{1}{\xi}\{(r_t^i w_1'^i + r_t^{i+1} w_2'^i)(1-2\xi+\xi^2) + (r_n^i\theta_1^i + r_n^{i+1}\theta_2^i)(1-\xi)\}\ \mathrm{d}\xi\ \Delta s \tag{4.265}$$

$$\begin{aligned} w_1'^i &= \mathrm{grad}^T w(\mathbf{P}^i)\boldsymbol{\tau}^i, \quad w_2'^i = \mathrm{grad}^T w(\mathbf{P}^i)\boldsymbol{\tau}^{i+1} \\ \theta_1^i &= -\mathrm{grad}^T w(\mathbf{P}^i)\boldsymbol{\nu}^i, \quad \theta_2^i = -\mathrm{grad}^T w(\mathbf{P}^i)\boldsymbol{\nu}^{i+1} \end{aligned} \tag{4.266}$$

and

$$\begin{aligned} r_t^i &= \mathrm{grad}_{\mathbf{p}}^T r\mathbf{t} = (\boldsymbol{\tau}^i)^T\mathbf{t}, \quad r_t^{i+1} = (-\boldsymbol{\tau}^{i+1})^T\mathbf{t} \\ r_n^i &= (\boldsymbol{\tau}^i)^T\mathbf{n}, \quad r_n^{i+1} = (-\boldsymbol{\tau}^{i+1})^T\mathbf{n} \end{aligned} \tag{4.267}$$

where $\boldsymbol{\tau}^k = [\tau_1^k, \tau_2^k] = [\nu_2^k, -\nu_1^k]$, $k = i, i+1$ is the tangent vector at a point $\mathbf{Q}$ on S_i resp. S_{i+1} and $\boldsymbol{\nu}^k = [\nu_1^k, \nu_2^k]$, $k = i, i+1$ is the outward normal at the same point. Therefore

$$r_t^i w_1'^i + r_t^{i+1} w_2'^i + r_n^i\theta_1^i + r_n^{i+1}\theta_2^i = 0 \tag{4.268}$$

(which could be proved by simple calculation) and the Cauchy principal value of the strongly singular term exists and is zero. Thus we obtain

$$\begin{aligned} J &= 2D(1+\bar{\nu})\int_0^1 \{(r_t^i w_1'^i + r_t^{i+1} w_2'^i)(-2+\xi) - (r_n^i\theta_1^i + r_n^{i+1}\theta_2^i)\}\ \mathrm{d}\xi\ \Delta s \\ &= 2D(1+\bar{\nu})\int_0^1 (r_t^i w_1'^i + r_t^{i+1} w_2'^i)(-1+\xi)\ \mathrm{d}\xi\ \Delta s \end{aligned} \tag{4.269}$$

which is a regular integral and we end with

$$A(\mathbf{P}^i) = 2D(1+\bar{\nu})\{r_t^i[-2w^{i-1} + 2w^i - \tfrac{1}{2}w_2'^{i-1} - \tfrac{1}{2}w_1'^i] + r_t^{i+1}[2w^{i+1} - 2w^i - \tfrac{1}{2}w_1'^{i+1} - \tfrac{1}{2}w_2'^i] + r_n^i\theta_2^{i-1} + r_n^{i+1}\theta_1^{i+1}\}\Delta s \tag{4.270}$$

We still require formulae for the integrals with the kernels $V_\nu(v_1)$ and $M_\nu(v_1)$ at the non-smooth point $\mathbf{P}$. On T_i

$$\begin{aligned} V_\nu(v_1) &= 0 \\ M_\nu(v_1) &= -D[2(1+\bar{\nu})\ \ln r + (1+3\bar{\nu})] \end{aligned} \tag{4.271}$$

so that we get

$$\int_{T_i} V_\nu(v_1)w(\mathbf{Q})\ \mathrm{d}s_{\mathbf{Q}}=0 \tag{4.272}$$

and after some simple calculations:

$$\int_{T_i} M_\nu(v_1)(\mathbf{P},\mathbf{Q})\theta_\nu(\mathbf{Q})\ \mathrm{d}s_{\mathbf{Q}}=D(1-\bar{\nu})(\theta_1^i+\theta_2^i)\Delta s- \\ D[\theta_2^{i-1}-\theta_1^i-\theta_2^i+\theta_1^{i+1}]\int_0^1[2(1+\bar{\nu})\xi\ \ln\ \xi+(1+3\bar{\nu})\xi]\ \mathrm{d}\xi\ \Delta s \tag{4.273}$$

which is a regular integral.

Note that $(-\ln r)$ is only a weakly singular kernel and therefore yields the non-vanishing contribution $D\Delta s(1-\bar{\nu})(\theta_1^i+\theta_2^i)$ contrary to the strongly singular kernels before. The same holds for the integral in Equation 4.261.

Our numerical scheme provides in general eight nodal parameters at a point $\mathbf{P}^i$:

$$w,\theta_1^i,\theta_2^i,M_1^i,M_2^i,V_1^i,V_2^i,F_i$$

Whether all these parameters are truely independent relies on the smoothness of the solution we want to approximate and whether $\mathbf{P}^i$ is a smooth point or not.

The simplest case is a smooth point with a smooth solution $w\in C^3(\bar{B})$ where we have four independent nodal values:

$$w,\theta_1^i=\theta_2^i,M_1^i=M_2^i,V_1^i=V_2^i$$

and it is $F_i=0$.

The most complicated case is a non-smooth point with a non-smooth solution. In between lie all possible combinations of w(smooth/non-smooth) with $\mathbf{P}^i$(smooth/non-smooth).

To determine the nodal parameters we have the boundary conditions at hand and two integral equations for w and $\partial w/\partial n$ at a smooth point and three such equations for w, $w_{,x}$ and $w_{,y}$ at a non-smooth point.

While at a smooth point these conditions and equations are usually equal to the number of unknown parameters this may not be so at a non-smooth point where sometimes we have to gather further information by studying the behaviour of the exact solution near the point.

A valuable contribution to this study is a paper by Melzer and Rannacher[64]. They are concerned with singularities of plate solutions that arise at non-smooth points of a plate or at points on the boundary where the boundary condition changes. They provide diagrams which allow one to determine the character of the singular terms ($r^{-\alpha}$, $\ln^{\beta} r$) in plate solutions and they calculate critical values of the internal angle. If the angle passes these values M_n and/or Q_n become singular. The paper is an extension of Williams'[65] well known paper and utilizes results of Kondratiev[66].

If the layers become singular as, e.g. M_n we should naturally provide appropriate interpolates to these layers by replacing the linear function $\xi, 1-\xi$ by $\xi^{-\alpha}$, $(1-\xi)^{-\alpha}$, $\ln^{\beta}\xi$ etc. In such cases the nodal parameters are no longer nodal values but should be understood as weights.

It seems that more than any other method the boundary element method depends on a good knowledge of the behaviour of the exact solution near critical points on the boundary and thus practical computational aspects revive the interest in such questions.

Concerning the bvp of classical elasticity, the Cauchy-Navier equations, we refer to Williams[65] and England[67] where further references regarding these questions may be found.

4.7.4 **Plates on elastic foundations and shells**

If the plate rests on an elastic foundation with stiffness k, the differential equation for the deflection becomes

$$\Delta\Delta w + \frac{k}{D} w = \frac{p}{D} \tag{4.274}$$

The principal part of this equation can be factorized and leads then to an equation whose fundamental solution is known. The further treatment is similar to the case $k=0$ and we refer to Tottenham[58] for a short discussion. Forbes and Robinson[55] do not use the fundamental solution of Equation 4.274 but employ the fundamental solution of plate bending. They start with the Rayleigh-Green identity and set

$$\Delta\Delta w = \frac{p}{D} - \frac{k}{D} w,$$

which leads to an integral extending over the plate containing, beside the load p, the term kw/D:

$$\int_B v_1(\mathbf{P},\mathbf{Q}) \frac{1}{D} [p(\mathbf{Q}) - kw(\mathbf{Q})] \, da_{\mathbf{Q}} \tag{4.275}$$

This way the identity 4.233 becomes an integral equation which must be solved on the whole plate B. With a similar approach they treat the problem of shallow shells, which is solved by Newton and Tottenham[68] by a true boundary integral equation method. The differential equations of shallow shells can under certain circumstances be reduced to differential equations whose principal part is the Laplacian and then can be treated in the framework of classical potential theory, see Tottenham[58]. The solution of general shell problems with boundary element methods seems still to be an open question but recently we learnt about a new publication by Lukasiewicz[69] which might eventually clear the way for boundary element methods in general shell problems.

4.8 FURTHER APPLICATIONS

We close with a list of those fields in elasticity where we know the fundamental solutions of the governing differential equation:

(1) Anisotropic elasticity:
  Plane problems: Kupradze[7], Chapter VIII
  3-D problems: Vogel and Rizzo[70]

(2) Harmonic oscillations of displacement fields: Kupradze[9]

(3) General dynamic problems: Gurtin[2], p. 239

(4) Thermoelasticity (steady state and harmonic oscillation): Kupradze[9]

(5) Cosserat continua: Kupradze[9]

(6) Saint-Venant's torsion: Christiansen[71]

The treatment of bvp in these fields with integral equation methods follows naturally the same pattern discussed in the preceding sections.

ACKNOWLEDGEMENT

Part of this work was completed while the author was visiting the School of Engineering at the University of California, Irvine. The financial support of the NATO Science Committee is gratefully acknowledged.

REFERENCES

1 Fichera, G., 'Il teorema del massimo modulo per l'equazione dell'elastostatica tri-dimensionale,' *Arch. Rational Mech. Anal.*, **7**, pp. 373–387, (1961)

2 Gurtin, M. E., 'The linear theory of elasticity,' in *Encyclopedia of Physics*, Vol. VIa/2, Flügge and Truesdell (Ed.), Springer Verlag (1972)

3 Adler, G., 'Majoration des tensions dans un corps elastique a l'aide des deplacements superficiels', *Arch. Rational Mech. Anal.*, **16**, pp. 345–372 (1964)

4 Stippes, M. and Rizzo, F. J., 'A note on the body force integral of classical elastostatics', *ZAMP*, **28**, pp. 339–341 (1977)

5 Hartmann, F., *Elastische Potentiale in Gebieten mit Ecken*, Thesis, University of Dortmund (1980)

6 Mikhlin, S. G., *Multidimensional Singular Integrals and Integral Equations*, Pergamon Press (1965)

7 Kupradze, V. D., *Potential Methods in the Theory of Elasticity*, Israel program for scientific translations, Jerusalem (1965)

8 Jaswon, M. A. and Symm, G. T., *Integral Equation Methods in Potential Theory and Elastostatics*, Academic Press, (1977)

9 Kupradze, V. D. (Ed.), *Three-dimensional Problems of the Mathematical Theory of Elasticity and Thermoelasticity*, North-Holland Publishing Co. (1979)

10 Fichera, G., 'Existence theorems in elasticity', in *Encyclopedia of Physics*, Vol. VIa/2, Flügge and Truesdell (Ed.), Springer-Verlag (1972)

11 Rizzonelli, P. C., 'On the first boundary value problem for the classical theory of elasticity in a three-dimensional domain with a singular boundary', *J. Elasticity*, **3**, No. 4, pp. 225–259 (1973)

12 Wendland, W., 'Die Behandlung von Randwertaufgaben im R_3 mit Hilfe von Einfach-und Doppelschichtpotentialen', *Numerische Mathematik*, **11**, pp. 380–404 (1968)

13 Burago, Yu. D., Maz'ya, V. G. and Sapozhwikova, V. D., 'On the theory of simple and dobule-layer potentials for domains with irregular boundaries', in *Problems in Mathematical Analysis 1*, pp. 1–30, Smirnov, V. I. (Ed.), New York: Consultants Bureau (1968)

14 Rieder, G., 'Mechanische Deutung und Klassifizierung einiger Integralverfahren der ebenen Elastizitatstheorie I, II', *Bulletin de l'academie polonaise des sciences, 5 and 6*, Vol. XVI, No. 2 (1968)

15 Heise, U., 'The spectra of some integral operators for plane elastostatic boundary value problems', *J. Elasticity*, **8**, No. 1, 47–49 (1978)

16 Heise, U., 'Numerical properties of integral equations in which the given boundary values and the sought solutions are defined on different curves', *Computers and Structures*, **8**, pp. 199–205 (1978)

17 Heise, U. and Müller, C. H., 'Numerical calculation of eigenvalues of integral operators for plane elastostatic boundary value problems', *Computer Methods in Appl. Mechanics and Engng.*, **21**, pp. 17–43 (1980)

18 Heise, U., 'The calculation of Cauchy principal values in integral equations for bvp of the plane and three-dimensional theory of elasticity', *J. Elasticity*, **5**, pp. 99–110 (1975)

19 Antes, H., 'Die Splineinterpolation zur lösung von Integralgleichungen und ihre Anwendung bei der Berechnung von Spannungen in krummlinig berandeten Scheiben, Dissertation', Aachen (1970)

20 Borgwardt, F., 'Berechnung von krummlinig berandeten Scheiben mit Hilfe eines Integralgleichungsverfahrens', Dissertation, Braunschweig (1965)

21 Pahnke, U., 'Berechnung von beliebig geformten Scheiben und Platten mit ausspringenden Ecken und geschlitzten Scheiben mit, Hilfe eines Integralgleichungsverfahrens', Dissertation, Aachen (1970)

22 Massonet, Ch., 'Solution generale du probleme aux tensions de l'elasticite tridimensionelle', *Proc. 9th Congr. Appl. Mechanics*, Bruxelles, pp. 168–180 (1965)

23 Heise, U., 'Systematic compilation of integral equations of the Rizzo type and of Kupradze's functional equations for boundary value problems of plane elastostatics', *J. Elasticity*, **10**, pp. 23–56 (1980)

24 Oliveira, E. R. A., 'Plane stress analysis by a general integral method', *J. Eng. Mechanics Div., Proc. Am. Soc. Civil Eng.*, pp. 79–101 (1968)

25 Friemann, H., 'Anwendung eines Integralverfahrens auf die Berechnung elastischer Scheiben mit statischen oder geometrischen Randbedingungen unter der Einwirkung beliebiger Rand-oder Innenlasten'. Habilitationssehrift, Darmstadt (1973)

26 Christiansen, S., 'On Kupradze's functional equations for plane harmonic problems', *ZAMM*, No. 44 (1975)

27 Rizzo, F. J., 'An integral equation approach to boundary value problems of classical elastostatic', *Quart. Appl. Math.*, **25**, pp. 83–95 (1967)

28 Cruse, T. A., 'Numerical solutions in three dimensional elastostatics', *Int. J. Solids Structures*, **5**, pp. 1259–1274 (1969)

29 Cruse, T. A., 'Application of the boundary integral equation method to three dimensional stress analysis', *Computers and Structures*, **3**, pp. 509–527 (1973)

30 Cruse, T. A., 'An improved boundary-integral equation method for three dimensional elastic stress analysis', *Computers and Structures*, **4**, pp. 741–754 (1974)

31 Turteltaub, M. J. and Sternberg, E., 'On concentrated loads and Green's functions in elastostatics', *Arch. Rational Mech. Anal.*, **29**, pp. 193–240 (1968)

32 Brebbia, C. A., *The Boundary Element Method for Engineers*, Pentech Press, Plymouth, London (1978)

33 Warlo, F., 'Behandlung des Rizzo-Integralgleichungsverfahrens der Elastostatik im Hinblick auf die Lösung technischer Probleme', Dissertation, TU München (1974)

34 Atkinson, K. A., 'A survey of numerical methods for the solution of Fredholm integral equations of the second kind', *SIAM*, Philadelphia (1976)

35 Ben Noble., 'Error analysis of collocation methods for solving Fredholm integral equations', in *Topics in Numerical Analysis*, J. H. Miller (Ed.), Academic Press, London (1972)

36 Ciarlet, P. G., *The Finite Element Method for Elliptic Problems*, North Holland, Amsterdam (1978)

37 Hsiao, G. C. and Wendland, W. L., 'A finite element method for some integral equations of the first kind', *J. Math. Anal. Appl.*, **58**, pp. 449–481 (1977)

38 Nedelec, J. C., 'Curved finite element methods for the solution of singular integral equations on surfaces in R^3', *Comp. Math. Appl. Mech. Engng.*, **8**, pp. 61–80 (1976)

39 Prössdorf, S. and Silbermann, B., *Projektionsverfahren und die näherungsweise Lösung singulärer Gleichungen*, Teubner, Leipzig (1977)

40 Wendland, W., 'The synthesis of the collocation and the Galerkin method applied to some integral equations of the first kind', in *New Developments in Boundary Element Methods*, C. A. Brebbia (Ed.), CML Publications, Southampton (1980)

41 Chaudonneret., 'On the discontinuity of the stress vector in the boundary integral equation method for elastic analysis', in *Recent Advances in Boundary Element Methods*, C. A. Brebbia (Ed.), Pentech Press (1978)

42 Wardle, L. J. and Crotty, J. M., 'Two-dimensional boundary integral equation analysis for non-homogeneous mining applications', in *Recent Advances in Boundary Element Methods*, C. A. Brebbia (Ed.), Pentech Press (1978)

43 Lachat, J. C. and Watson, J. O., 'Effective numerical treatment of boundary integral equations: a formulation for three-dimensional elastostatics', *Int. J. Num. Meth. Engng.*, **10**, pp. 991–1005 (1976)

44 Rizzo, F. J. and Shippy, D. J., 'An advanced boundary integral equation method for three-dimensional thermoelasticity', *Int. J. Num. Meth. Engng.*, **11**, pp. 1753–1768 (1977)

45 Cristescu, M. and Loubignac, G., 'Gaussian quadrature formulas for functions with singularities in 1/R over triangles and quadrangles', in *Recent Advances in Boundary Element Methods*, C. A. Brebbia (Ed.), Pentech Press (1978)

46 Mayr, M., 'Ein Integralgleichungsverfahren zur Lösung rotationssymmetrischer Elastizitätsprobleme', Dissertation, TU München (1975)

47 Mayr, M. and Neureiter, W., 'Ein numerisches Verfahren zur Lösung des axialsymmetrischen Torsionproblems', *Ingenieur Archiv.*, **46**, pp. 137–142 (1977)

48 Sneddon, I. N., 'Application of integral transforms in the theory of elasticity', *CISM, Courses and Lectures*, No. 220, Springer Verlag, Wien–New York (1976)

49 Maiti, M., Das, B. and Palit, S. S., 'Somigliana's method applied to plane problems of elastic halfspaces', *J. Elasticity*, **6**, No. 4 (1976)

50 Jaswon, M. A., Maiti, M. and Symm, G. T., 'Numerical biharmonic analysis and some applications', *Int. J. Solids Struct.*, **3**, pp. 309–312 (1967)

51 Jaswon, M. A. and Maiti, M., 'An integral equation formulation of plate bending problems', *J. Engng. Math.*, **2**(1), pp. 83–93 (1968)

52 Maiti, M. and Chakrabart, S. K., 'Integral equation solution for simply supported polygonal plates', *Int. J. Engng. Sci.*, **12**, pp. 793–806 (1974)

53 Holzemer, K., 'Integralgleichungsverfahren für Platten mit verschiedenen Randbedingungen auf nicht zusammenhängenden Randkurven', *Ingenieur Archiv*, **48**, pp. 1–11 (1979)

54 Wendland, W., *Elliptic Systems in the Plane*, Pitman, London (1979)

55 Forbes and Robinson., 'Numerical analysis of elastic plates and shallow shells by an integral equation method', University of Illinois, Structural Research Series Report 345 (1969)

56 Bezine, G. and Gamby, D. A., 'A new integral equation for plate bending problems', in *Recent Advances in Boundary Elements Methods*, C. A. Brebbia (Ed.), Pentech Press (1978)

57 Stern, M., 'A general boundary integral formulation for the numerical solution of plate bending problems', *Int. J. Solids Structures*, **15**, pp. 769–782 (1979)

58 Tottenham, H., 'The boundary element method for plates and shells', in *Developments in Boundary Element Methods*, Banerjee and Butterfield (Ed.), Applied Science Publishers Ltd., London (1979)

59 Altiero, N. and Sikarskie, D., 'A boundary integral method applied to plates of arbitrary plane form', *Computers and Structures*, **9**, pp. 163–168 (1978)

60 Freitag, H., 'Berechnung von isotropen, elastischen Platten mit beliebigen Randbedingungen unter der Einwirkung von Rand- und Innenlasten nach der Integralmethode mit Hilfe der Scheiben-Platten-Analogie', Veröffentlichung des Instituts für Statik und Stahlbau der TH Darmstadt, Darmstadt (1979)

61 Bergman, S. and Schiffer, M., *Kernel Functions and Elliptic Differential Equations in Mathematical Physics*, Academic Press, New York, London (1953)

62 Timoshenko, S. and Woinowsky-Krieger, W., *Theory of Plates and Shells*, 2nd ed., McGraw Hill, New York (1959)

63 Davis, P. J. and Rabinowitz, P., *Methods of Numerical Integration*, Academic Press, New York (1975)

64 Melzer, H. and Rannacher, R., 'Spannungskonzentration in Eckpunkten der Kirchhoffschen Platte', *Bauingenieur*, **55**, p. 181–184 (1980)

65 Williams, M. L., 'Surface stress singularities resulting from various boundary conditions in angular corners of plates under bending', *Proc. 1st U.S. Nat. Congr. of Applied Mechanics*, pp. 325–329 (1951)

66 Kondratiev, V. A., 'Boundary problems for elliptic equations with conical or angular points', *Trans. Moscow Math. Soc.*, **16**, pp. 227–313 (1967)

67 England, H., 'On stress singularities in linear elasticity', *Int. J. Engrg. Sci.*, **9**, pp. 571–585 (1965)

68 Newton, D. A. and Tottenham, H., 'Boundary value problems in thin shallow shells of arbitrary plan form', *J. Engng. Maths.*, **2**, pp. 211–224 (1968)

69 Lukasiewicz, S., *Local Loads in Plates and Shells*, Sijthoff and Noordhoff, Alphen aan den Rijn and PWN, Polish Scientific Publishers (1979)

70 Vogel, S. M. and Rizzo, F. J., 'An integral equation formulation of three-dimensional anisotropic elastostatic boundary value problems', *J. Elasticity*, **3**, No. 3, (1973)

71 Christiansen, S., 'A review of some integral equations for solving the Saint-Venant torsion problem', *J. Elasticity* (1978)

72 Kuhn, Lehrstuhl A für Mechanik, TU München, Arcisstr. 21, D-8000 München 2.

BIBLIOGRAPHY

Altiero, N. J. and Gavazza, S. C., 'On a unified boundary-integral equation method', *J. Elasticity*, **10**, No. 1 (1980)

Antes, H., 'Die Spline-Funktionen bei der Lösung von Integralgleichungen', *Numerische Mathematik*, **19**, pp. 116–126 (1972)

Antes, H., 'Über die Integralgleichungen von Massonet und Rieder', *ZAMM*, **53**, pp. 64–66 (1973)

Bergmann, S., *Integral Operators in the Theory of Linear Partial Differential Equations*, (3rd ed.) Springer Verlag (1971)

Bernal, M. J. M. and Whiteman, J. R., 'Numerical treatment of biharmonic problems with reentrant boundaries', *Computer J.*, **13**(1), pp. 87–91 (1970)

Brebbia, C. A. (Ed.), *Recent Advances in Boundary Element Methods*, Pentech Press, Plymouth, London (1978)

Brebbia, C. A. (Ed.), *New Developments in Boundary Element Methods*, CML Publications, Southampton (1980)

Brickell, D. G. A., 'Integral equation method for influence surfaces of a corner plate', *Proc. Inst. Civ. Engrs.*, **46**, pp. 185–194 (1970)

van Buren, W., 'The indirect potential method for three dimensional boundary value problems of classical elastic equilibrium', Westinghouse Research Lab. Pittsburgh, PA., Research Report 68–1D7–MEKMA–R2 (1968)

Cruse, T. A. and Rizzo, F. J., 'A direct formulation and numerical solution of the general transient elastodynamic problem', *J. Math. Analysis and Appl.*, **22**, pp. 244–259 (1968)

Cruse, T. A., Snow, D. W. and Wilson, R. B., 'Numerical solutions in axisymmetric elasticity', *Computer Struct.*, **7**, pp. 445–451 (1977)

Drexler, W., Kuhn, G. and Mayr, M., 'Behandlung der axialsymmetrischen Torsion scharf gekerbter Bauteile mittels Randintegralgleichungsmethoden', *ZAMM*, **59**, pp. 177–179 (1979)

Fichera, G., 'Linear elliptic equations of higher order in two independent variables and singular integral equations', in *Proc. Conf. Partial Differential Equations and Continuum Mechanics*, University of Wisconsin Press (1961)

Fichera, G. and Ricci, P., 'The single layer potential approach in the theory of boundary value problems for elliptic equations', in *Lecture Notes in Math.*, No. 561, pp. 39–50, Springer Verlag, Berlin (1976)

Fischer, D., 'Über die singulären Grundlösungen der partiellen DGL für die Plattenbiegung mit veränderlicher Dicke', (On the fundamental sol. of plates with variable thickness), Dissertation, TH Wien (1965)

Hansen, E. B., 'Numerical solution of integrodifferential and singular integral equations for plate bending problems', *J. Elasticity*, **6**(1), pp. 39–56 (1976)

Hein, J. C., 'Eine Integralgleichungsmethode zur Lösung eines gemischten Randwertproblems der beliebig belasteten Kreiszylinderschale mit Ausschnitten glatter Berandung', Dissertation, TH Darmstadt (1979)

Heise, U., 'Integral equations for the mixed boundary value problem in plane elastostatics', *Appl. Math. Modelling*, **4**, February, p. 63–66 (1980)

Massonet, Ch., 'Numerical use of integral procedures', in *Stress Analysis*, O. C. Zienkiewicz and S. G. Holister (Ed.), John Wiley (1965)

Mayr, M., 'On the numerical solution of axisymmetric elasticity problems using an integral equation approach', *Mech. Res. Comm.*, **3**, pp. 393–398 (1976)

Mayr, M., Drexler, W. and Kuhn, G., 'A semianalytical boundary integral approach for axisymmetric elastic bodies with arbitrary boundary conditions', *Int. J. Solids Structures*, **16**, pp. 863–871 (1980)

Nemenyi, P., 'Über die Singularitäten der Elastizitätstheorie', *ZAMM*, **10**, pp. 383–399 (1930)

Pahnke, U., 'Über die Hebung von Ecksingularitäten bei Integralgleichungsverfahren', *ZAMM*, **56**, T 138–T 140 (1976)

Pahnke, U., 'Genauigkeits-und Wirtschaftlichkeitsbetrachtungen bei Integralgleichungsverfahren der ebenen Elastizitätstheorie', *ZAMM*, **55**, pp. 105–107 (1975)

Pucher, A., (Über die Singularitätenmethode an elastischen Platten', *Ing. Archiv*, **12**, pp. 76–100 (1941)

Rizzo, F. J. and Shippy, D. J., 'A boundary integral approach to potential and elasticity problems for axisymmetric bodies with arbitrary boundary conditions', *Mech. Res. Com.*, **6**(2), pp. 99–103 (1979)

Schweiger, W. and Mayr, M., 'On the solution of a certain fluid-structure-coupling problem using the boundary-integral equation method', *Mech. Res. Comm.*, **3**, pp. 495–500 (1976)

Segedin, C. M. and Brickell, D. G.A., 'Integral equation method for a corner plate', *J. Struct. Div. Proc. ASCE*, **94**, pp. 41–52 (1968)

Villagio, P., *Qualitative Methods in Elasticity*, Noordhoff International Publishing, Leyden (1977)

Watson, J. O., 'The analysis of thick shells with holes, by integral representation of displacement', Ph.D. Thesis, University of Southampton (1972)

Weinel, E., 'Die Integralgleichungen des ebenen Spannungszustandes und der Plattentheorie', *J. Appl. Math. Mech.*, (*ZAMM*), **11**(5), pp. 349–360 (1931)

Zabreyko, P. P., *Integral Equations — a Reference Text*, Noordhoff International Publishing, Leyden (1975)

5

Plasticity

by J. C. F. Telles and C. A. Brebbia

5.1 INTRODUCTION

In recent years researchers have become increasingly interested in solving complex stress analysis problems using boundary methods rather than domain techniques. This has resulted in a reassessment of classical boundary integral equation techniques and in their interpretation as an approximate method of analysis[1]. The concept of boundary elements, i.e. elements only on the surface of the domain but with interpolation functions similar to those used in finite elements, was introduced at the beginning of the 1970s. The combination of integral equations accuracy with finite elements versatility resulted in what is now called the boundary element method. This technique offers several advantages over finite elements such as, the possibility of working with boundaries extending to infinity; having smaller systems of equations and more accurate results for stress concentration regions and requiring much less data than finite elements to run the same problem. This last characteristic is, without doubt, the most interesting feature of boundary elements for the practicing engineer or user.

Several important papers have been published on the application of integral equations for solving plasticity problems[2,3,4] but none of them presented the complete expressions for strains or stresses at internal points. These expressions were given by Bui[5] in 1978 for three dimensional bodies and then Telles and Brebbia[6] in 1979 extended them to two dimensional problems. The latter authors presented the complete formulation for the boundary element method applied to two- and three-dimensional plasticity problems.

In this chapter, the application of boundary elements to plasticity is presented, employing linear interpolation functions for the boundary elements as well as for the internal cells. The problem of accurately integrating the domain integrals has been solved using a semi-analytical approach, which can also be applied to compute principal values. Three alternative boundary element formulations are discussed; (1) the initial strain, (2) the initial stress and (3) the fictitious traction and body forces approach. The chapter explains how different yield criteria can be introduced into the formulation in conjunction with the normality principle.

The required stepwise procedure is formulated in matrix form and a simple recursive relation is presented relating the stresses to the initial elastic solution and the plastic strains. This approach is only possible due to the reduced number of unknowns originated by the boundary element method, but would not be suitable for finite element programs for instance.

The results found using boundary elements for a series of examples are compared with analytical solutions and finite element results. Some of the applications emphasize the advantages of using boundary elements over domain type techniques to solve infinite medium problems. They also show how one can take advantage of symmetry conditions to significantly reduce the required number of unknowns. In all the cases the agreement between boundary elements and other solutions is satisfactory. The examples illustrate the considerable reductions achieved in the amount of data required to run a problem using boundary elements. Internal cells, for instance, are only needed in the region of plasticity, but they do not increase the problem unknowns which are still those defined on the boundary of the body.

5.2 GOVERNING EQUATIONS

The total strain rate is assumed to be represented by,

$$\dot{\varepsilon}_{ij}=\frac{1}{2}(\dot{u}_{i,j}+\dot{u}_{j,i})=\dot{\varepsilon}^{e}_{ij}+\dot{\varepsilon}^{p}_{ij} \tag{5.1}$$

where $\dot{u}_i$ are the displacement rate components, $\dot{\varepsilon}^e_{ij}$ and $\dot{\varepsilon}^p_{ij}$ are respectively the elastic and plastic components of the strain rate tensor. Time derivatives or increments are indicated by a dot and space derivatives by a comma.

The equilibrium conditions are,

$$\dot{\sigma}_{ij,i}+\dot{b}_j=0 \tag{5.2}$$

in the interior Ω of the body, and

$$\dot{p}_i-\dot{\sigma}_{ij}n_j=0 \tag{5.3}$$

on the boundary Γ, being $\dot{b}_j$ the body force rate components per unit volume, $\dot{\sigma}_{ij}$ the stress rate components, $\dot{p}_i$ the traction rate components per unit area and n is the outward normal to the boundary of the body.

If we consider the plastic strain rates as initial strains and apply Hooke's law to the elastic part of the three-dimensional strain rate tensor, the following expression arises,

$$\dot{\sigma}_{ij}=2G(\dot{\varepsilon}_{ij}-\dot{\varepsilon}^{p}_{ij})+\frac{2G\nu}{1-2\nu}(\dot{\varepsilon}_{ll}-\dot{\varepsilon}^{p}_{kk})\delta_{ij} \tag{5.4}$$

where G is the shear modulus and ν is Poisson's ratio.

The above expression can be rewritten in terms of initial stresses,

$$\dot{\sigma}_{ij} = 2G\dot{\varepsilon}_{ij} + \frac{2G\nu}{1-2\nu}\dot{\varepsilon}_{ll}\delta_{ij} - \dot{\sigma}^{p}_{ij} \tag{5.5}$$

where $\dot{\sigma}^{p}_{ij}$ represents the components of the 'initial stresses' given by,

$$\dot{\sigma}^{p}_{ij} = 2G\dot{\varepsilon}^{p}_{ij} + \frac{2G\nu}{1-2\nu}\dot{\varepsilon}^{p}_{kk}\delta_{ij} \tag{5.6}$$

The substitution of Equation 5.4 into 5.2 and 5.3, together with Equation 5.1 gives[7],

$$\dot{u}_{j,ll} + \frac{1}{1-2\nu}\dot{u}_{l,lj} = 2\left(\dot{\varepsilon}^{p}_{ij,i} + \frac{\nu}{1-2\nu}\dot{\varepsilon}^{p}_{kk,j}\right) - \frac{\dot{b}_j}{G} \tag{5.7}$$

and

$$\dot{p}_i + 2G\left(\dot{\varepsilon}^{p}_{ij}n_j + \frac{\nu}{1-2\nu}\dot{\varepsilon}^{p}_{kk}n_i\right) = \frac{2G}{1-2\nu}\dot{u}_{l,l}n_i + G(\dot{u}_{i,j} + \dot{u}_{j,i})n_j \tag{5.8}$$

Equation 5.7 is an extended form of Navier's equation and 5.8 represents its boundary conditions. The above expressions can be written in the following form,

$$\dot{u}_{j,ll} + \frac{1}{1-2\nu}\dot{u}_{l,lj} = -\dot{\bar{b}}_j/G \tag{5.9}$$

and

$$\dot{\bar{p}}_i = \frac{2G\nu}{1-2\nu}\dot{u}_{l,l}n_i + G(\dot{u}_{i,j} + \dot{u}_{j,i})n_j \tag{5.10}$$

where $\dot{\bar{b}}_j$ and $\dot{\bar{p}}_i$ are pseudo body forces and pseudo tractions given by,

$$\dot{\bar{b}}_j = \dot{b}_j - 2G\left(\dot{\varepsilon}^{p}_{ij,i} + \frac{\nu}{1-2\nu}\dot{\varepsilon}^{p}_{kk,j}\right) \tag{5.11}$$

and

$$\dot{\bar{p}}_i = \dot{p}_i + 2G\left(\dot{\varepsilon}^{p}_{ij}n_j + \frac{\nu}{1-2\nu}\dot{\varepsilon}^{p}_{kk}\mathrm{n}_i\right) \tag{5.12}$$

One can notice that Equation 5.9 represents a set of three quasi-linear partial differential equations for the displacements (plasticity terms appear on the right-hand side). This enables us to apply as a fundamental solution for the boundary element formulation the singular solution of the Navier equations due to Kelvin[1].

Expressions 5.1 to 5.12 are valid for plane strain ($i,j,l = 1,2$; $k = 1,2,3$) and also for plane stress ($i,j,k,l = 1,2$) if ν is replaced by $\bar{\nu} = \nu/(1+\nu)$.

5.2.1 Fundamental solutions

The singular fundamental solutions corresponding to unit point loads can be written as,

$$u^*_{ij} = \frac{1}{16\pi(1-\nu)Gr}\{(3-4\nu)\delta_{ij} + r_{,i}r_{,j}\} \tag{5.13}$$

for 3-D problems

$$u^*_{ij} = \frac{-1}{8\pi(1-\nu)G}\{(3-4\nu)\ \ln\ (r)\delta_{ij} - r_{,i}r_{,j}\} \tag{5.14}$$

and for 2-D plane strain problems

$$p^*_{ij} = \frac{1}{4\alpha\pi(1-\nu)r^\alpha}\left\{[(1-2\nu)\delta_{ij} + \beta r_{,i}r_{,j}]\frac{\partial r}{\partial n} - (1-2\nu)(r_{,i}n_j - r_{,j}n_i)\right\} \tag{5.15}$$

$\alpha = 2,1$; $\beta = 3,2$ for 3-D and 2-D plane strain respectively.

u^*_{ij} and p^*_{ij} represent the displacements and tractions in the j direction due to a unit force acting in i direction in which r is the distance from the point of application of the load to the point under consideration. The derivatives of r are taken with reference to the coordinates of the latter point and δ_{ij} is the Kronecker delta.

The strains (ε_{jk}) at any point within the body are given by,

$$\varepsilon^*_{jki} = \frac{-1}{8\alpha\pi(1-\nu)Gr^\alpha}\left\{(1-2\nu)(r_{,k}\delta_{ij} + r_{,j}\delta_{ik}) - r_{,i}\delta_{jk} + \beta r_{,i}r_{,j}r_{,k}\right\} \tag{5.16}$$

and the stresses,

$$\sigma^*_{jki} = C_{jkrs}\varepsilon^*_{rsi} = \frac{-1}{4\alpha\pi(1-\nu)r^\alpha}\{(1-2\nu)(r_{,k}\delta_{ij} + r_{,j}\delta_{ki} - r_{,i}\delta_{jk}) + \beta r_{,i}r_{,j}r_{,k}\} \tag{5.17}$$

where C_{jkrs} is the isotropic tensor of elastic constants which can be concisely written as,

$$C_{ijrs} = \frac{2G\nu}{1-2\nu}\delta_{ij}\delta_{rs} + G(\delta_{ir}\delta_{js} + \delta_{is}\delta_{jr}) \tag{5.18}$$

The plane strain expressions are valid for plane stress if ν is replaced by $\bar{\nu}$.

5.3 ALTERNATIVE BOUNDARY ELEMENT FORMULATIONS

5.3.1 Initial strain

We start this section by introducing an initial strain formulation which is particularly simplified by considering the plastic strains to be incompressible.

This is not restrictive but is justified by the fact that the implementation of the formulation is only intended for the application of the von Mises yield criterion.

The adoption of an initial strain formulation for three dimensional plasticity problems leads to the following equation[1],

$$C_{ij}\dot{u}_j = \int_\Gamma u^*_{ij}\dot{p}_j \, d\Gamma - \int_\Gamma p^*_{ij}\dot{u}_j \, d\Gamma + \int_\Omega u^*_{ij}\dot{b}_j \, d\Omega + \int_\Omega \sigma^*_{jki}\dot{\varepsilon}^p_{jk} \, d\Omega \qquad (5.19)$$

Equation 5.19 is valid for boundary points ($C_{ij}=\delta_{ij}/2$ on smooth surfaces, see Lachat and Watson[19] otherwise) and also for interior points with $C_{ij}=\delta_{ij}$. If metal plasticity is to be considered, plastic strains are deviatoric, i.e.

$$\dot{\varepsilon}^p_{kk} = \dot{\varepsilon}^p_{11} + \dot{\varepsilon}^p_{22} + \dot{\varepsilon}^p_{33} = 0 \qquad (5.20)$$

Under this assumption, stress rates at interior points can be computed by use of expressions 5.1 and 5.4. The derivative of Equation 5.19 yields,

$$\frac{\partial u_i}{\partial x_m} = \int_\Gamma \frac{\partial u^*_{ij}}{\partial x_m}\dot{p}_j \, d\Gamma - \int_\Gamma \frac{\partial p^*_{ij}}{\partial x_m}\dot{u}_j \, d\Gamma + \int_\Omega \frac{\partial u^*_{ij}}{\partial x_m}\dot{b}_j \, d\Omega + \frac{\partial}{\partial x_m}\left\{\int_\Omega \sigma^*_{jki}\dot{\varepsilon}^p_{jk} \, d\Omega\right\} \qquad (5.21)$$

Where the derivatives are taken with reference to the load point (these derivatives are written explicitly to differentiate from the previous ones indicated by a comma). Notice that we need to perform the derivative of the plastic strain rate integral, which is strongly singular. The proper expression for this derivative has been presented elsewhere, Telles and Brebbia[6], and is repeated here for completeness,

$$\frac{\partial}{\partial x_m}\left\{\int_\Omega \sigma^*_{jki}\dot{\varepsilon}^p_{jk} \, d\Omega\right\} = \int_\Omega \frac{\partial \sigma^*_{jki}}{\partial x_m}\dot{\varepsilon}^p_{jk} \, d\Omega + \frac{(8-10\nu)}{15(1-\nu)}\dot{\varepsilon}^p_{im} \qquad (5.22)$$

where the integral on the right-hand side is to be interpreted in the Cauchy principal value sense and the last term comes from a boundary integral over the surface of a sphere of unit radius centred at the singular point.

From now on the reader is referred to the appendix for all components of the tensors related to the fundamental solutions.

The above expressions together with Equations 5.1 and 5.4 allow for the determination of the internal stresses,

$$\dot{\sigma}_{ij} = \int_\Gamma (-\sigma^*_{ijk})\dot{p}_k \, d\Gamma - \int_\Gamma p^*_{ijk}\dot{u}_k \, d\Gamma$$

$$+ \int_\Omega (-\sigma^*_{ijk})\dot{b}_k \, d\Omega + \int_\Omega \sigma^*_{ijkl}\dot{\varepsilon}^p_{kl} \, d\Omega - \frac{2G(7-5\nu)}{15(1-\nu)}\dot{\varepsilon}^p_{ij} \tag{5.23}$$

where the last two terms represent the influence of the plastic strains.

For plane strains the procedure is analogous, the only difference being the fact that the plastic strain rate integrals still have to take into consideration the work performed in the third direction $(\sigma^*_{33i}\dot{\varepsilon}^p_{33})$. This effect is easily incorporated through the assumption of incompressible plastic strains[4] leading to what follows;

$$C_{ij}u_j = \int_\Gamma u^*_{ij}\dot{p}_j \, d\Gamma - \int_\Gamma p^*_{ij}\dot{u}_j \, d\Gamma + \int_\Omega u^*_{ij}\dot{b}_j \, d\Omega + \int_\Omega \hat{\sigma}^*_{jki}\dot{\varepsilon}^p_{jk} \, d\Omega \tag{5.24}$$

where

$$\hat{\sigma}^*_{jki} = \sigma^*_{jki} + \frac{2\nu\delta_{jk}r_{,i}}{4\pi(1-\nu)r} \tag{5.25}$$

and the internal stress rates,

$$\dot{\sigma}_{ij} = \int_\Gamma (-\sigma^*_{ijk})\dot{p}_k \, d\Gamma - \int_\Gamma p^*_{ijk}\dot{u}_k \, d\Gamma + \int_\Omega (-\sigma^*_{ijk})\dot{b}_k \, d\Omega + \int_\Omega \hat{\sigma}^*_{ijkl}\dot{\varepsilon}^p_{kl} \, d\Omega$$

$$- \frac{G}{4(1-\nu)}[2\dot{\varepsilon}^p_{ij} + (1-4\nu)\dot{\varepsilon}^p_{ll}\delta_{ij}] \tag{5.26}$$

in which

$$\hat{\sigma}^*_{ijkl} = \sigma^*_{ijkl} + \frac{G}{2\pi(1-\nu)r^2}[4\nu r_{,i}r_{,j}\delta_{kl} - 2\nu\delta_{ij}\delta_{kl}] \tag{5.27}$$

and the last integral is to be taken in the Cauchy principal value sense.

Plane stress problems can also be solved by using Equations 5.24 and 5.26 with $\hat{\sigma}^*_{jki} = \sigma^*_{jki}$, $\hat{\sigma}^*_{ijkl} = \sigma^*_{ijkl}$, ν replaced by $\bar{\nu}$ in all ( )* tensors and the last term in 5.26 being substituted by,

$$- \frac{G}{4(1-\bar{\nu})}[2\dot{\varepsilon}^p_{ij} + \dot{\varepsilon}^p_{ll}\delta_{ij}] \tag{5.28}$$

5.3.2 Initial stress

In order to introduce the initial stress formulation let us merely study the plastic strain rate integral presented in Equation 5.19. Recalling expression 5.17 we see that,

$$\int_{\Omega} \sigma^*_{jki} \dot{\varepsilon}^p_{jk} \, d\Omega = \int_{\Omega} C_{jkrs} \varepsilon^*_{rsi} \dot{\varepsilon}^p_{jk} \, d\Omega \tag{5.29}$$

By simple inspection of Equation 5.18 we can make,

$$C_{jkrs} = C_{rsjk} \tag{5.30}$$

moreover

$$C_{rsjk} \dot{\varepsilon}^p_{jk} = \dot{\sigma}^p_{rs} \tag{5.31}$$

where $\dot{\sigma}^p_{rs}$ was given in Equation 5.6. Thus,

$$\int_{\Omega} \sigma^*_{jki} \dot{\varepsilon}^p_{jk} \, d\Omega = \int_{\Omega} \varepsilon^*_{jki} \dot{\sigma}^p_{jk} \, d\Omega \tag{5.32}$$

Hence the initial stress formulation can be applied with equal ease. For plane strain problems the formulation follows the same pattern and the equivalent expression is,

$$\int_{\Omega} \hat{\sigma}^*_{jki} \dot{\varepsilon}^p_{jk} \, d\Omega = \int_{\Omega} \varepsilon^*_{jki} \dot{\sigma}^p_{jk} \, d\Omega \tag{5.33}$$

and internal stresses can be computed by,

$$\begin{aligned} \dot{\sigma}_{ij} = & \int_{\Gamma} (-\sigma^*_{ijk}) \dot{p}_k \, d\Gamma - \int_{\Gamma} p^*_{ijk} \dot{u}_k \, d\Gamma \\ & + \int_{\Omega} (-\sigma^*_{ijk}) \dot{b}_k \, d\Omega + \int_{\Omega} \varepsilon^*_{ijkl} \dot{\sigma}^p_{kl} \, d\Omega \\ & - \frac{1}{8(1-\nu)} [2\dot{\sigma}^p_{ij} + (1-4\nu) \dot{\sigma}^p_{ll} \delta_{ij}] \end{aligned} \tag{5.34}$$

where the initial stress integral is to be interpreted in the principal value sense.

It is worth noting that here the initial stress integrals do not require the contribution of the work performed in the third direction, nor the assumption of incompressibility of the plastic strains needs to be made. This is because $\varepsilon^*_{33i}=0$ and the effect of $\dot{\varepsilon}^p_{33}$ is already included into the components of $\dot{\sigma}^p_{ij}$. As a consequence, plane stress problems can be handled by the above expressions with the replacement of v by $\bar{v}$ being the only modification.

5.3.3 Fictitious tractions and body forces

The last integral presented in Equation 5.19 can be written in terms of the derivatives of u^*_{ij} as follows,

$$\int_\Omega \sigma^*_{jki}\dot{\varepsilon}^p_{jk}\,\mathrm{d}\Omega=\int_\Omega \left\{G(u^*_{ij,k}+u^*_{ik,j})+\frac{2Gv}{1-2v}u^*_{il,l}\delta_{jk}\right\}\dot{\varepsilon}^p_{jk}\,\mathrm{d}\Omega \tag{5.35}$$

which after integrating by parts gives the identity,

$$\begin{aligned}\int_\Omega \sigma^*_{jki}\dot{\varepsilon}^p_{jk}\,\mathrm{d}\Omega=&\int_\Gamma u^*_{ij}2G\left(\dot{\varepsilon}^p_{jk}n_k+\frac{v}{1-2v}\dot{\varepsilon}^p_{ll}n_j\right)\mathrm{d}\Gamma\\&-\int_\Omega u^*_{ij}2G\left(\dot{\varepsilon}^p_{jk,k}+\frac{v}{1-2v}\dot{\varepsilon}^p_{ll,j}\right)\mathrm{d}\Omega\end{aligned} \tag{5.36}$$

The substitution of Equation 5.36 in 5.19 gives as a result,

$$C_{ij}\dot{u}_j=\int_\Gamma u^*_{ij}\dot{\bar{p}}_j\,\mathrm{d}\Gamma-\int_\Gamma p^*_{ij}\dot{u}_j\,\mathrm{d}\Gamma+\int_\Omega u^*_{ij}\dot{\bar{b}}_j\,\mathrm{d}\Omega \tag{5.37}$$

where $\dot{\bar{b}}_j$ and $\dot{\bar{p}}_j$ were given in Equation 5.11 and 5.12 respectively.

We have arrived, therefore, at a plasticity formulation in which tractions and body force rates are fictitious (depend on the plastic strains), but the displacements are correct. In order to apply Equation 5.37 one has to be aware that although it looks like the elastic application of the boundary element method, the internal stresses still have to be computed by use of Equations 5.1 and 5.4, i.e.

$$\dot{\sigma}_{ij}=\int_\Gamma(-\sigma^*_{ijk})\dot{\bar{p}}_k\,\mathrm{d}\Gamma-\int_\Gamma p^*_{ijk}\dot{u}_k\,\mathrm{d}\Gamma+\int_\Omega(-\sigma^*_{ijk})\dot{\bar{b}}_k\,\mathrm{d}\Omega-C_{ijkl}\dot{\varepsilon}^p_{kl} \tag{5.38}$$

Another feature of this formulation is that in contrast with the two previous approaches, it needs computation of space derivatives of the plastic strains (see Equation 5.11). This may be considered as a disadvantage for numerical implementation, but, nevertheless, it is a valid procedure for formulating the BEM to plasticity and still remains to be properly attempted.

5.4 BOUNDARY ELEMENT DISCRETIZATION

The numerical implementation of the initial strain formulation presented in Equations 5.24 and 5.26 was given in Telles and Brebbia[9]. It is the aim of this section to illustrate the initial stress formulation by an entirely similar procedure.

Let us suppose that the boundary of the body is represented by surface elements while the part (or parts) of its interior where plasticity is likely to occur is discretized into a number of internal cells. Under these assumptions, linear piecewise functions are chosen to interpolate tractions, displacements and 'initial stresses', the two former over boundary elements and the latter over internal cells (body forces are not considered in this text). The unknowns produced by this discretization are interrelated by a series of coefficients found by integrating over boundary elements and internal cells.

From the substitution of Equation 5.33 into 5.24 comes the following matrix relationship,

$$\mathbf{H}\dot{\mathbf{u}} = \mathbf{G}\dot{\mathbf{p}} + \mathbf{Q}\dot{\boldsymbol{\sigma}}^p \tag{5.39}$$

where matrices $\mathbf{H}$ and $\mathbf{G}$ are the same as obtained for elastic analysis and matrix $\mathbf{Q}$ is due to the initial stress integral. Similarly, Equation 5.34 gives,

$$\dot{\boldsymbol{\sigma}} = \mathbf{G}'\dot{\mathbf{p}} - \mathbf{H}'\dot{\mathbf{u}} + (\mathbf{Q}' + \mathbf{E}')\dot{\boldsymbol{\sigma}}^p \tag{5.40}$$

in which $\mathbf{E}'$ is a well defined matrix that represents the independent terms and $\mathbf{Q}'$ stands for the initial stress rate integral. Matrices $\mathbf{H}'$ and $\mathbf{G}'$ are the same as those obtained for elastic analysis.

In order to extend the validity of Equation 5.40 to stresses at boundary nodes, expressions computed by means of strain displacement relationships and traction rate values along each boundary element have to be employed. These expressions do not require any integration and in the local coordinate system $(\eta_1;\eta_2)$ of a linear boundary elment located between nodes i and j are given by,

$$\dot{\sigma}_{11} = \frac{1}{1-\nu}\left\{\frac{2G}{l}({}^j\dot{u}_1 - {}^i\dot{u}_1) + \nu\dot{p}_2\right\} + \frac{\nu}{1-\nu}\dot{\sigma}^p_{22} - \dot{\sigma}^p_{11} \tag{5.41}$$

$$\dot{\sigma}_{12} = \dot{p}_1 \tag{5.42}$$

$$\dot{\sigma}_{22} = \dot{p}_2 \tag{5.43}$$

where ${}^i\dot{u}_1$ and ${}^j\dot{u}_1$ are the displacement rate values at nodes i and j and l is the element size. For plane stress ν is replaced by $\bar{\nu}$.

5.5 NUMERICAL INTEGRATION

The process of integration over boundary elements is well-known and the procedure presented in Brebbia and Walker[1] was followed. For the initial stress rate integrals the semi-analytical integration scheme fully described in Telles and Brebbia[9] was adopted. The technique is formulated with reference to a cylindrical coordinate system (r,φ) based at the singular point. Due to the very nature of the fundamental solution, analytical integration with respect to r turns to be very simple and removes the singularities. Integration with respect to the angle φ is then performed using one dimensional Gaussian quadrature formulae, five integration points being sufficient.

Notice that the procedure is also applied to compute the principal value of the domain integral presented in Equation 5.34. But if higher order interpolation functions as well as different cell shapes were chosen, the principal value could still be computed by applying Equation 5.40 to represent a state of constant initial stresses, in much the same way as it was shown in Telles and Brebbia[6].

5.6 ELASTOPLASTIC STRESS STRAIN RELATIONS

For the formulation of a theory which models elastoplastic material deformation, three requirements have to be met, these are:

(1) Explicit elastic relationship between stress and strain before the onset of plastic deformation.

(2) A yield criterion indicating the stress level at which plastic flow commences.

(3) Relationship between stress and strain for post yield behaviour.

The yield criterion for isotropic hardening can be written in general form as,

$$F(\sigma_{ij},k)=0 \tag{5.44}$$

where k is a hardening parameter that gives the instantaneous position of the yield surface in the n-dimensional stress space. Here, the work hardening hypothesis has been adopted, hence

$$k=w^p=\int\sigma_{ij}\,\mathrm{d}\varepsilon_{ij}^p \tag{5.45}$$

On physical grounds one can notice that the yield criterion, to be independent of the orientation of the coordinate system employed, should be a function of the three stress invariants only. It is common to represent two of these invariants as functions of the deviatoic stresses,

$$\begin{aligned}J_1&=\sigma_{ii}\\ J_2&=\tfrac{1}{2}S_{ij}S_{ij}\\ J_3&=\tfrac{1}{3}S_{ij}S_{jk}S_{ki}\end{aligned} \tag{5.46}$$

where

$$S_{ij}=\sigma_{ij}-\tfrac{1}{3}J_1\delta_{ij} \tag{5.47}$$

In the present work, instead of J_3 the alternative stress invariant θ, known as the Lode angle[10] was used.

$$-\frac{\pi}{6}\leqslant\theta=\frac{1}{3}\sin^{-1}\left(-\frac{3\sqrt{3}}{2}\frac{J_3}{J_2^{3/2}}\right)\leqslant\frac{\pi}{6} \tag{5.48}$$

By using these stress invariants, different yield criteria can be applied, such as

Tresca: $$2\sqrt{J_2}\cos\theta-\sigma_0(k)=0 \tag{5.49}$$

where $\sigma_0(k)$ is the uniaxial yield stress.

Von Mises: $$\sqrt{(3J_2)}-\sigma_0(k)=0 \tag{5.50}$$

Mohr-Coulomb: $$\frac{J_1}{3}\sin\varphi+\sqrt{J_2}\left(\cos\theta-\frac{1}{\sqrt{3}}\sin\theta\sin\varphi\right)-c\cos\varphi=0 \tag{5.51}$$

in which φ is the angle of internal friction and c is the cohesion of the material.

Drucker-Prager: $$\alpha J_1+\sqrt{J_2}-K=0 \tag{5.52}$$

where,

$$\alpha=\frac{2\sin\varphi}{\sqrt{3}(3-\sin\varphi)};\quad K=\frac{6c\cos\varphi}{\sqrt{3}(3-\sin\varphi)} \tag{5.53}$$

Mohr-Coulomb hypothesis can be simulated by the Drucker-Prager criterion in plane strains if α and K are written as[11],

$$\alpha=\frac{\tan\varphi}{(9+12\tan^2\varphi)^{1/2}};\quad K=\frac{3c}{(9+12\tan^2\varphi)^{1/2}} \tag{5.54}$$

For our practical purposes, Equation 5.44 can then be written as,

$$F(\sigma_{ij},k)=f(\sigma_{ij})-\psi(k)=0 \tag{5.55}$$

where one can notice that $f(\sigma_{ij})$ is a scalar function of σ_{ij} which plays the role of an equivalent stress here designated by σ_e. As a consequence, we can define an equivalent plastic strain ε_e^p whose increment produces an increment in the plastic strain energy as follows,

$$\sigma_e\,\mathrm{d}\varepsilon_e^p=\sigma_{ij}\,\mathrm{d}\varepsilon_{ij}^p=\mathrm{d}k \tag{5.56}$$

In order to obtain the stress-strain relations for post yield behaviour, let us first rewrite Equation 5.4,

$$\dot{\sigma}_{ij} = C_{ijkl}(\dot{\varepsilon}_{kl} - \dot{\varepsilon}^{p}_{kl}) \tag{5.57}$$

Within the context of associated plasticity, the flow rule also known as normality principle, can be described by,

$$\dot{\varepsilon}^{p}_{ij} = \mathrm{d}\lambda \frac{\partial F}{\partial \sigma_{ij}} \tag{5.58}$$

where $\mathrm{d}\lambda$ is a proportionality factor, termed the plastic multiplier. The substitution of Equation 5.58 into 5.57 gives,

$$\dot{\sigma}_{ij} = C_{ijkl}(\dot{\varepsilon}_{kl} - a_{kl}\,\mathrm{d}\lambda) \tag{5.59}$$

in which

$$a_{kl} = \frac{\partial F}{\partial \sigma_{kl}} = \frac{\partial f}{\partial \sigma_{kl}} \tag{5.60}$$

When plastic yielding is occurring, the stresses satisfy Equation 5.55 which by differentiating gives,

$$\dot{F} = a_{ij}\dot{\sigma}_{ij} - \frac{\partial \psi}{\partial \mathrm{k}}\dot{k} = 0 \tag{5.61}$$

Or, according to Equation 5.45,

$$a_{ij}\dot{\sigma}_{ij} - \frac{\mathrm{d}\psi}{\mathrm{d}k}\sigma_{ij}\dot{\varepsilon}^{p}_{ij} = 0 \tag{5.62}$$

From the application of the normality principle to Equation 5.62 results

$$a_{ij}\dot{\sigma}_{ij} - \frac{\mathrm{d}\psi}{\mathrm{d}k}\sigma_{ij}a_{ij}\,\mathrm{d}\lambda = 0 \tag{5.63}$$

If we substitute Equation 5.59 into 5.63 and solve for $\mathrm{d}\lambda$,

$$\mathrm{d}\lambda = \frac{1}{\gamma}a_{ij}C_{ijkl}\dot{\varepsilon}_{kl} \tag{5.64}$$

where

$$\gamma = a_{ij}C_{ijkl}a_{kl} + \frac{\mathrm{d}\psi}{\mathrm{d}k}\sigma_{ij}a_{ij} \tag{5.65}$$

Before we go further the last term in Equation 5.65 can be examined. It can be shown that $f(\sigma_{ij})$ is homogeneous of degree one and this allows for the application of Euler's theroem as follows,

$$\sigma_{ij}\frac{\partial f}{\partial \sigma_{ij}}=f(\sigma_{ij})=\sigma_e \tag{5.66}$$

The substitution of Equations 5.66 and 5.56 into Equation 5.65 gives,

$$\gamma=a_{ij}C_{ijkl}a_{kl}+H' \tag{5.67}$$

where $H'=\mathrm{d}\psi/\mathrm{d}\varepsilon_e^p$ can be interpreted as the slope of the uniaxial curve plotted as true stress versus true plastic strain.

Equation 5.64 can now be used to substitute $\mathrm{d}\lambda$ in Equation 5.59 providing the required incremental stress-strain relations

$$\dot{\sigma}_{ij}=C_{ijkl}^{ep}\dot{\varepsilon}_{kl} \tag{5.68}$$

in which

$$C_{ijkl}^{ep}=C_{ijkl}-\frac{1}{\gamma}C_{ijmn}a_{mn}a_{op}C_{opkl} \tag{5.69}$$

For the application of the above relations to the initial stress formulation, a further modification has proved to be convenient. Let us adopt the following notation,

$$\dot{\sigma}_{ij}^e=C_{ijkl}\dot{\varepsilon}_{kl} \tag{5.70}$$

where $\dot{\sigma}_{ij}^e$ stands for the components of the elastic stress rate tensor (i.e. these represent the stress values as if a pure elastic problem were being solved). Equation 5.68 can then be rewritten in the following form

$$\dot{\sigma}_{ij}=\dot{\sigma}_{ij}^e-\frac{1}{\gamma}C_{ijmn}a_{mn}a_{kl}\dot{\sigma}_{kl}^e \tag{5.71}$$

which means that the true stresses can be computed from the corresponding elastic stresses. In addition to this, the initial stress rates presented in Equation 5.6 can also be calculated by the relation,

$$\dot{\sigma}_{ij}^p=\dot{\sigma}_{ij}^e-\dot{\sigma}_{ij}=\frac{1}{\gamma}C_{ijmn}a_{mn}a_{kl}\dot{\sigma}_{kl}^e \tag{5.72}$$

By simply examining Equation 5.5 we notice that Equation 5.40 can be applied for the computation of $\dot{\sigma}_{ij}^e$ if matrix $\mathbf{E}'$ is replaced by

$$\bar{\mathbf{E}}=\mathbf{E}'+\mathbf{I} \tag{5.73}$$

where **I** is the identity matrix. This gives

$$\dot{\boldsymbol{\sigma}}^e = \mathbf{G}'\dot{\mathbf{p}} - \mathbf{H}'\dot{\mathbf{u}} + \mathbf{Q}^*\dot{\boldsymbol{\sigma}}^p \tag{5.74}$$

where

$$\mathbf{Q}^* = \mathbf{Q}' + \bar{\mathbf{E}} \tag{5.75}$$

Before the application of the above expressions to solve plasticity problems, it should be pointed out that although solution algorithms are incremental, always finite sized load increments are prescribed and this may create some drifts of the stress level beyond the yield surface. If load increments are kept sufficiently small this problem is practically eliminated, but if relatively large load increment sizes are to be permiited, special techniques of the type presented in Nayak and Zienkiewicz[12] have been found necessary to maintain the stresses on the yield surface.

5.7 OUTLINE OF SOLUTION TECHNIQUE

In order to minimise the computer effort for the initial stress formulation, let us re-examine Equations 5.39 and 5.74. For a well-posed problem, a sufficient number of tractions and boundary displacements needs to be prescribed. The unknowns are then reordered leading to

$$\mathbf{A}\dot{\mathbf{x}} = \dot{\mathbf{f}} + \mathbf{Q}\dot{\boldsymbol{\sigma}}^p \tag{5.76}$$

and similarly,

$$\dot{\boldsymbol{\sigma}}^e = -\mathbf{A}'\dot{\mathbf{x}} + \dot{\mathbf{f}}' + \mathbf{Q}^*\dot{\boldsymbol{\sigma}}^p \tag{5.77}$$

where vector $\dot{\mathbf{x}}$ contains the unknown tractions and boundary displacements. From the multiplication of Equation 5.76 by $\mathbf{A}^{-1}$ we get

$$\dot{\mathbf{x}} = \mathbf{R}\dot{\boldsymbol{\sigma}}^p + \dot{\mathbf{m}} \tag{5.78}$$

where

$$\mathbf{R} = \mathbf{A}^{-1}\mathbf{Q} \tag{5.79}$$

and

$$\dot{\mathbf{m}} = \mathbf{A}^{-1}\dot{\mathbf{f}} \tag{5.80}$$

Substituting Equation 5.78 into 5.77 gives,

$$\dot{\boldsymbol{\sigma}}^e = \mathbf{S}\dot{\boldsymbol{\sigma}}^p + \dot{\mathbf{n}} \tag{5.81}$$

in which

$$\mathbf{S} = \mathbf{Q}^* - \mathbf{A}'\mathbf{R} \tag{5.82}$$

and

$$\dot{\mathbf{n}} = \dot{\mathbf{f}}' - \mathbf{A}'\dot{\mathbf{m}} \tag{5.83}$$

Note that vectors **m** and **n** represent the elastic solution to the incremental problem (actual solution in absence of plasticity). Furthermore, equations 5.78 and 5.81 remain valid if instead of incremental loading, the total load is applied. The only reason to proceed incrementally being the constitutive relations presented in Equation 5.68. This enables us to compute the load at first yield by simply scaling down the total elastic solution by a load factor λ_0. The incremental process starts at this load level and further values of the load factor are given by,

$$\lambda_i = \lambda_{i-1} + \beta \tag{5.84}$$

where $\beta = \lambda_0 \omega$; ω is the given value of the load increment with reference to load at first yield.

For elasto-plastic solutions, Equations 5.78 and 5.81 can therefore be applied as,

$$\mathbf{x} = \mathbf{R}(\boldsymbol{\sigma}^p + \dot{\boldsymbol{\sigma}}^p) + \lambda_i \mathbf{m} \tag{5.85}$$

and

$$\boldsymbol{\sigma}^e = \mathbf{S}(\boldsymbol{\sigma}^p + \dot{\boldsymbol{\sigma}}^p) + \lambda_i \mathbf{n} \tag{5.86}$$

or alternatively for pure incremental relations,

$$\dot{\mathbf{x}} = \mathbf{R}\dot{\boldsymbol{\sigma}}^p + \beta \mathbf{m} \tag{5.87}$$

and

$$\dot{\boldsymbol{\sigma}}^e = \mathbf{S}\dot{\boldsymbol{\sigma}}^p + \beta \mathbf{n} \tag{5.88}$$

where in both cases vectors **m** and **n** correspond to the application of the total load and $\dot{\boldsymbol{\sigma}}^p$ stands for the current initial stress increment.

For a typical load increment (i.e., a given value of λ_i), the initial stress increment can be determined iteratively at each boundary node and internal point exhibiting elastoplastic behaviour by two different processes. The former is in fact a pure incremental procedure. Once the load increment $\beta\mathbf{n}$ is applied, the initial stress increment corresponding to the solution of the elastic problem is computed and has to be applied back into the body, providing an elastic stress redistribution. This operation, again generates a new initial stress field to be redistributed elastically and so on. Iteration is halted when the contribution of the last initial stress increment can be neglected.

The process is in essence comparable to what was presented in Zienhiewicz[13] for the finite element method and is summarized as follows;

(1) Compute elastic stress increment by Equation 5.88 if first iteration is being performed or $\dot{\boldsymbol{\sigma}}^e = \mathbf{S}\dot{\boldsymbol{\sigma}}^p$ otherwise.

(2) Find true stress increment $\dot{\sigma}_{ij}$ (Equation 5.71).

(3) Verify convergence, i.e. compare $d\varepsilon_e^p$ calculated with its accumulated value obtained during the current load increment to see if it can be neglected.

(4) Calculate initial stress increment by $\dot{\sigma}_{ij}^p = \dot{\sigma}_{ij}^e - \dot{\sigma}_{ij}$.

(5) Accumulate values of initial stress and true stress,

$\sigma_{ij}^p = \sigma_{ij}^p + \dot{\sigma}_{ij}^p$

$\sigma_{ij} = \sigma_{ij} + \dot{\sigma}_{ij}$

(6) Continue with next node or point and start with (2) until all nodes and points have been considered.

(7) Go to (1) for a new iteration.

Iterations are performed until convergence is obtained (within prescribed tolerance) at every node or point.

It is interesting to note that in order to avoid cumulative errors, $\dot{\boldsymbol{\sigma}}^p$ obtained at the end of iterations is applied together with $\beta\mathbf{n}$ in Equation 5.88 for the first iteration of the next load increment.

The second process, which proved to be less dependent on the load increment size but not always more economical, deals with accumulated values of the elastic stress in a similar fashion to the procedure adopted for the initial strain implementation[9].

The initial stress increment is kept separate from its accumulated values until convergence is obtained as follows,

(1) Compute elastic stress (Equation 5.86).

(2) Calculate elastic stress increment by,

$\dot{\sigma}_{ij}^e = \sigma_{ij}^e - \sigma_{ij} - \sigma_{ij}^p$

(3) Find true stress increment $\dot{\sigma}_{ij}$ (Equation 5.71).

(4) Verify convergence, i.e. compare $d\varepsilon_e^p$ calculated with its previous value.

(5) Calculate new estimate of initial stress increment by,

$\dot{\sigma}_{ij}^p = \dot{\sigma}_{ij}^e - \dot{\sigma}_{ij}$

(6) Continue with next node or point and start with (2) until all nodes and points have been considered.

(7) Go to (1) for a new iteration.

Once that convergence is obtained for all nodes and points, the initial stress increment is accumulated and its value is also used as an initial guess for the next load increment.

Note that neither procedure requires computation of the boundary unknowns. Consequently Equation 5.85 need only be used to print the boundary unknowns once convergence is achieved. In addition we should emphasize that matrices **R** and **S** as well as vectors **m** and **n** remain unchanged throughout the incremental process. These features, also common with the initial strain procedure, represent a great saving in computer time.

Finally, a question is posed; Which procedure is more efficient? The answer can only be obtained in the light of experience. Regarding precision, the examples solved show that if load increment is kept small, the initial strain and both initial stress processes lead to entirely equivalent results (difference within the prescribed tolerance). Nevertheless with relation to computer time, the initial strain version has proved to be faster than the initial stress. This difference however, seems to be due to the fact that in contrast with the initial strain program which is restricted to von Mises yield criterion, both versions of the initial stress routine are more general and can handle four different yield criteria.

As load increment was increased, the initial strain and the second initial stress versions exhibited more stable results than the former initial stress procedure, despite special techniques to avoid stress drifts beyond the yield surface. But in some of the examples run with relatively large load increments, the pure incremental form of the initial stress program produced equally reliable results, requiring less computer time than its second version.

In conclusion, the above discussion demonstrates the equivalence of the three processes and furthermore shows that the efficiency of the procedure is problem dependent. Ideally, all three options should be available for a general case.

5.8 EXAMPLES

Following the discussion presented in the end of the last section, the results for a series of examples solved by the boundary element technique are now compared with analytical solutions where such solutions are available and with finite element results.

5.8.1 **Notched tensile specimen**

This first example is one of the very early plasticity problems solved by using the finite element technique. Plane stress and plane strain results have been presented in several papers, creating a good opportunity to compare the boundary element computations.

Material parameters are as follows:

$$
\begin{aligned}
E &= 7000 \text{ kg/mm}^2 \\
\sigma_0 &= 24.3 \text{ kg/mm}^2 \\
\nu &= 0.2 \\
H' &= 0 \text{ (von Mises yield criterion)}
\end{aligned}
$$

Plane stress analysis was carried out using the discretization shown in Figure 5.1. Note that symmetry was taken into consideration without boundary discretization of the symmetry axes, this is due to a direct condensation process which integrates automatically over reflected elements and cells in such a way that the size of the final matrices correspond to the reduced number of boundary nodes and internal points presented.

Figure 5.2 gives the load displacement curve for this case. It is seen that the curve remains nearly straight until very close the the limit load, when a sharp bend then occurs. Such behaviour was also observed by Yamada *et al.*[14] in an

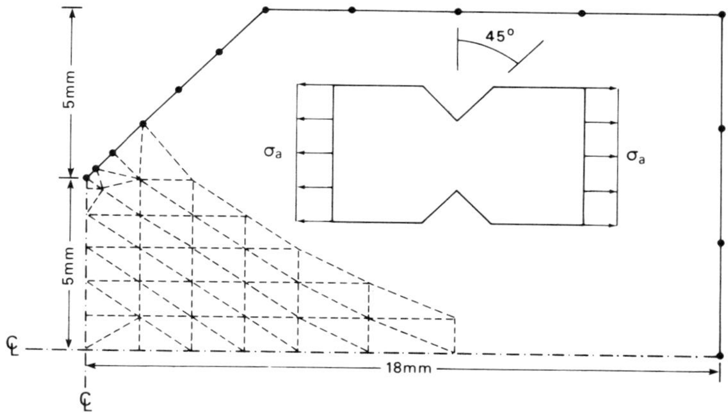

Figure 5.1 Notched tensile specimen. Boundary element and internal cell discretization (plane stress case)

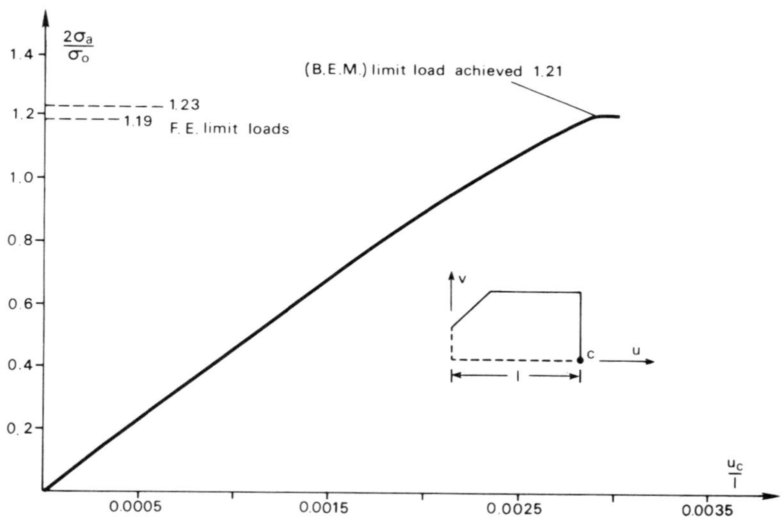

Figure 5.2 Load-displacement curve for notched specimen in plane stress

entirely similar problem. The limit load achieved by the boundary elements ($2\sigma_a/\sigma_0 = 1.21$) coincides with the results presented by Nayak and Zienkiewicz[12] using four different finite elements to analyse the same problem. Their limit load was found to vary between $2\sigma_a/\sigma_0 = 1.19$–$1.23$ and simple triangular, isoparametric linear, quadratic and cubic elements were used, all four meshes with approximately 97 nodes.

For the plane strain case, because of a large spread of plastic zones before limit load is achieved, the number of internal points and cells was increased

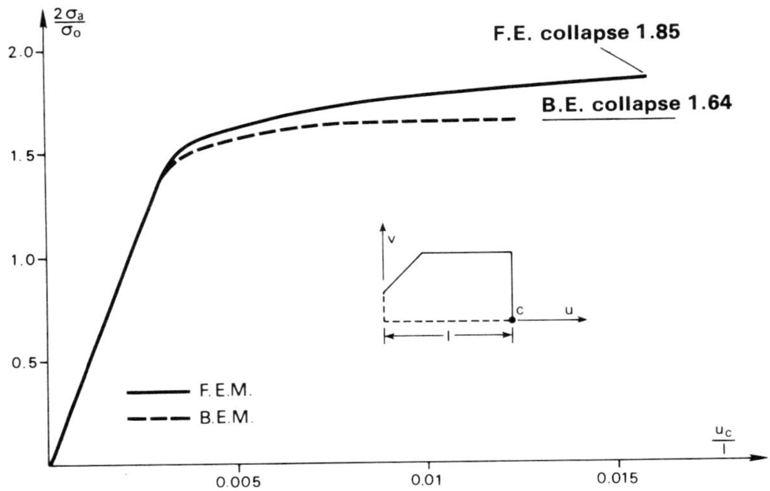

Figure 5.3 Load-displacement curves for notched specimen in plane strain

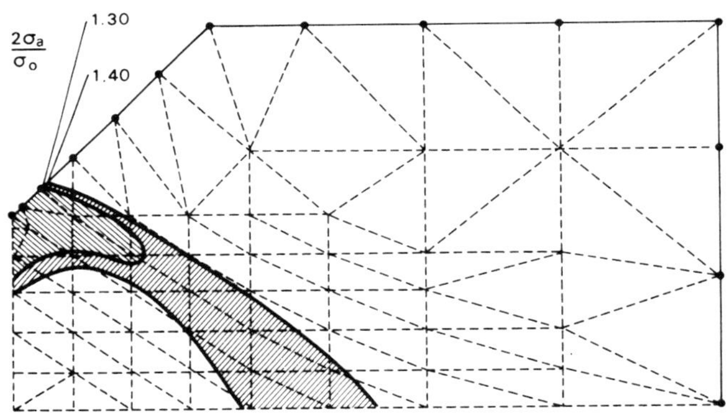

Figure 5.4 Plastic zones obtained by BEM for different load levels (plane strain)

from 33 and 51 to 59 and 97 respectively. The load displacement curve is shown in Figure 5.3 where equivalent finite element results presented by Chen[15] are also given. The limit load obtained by BEM ($2\sigma_a/\sigma_0 = 1.64$) is below the value given by the finite element method ($2\sigma_a/\sigma_0 = 1.85$). But, as stated by Chen, bound theorems demonstrate that the maximum load should lie between 1.52 and 1.73, which supports the boundary element results.

The spread of plastic zones at lower load levels presented in Figure 5.4 exhibits good agreement with finite element computations[13,15] for the same problem.

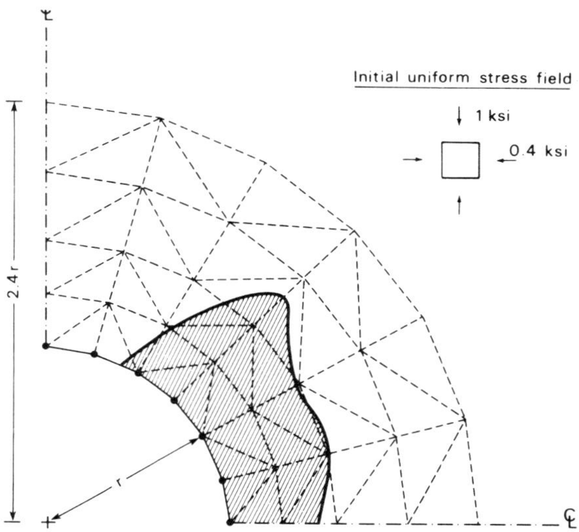

Figure 5.5 *Deep circular tunnel problem. Discretization used for BE results and spread of plastic zone at the end of relaxation process*

5.8.2 **Deep circular tunnel**

The second example was selected to emphasize the advantages of boundary elements over 'domain' type techniques to solve infinite medium problems.

A circular excavation studied by Reyes[16] and later by Baker *et al.*[17] with linear displacement triangular and quadrilateral finite elements respectively is here compared with boundary element results. The plane strain problem was analysed under the Drucker-Prager simulation of Mohr-Coulomb yield criterion (α and K given by Equation 5.54) and by assuming the infinite domain to be initially subjected to a uniform stress field of 1 ksi vertical and 0.4 ksi horizontal ($K_0 = 0.4$). For the present study, external loads corresponding to the relaxation of this free stress field were applied along the surface of the opening.

The material (Rock) was assumed to be perfectly plastic with,

$$E = 500 \text{ ksi}$$
$$c = 0.28 \text{ ksi}$$
$$\nu = 0.2$$
$$\varphi = 30^\circ$$

Boundary element and internal cell discretization is presented in Figure 5.5 where the plastic zone on complete removal of the free stress from the boundary of the cavity is also given.

Stresses along the horizontal section computed at the end of the relaxation process are compared with the corresponding results presented by Reyes and Baker in Figure 5.6. Here, internal stresses outside the discretized region were calculated at simple internal points not connected to any internal cells.

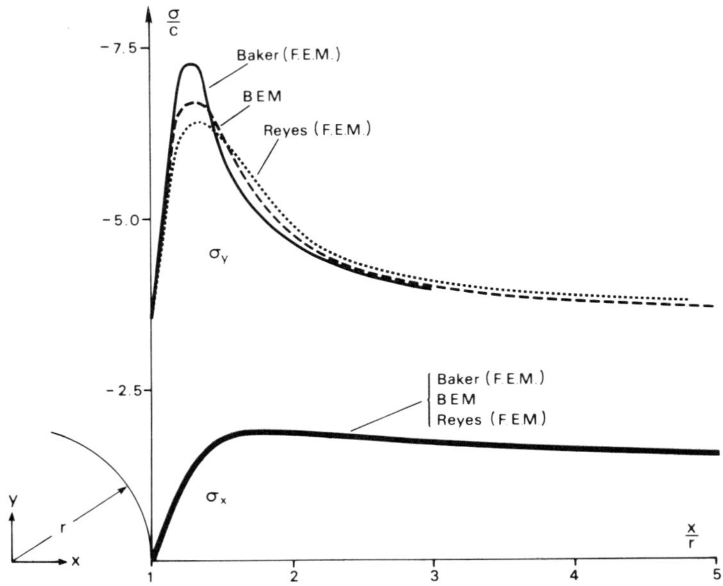

Figure 5.6 Final stresses along the horizontal section through the infinite medium

It is important to note that the refinement of the two finite element meshes (about 253 nodes) should not lead to the differences in the σ_y values in Figure 5.6. Although no reference was made by the authors, this discrepancy is probably due to the outer boundary conditions considered in the two analyses. The boundary element technique does not require any outer boundary discretization, but in order to study its influence in the results a quarter of a circle with radius equal to nine times the radius of the cavity was discretized into six boundary elements, this is approximately the extent of the finite element meshes. The outer circle was then considered to be free to displace, giving as a result a better agreement with Baker computations. The second alternative was carried out by prescribing zero displacements over the outer boundary, leading now to improved agreement with Reyes results.

5.8.3 **Rough punch**

In the third example the elastic-plastic behaviour of a square block compressed by two opposite perfectly rough rigid punches is studied. The problem is analysed under plane strain condition and material is considered to be perfectly plastic obeying von Mises yield criterion.

By using a very refined mesh of 274 linear displacement triangular finite elements and 173 nodal points, Figure 5.7(a), results to this problem were presented by Chen[15,18]. The boundary element analysis was performed with the discretization shown in Figure 5.7(b), requiring less than one third of the FE data to run the problem.

The indentation process was developed by prescribing the flat punch

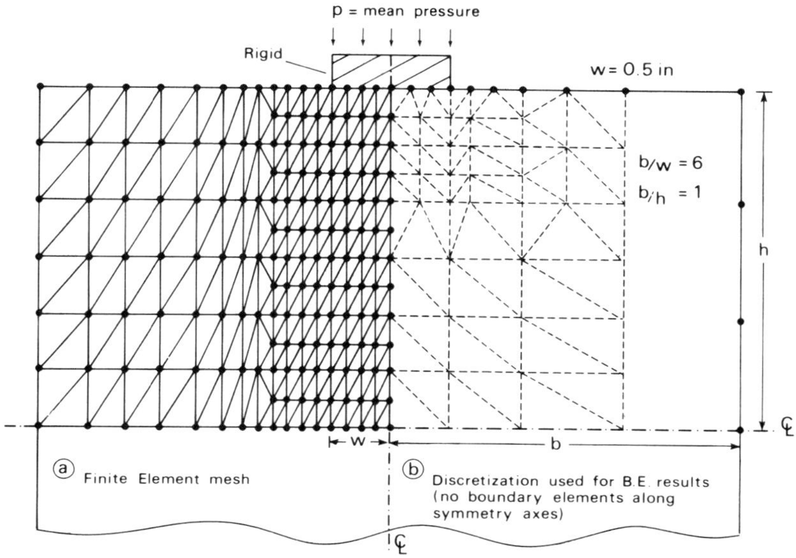

Figure 5.7 Geometry of rough punch problem

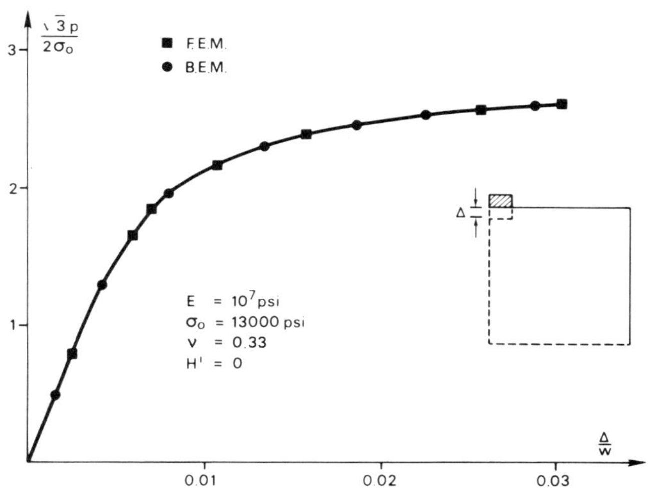

Figure 5.8 Mean pressure — applied displacement curve for rough punch problem

displacements leading to the average pressure-applied displacement curve presented in Figure 5.8. As can be seen, agreement between the two analyses has been thoroughly obtained, both methods slightly exceeding (4%) the theoretical limit load $\sqrt{3}p/2\sigma_0 = 2.5$.

5.9 CONCLUSIONS

In this chapter the correct and complete formulation for the boundary element method applied to plasticity has been presented. Three alternative procedures — initial stress, initial strain and fictitious forces — were discussed together with the possibility of applying different yield criteria. The required stepwise procedure has been formulated in matrix form using a simple recursive relation relating stresses to initial elastic solution and plastic strains.

The examples presented show the advantages of using boundary elements in plasticity in preference to finite elements. Considerable reductions in the data required to run a problem can be achieved for the same degree of accuracy. In addition, problems with boundaries at infinity can be properly modelled using boundary elements and the resulting matrices are smaller than those corresponding to finite element or finite difference techniques.

REFERENCES

1. Brebbia, C. A. and Walker, S., *The Boundary Element Techniques in Engineering*, Butterworths, London (1979)
2. Swedlow, J. L. and Cruse, T. A., 'Formulation of Boundary Integral Equations for Three-Dimensional Elasto-Plastic Flow', *Int. J. Solids Structures*, **7**, pp 1673–1683 (1971)
3. Mendelson, A., 'Boundary Integral Methods in Elasticity and Plasticity', Report No. NASA–TN–D–7418 (1973)
4. Mukherjee, S., 'Corrected Boundary Integral Equation in Planar Thermoelastoplasticity', *Int. J. Solids Structures*, **13**, pp. 331–335 (1977)
5. Bui, H. D., 'Some Remarks about the Formulation of Three-Dimensional Thermoelastoplastic Problems by Integral Equations', *Int. J. Solids Structures*, **14**, pp. 935–939 (1978)
6. Telles, J. C. F. and Brebbia, C. A., 'On the Application of the Boundary Element Method to Plasticity, Appl. Math. Modelling', **3**, pp. 466–470 (1979)
7. Lin, T. H., *Theory of Inelastic Structures*, Wiley, New York (1968)
8. Riccardella, P. C., 'An Implementation of the Boundary-Integral Technique for Planar Problems in Elasticity and Elasto-Plasticity', Report No. SM–73–10, Dept. Mech. Eng., Carnegie Mellon University, Pittsburgh (1973)
9. Telles, J. C. F. and Brebbia, C. A., 'The Boundary Element Method in Plasticity', *Proc. 2nd Int. Seminar on Recent Advances in Boundary Element Methods*, C. A. Brebbia (Ed.), University of Southampton, pp. 295–317 (1980)
10. Nayak, G. C. and Zienkiewicz, O. C., 'Convenient Form of Stress Invariants for Plasticity', *Proc. Am. Soc. Civ. Engrs., J. Struc. Div.*, **98**, pp. 949–954 (1972)
11. Drucker, D. C. and Prager, W., 'Soil Mechanics and Plastic Analysis or Limit Design', *Q. Appl. Math.*, **10**, pp. 157–165 (1952)
12. Nayak, G. C. and Zienkiewicz, O. C., 'Elasto-plastic Stress Analysis. A Generalization for Various Constitutive Relations including Strain Softening', *Int. J. Num. Meth. Engng*, **5**, pp. 113–135 (1972)
13. Zienkiewicz, O. C., Valliappan, S. and King, I. P., 'Elastoplastic Solutions of Engineering Problems; Initial Stress Finite Element Approach', *Int. J. Num. Meth. Engng*, **1**, pp. 75–100 (1969)
14. Yamada, Y., Yoshimura, N. and Sakurai, T., 'Plastic Stress-strain Matrix and its Application for the Solution of Elastic-plastic Problems by the Finite Element Method', *Int. J. Mech. Sci.*, **10**, pp. 343–354 (1968)
15. Chen, W. F., *Limit Analysis and Soil Plasticity*, Elsevier Scientific Publishing Co., Amsterdam, (1975)
16. Reyes, S. F. and Deere, D. U., 'Elastic-plastic Analysis of Underground Openings by the Finite Element Method', *Proc. 1st Int. Congr. Rock Mechanics*, Lisbon, pp. 477–483 (1966)

17 Baker, L. E., Sandhu, R. S. and Shieh, W. Y., 'Application of Elasto-plastic Analysis in Rock Mechanics by Finite Element Method', *Proc. 11th Symp. Rock Mechanics*, W. H. Somerton (Ed.), University of California, Berkeley, pp. 237–251 (1969)

18 Chen, A. C. T. and Chen, W. F., 'Constitutive Equations and Punch-indentation of Concrete', *Proc. Am. Soc. Civ. Engrs, J. Engng Mech. Div.*, **101**, pp. 889–906 (1975)

19 Lachat, J. C. and Watson, J. O., 'Effective Numerical Treatment of Boundary Integral Equations, a Formulation for Three-dimensional Elastostatics', *Int. J. Num. Meth. Engng*, **10**, pp. 991–1005 (1976)

APPENDIX: SOME TENSOR FORMS

Tensors related to the fundamental solutions that appear in the text are of the following form:

$$p_{ijk}^{*}=\frac{G}{2\alpha\pi(1-v)r^{\beta}}\left\{\beta\frac{\partial r}{\partial n}[(1-2v)\delta_{ij}r_{,k}+v(\delta_{ik}r_{,j}+\delta_{jk}r_{,i})-\gamma r_{,i}r_{,j}r_{,k}]\right.$$

$$\left.+\beta v(n_i r_{,j}r_{,k}+n_j r_{,i}r_{,k})+(1-2v)(\beta n_k r_{,i}r_{,j}+n_j\delta_{ik}+n_i\delta_{jk})-(1-4v)n_k\delta_{ij}\right\} \tag{A5.1}$$

$$\begin{aligned}\sigma_{ijkl}^{*}=\frac{G}{2\alpha\pi(1-v)r^{\beta}}\{&\beta(1-2v)(\delta_{ij}r_{,k}r_{,l}+\delta_{kl}r_{,i}r_{,j})\\&+\beta v(\delta_{li}r_{,j}r_{,k}+\delta_{jk}r_{,l}r_{,i}+\delta_{ik}r_{,l}r_{,j}\\&+\delta_{jl}r_{,i}r_{,k})-\beta\gamma r_{,i}r_{,j}r_{,k}r_{,l}+\\&+(1-2v)(\delta_{ik}\delta_{lj}+\delta_{jk}\delta_{li})-(1-4v)\delta_{ij}\delta_{kl}\}\end{aligned} \tag{A5.2}$$

where $\alpha=2,1$, $\beta=3,2$, $\gamma=5,4$ for 3-D and plane strain respectively.
Also

$$\begin{aligned}\varepsilon_{ijkl}^{*}=\frac{1}{4\pi(1-v)r^{2}}\{&(1-2v)[\delta_{ik}\delta_{lj}+\delta_{jk}\delta_{li}-\delta_{ij}\delta_{kl}\\&+2\delta_{ij}r_{,k}r_{,l}]+2v[\delta_{li}r_{,j}r_{,k}+\delta_{jk}r_{,l}r_{,i}\\&+\delta_{ik}r_{,l}r_{,j}+\delta_{jl}r_{,i}r_{,k}]\\&+2\delta_{kl}r_{,i}r_{,j}-8r_{,i}r_{,j}r_{,k}r_{,l}\}\end{aligned} \tag{A5.3}$$

for the plane strain initial stress formulation.

6

Time dependent potential problems

by L. C. Wrobel and C. Brebbia

6.1 INTRODUCTION

Several formulations of the boundary element method (BEM) as applied to time-dependent problems are discussed in this chapter. Emphasis is given to the solution of the two-dimensional diffusion equation but the ideas presented can be readily extended to three-dimensional and axisymmetric analyses. Also more complex time-dependent problems such as the ones governed by the wave equation, the Navier-Stokes equations, the elastodynamics equations, etc., can be dealt with in a similar way.

The first boundary integral solution for the diffusion equation was proposed by Rizzo and Shippy[1]. Their approach removed the time dependence of the problem by means of Laplace transforms. The boundary integral equation in the transformed space was solved numerically and a transform inversion was then performed in order to evaluate the physical variables. This approach was later discussed in Lachat and Combescure[2] and Liggett and Liu[3], where some of its disadvantages were pointed out.

Chang *et al.*[4] solved the problem in the time domain but using only space and time piecewise constant values for the variables and presenting their formulation in an analytical rather than numerical way. This approach was extended by Wrobel and Brebbia[5]. Brebbia and Walker[6] also proposed a coupled BEM-finite difference scheme, where a step-by-step finite difference type procedure is employed to advance the solution in time.

The first part of the chapter presents a brief description of all the above-mentioned formulations. The second approach, which employs time-dependent fundamental solutions, is then described in detail and the time integrals that appear in its development are evaluated analytically, for constant and higher order time interpolation functions. It is also pointed out how the singular space integrals can be computed analytically and how complex boundary conditions that occur in practical problems can be introduced into the analysis. The latter feature is achieved by using an original formulation based on higher order time interpolation functions.

Finally, several applications are presented and their results compared with the ones obtained by the finite element method (FEM) and with known analytical solutions. They show the advantages of using boundary elements for time-dependent problems in preference to more classical numerical techniques such as finite differences and finite elements.

6.2 BEM FORMULATIONS AND FUNDAMENTAL SOLUTIONS

In this chapter, we discuss BEM solutions for the two-dimensional diffusion equation

$$k\nabla^2 u = \frac{\partial u}{\partial t} \text{ in } \Omega \tag{6.1}$$

with boundary conditions of the following types,

$$\begin{aligned} u &= \bar{u}, && \text{on } \Gamma_1 \\ q &= \frac{\partial u}{\partial n} = \bar{q}, && \text{on } \Gamma_2 \end{aligned} \tag{6.2}$$

and some initial conditions at $t = t_0$,

$$u = u_0 \quad \text{in } \Omega \tag{6.3}$$

Three alternative formulations are considered in this section. The first removes the time dependence of the problem by means of a Laplace transform; the second integrates Equation 6.1 numerically in time in a finite difference fashion while the third approach makes use of time-dependent fundamental solutions.

6.2.1 Laplace transforms

Let us denote the Laplace transform of a function u as,

$$U = \int_0^\infty u e^{-st} \, dt \tag{6.4}$$

Equation 6.1 in the transform space becomes,

$$k\nabla^2 U - sU + u_0 = 0 \tag{6.5}$$

The boundary conditions, Equation 6.2, must also be transformed and we assume, for simplicity, that they are constant in time, which gives

$$\begin{aligned} U &= \bar{U} = \frac{\bar{u}}{s} && \text{on } \Gamma_1 \\ Q &= \bar{Q} = \frac{\bar{q}}{s} && \text{on } \Gamma_2 \end{aligned} \tag{6.6}$$

The boundary integral equation in the transformed space, obtained by applying either the method of weighted residuals and integration by parts[7] or Green's reciprocal identity[1] is,

$$c_i U_i + k \int_\Gamma Q^* U \, d\Gamma = k \int_\Gamma U^* Q \, d\Gamma + \int_\Omega U^* u_0 \, d\Omega \tag{6.7}$$

in which c_i is a coefficient that depends on the solid angle of the boundary at point i[7] ($c_i = 1$ if i is an internal point) and $Q^* = \dfrac{\partial U^*}{\partial n}$, being n the outward normal to surface Γ and U^* the fundamental solution of Equation 6.5[1], i.e.

$$U^* = \frac{1}{2\pi k} K_0 \left[\left(\frac{s}{k} \right)^{1/2} r \right] \tag{6.8}$$

where K_0 is the modified Bessel function of the second kind of order zero and $r = [(x - x_i)^2 + (y - y_i)^2]^{1/2}$ is the distance from the point i to the point under consideration.

After discretizing the Γ boundary into a series of elements, Equation 6.7 can be solved numerically[7]. If u_0 is the solution of a steady-state Laplace equation, the domain integral in Equation 6.7 can be transformed into a boundary integral[8] and no domain discretization is needed for its solution.

The remaining step is the transform inversion of the solution, which is carried out numerically. Following the method of Schapery[9], we assume that the potential at each point can be represented as a finite series by

$$u_i = u_i^s + \sum_{n=1}^{N} a_n \exp(-b_n t) \tag{6.9}$$

where u^s is the steady-state solution (for $t \to \infty$), a_n and b_n are constants. The transform of Equation 6.9 gives,

$$U_i = \frac{u_i^s}{s} + \sum_{n=1}^{N} \frac{a_n}{s + b_n} \tag{6.10}$$

We now have to choose, somewhat arbitrarily, N values of s. The values of b_n are assumed to be equal to the selected s_n. Thus there remain the N values of the coefficient a to be computed at each boundary point (plus the internal points where the solution is required). By solving Equation 6.7 N times, for the selected values of s, we find N values of U at each point, which allows the evaluation of the a_n using Equation 6.10 and, consequently, the evaluation of the physical variable u using Equation 6.9. A similar calculation is also required in order to obtain the real boundary (and internal) fluxes.

The main advantage of this formulation is that the dimensionality of the problem under consideration is effectively reduced by one, provided u_0

satisfies Laplace's equation. Thus, for a two-dimensional problem, only a line integration over the boundary is required. However, the transform inversion is essentially a curve fitting process and as such, it is important for the analyst to have an idea of the expected behaviour of the solution in order to select the time function, Equation 6.9, since choosing too many values of s would quickly make Equation 6.10 unstable while too few values would not represent the curve adequately[3]. Furthermore, as pointed out in Lachat and Combescure[2], the method is not efficient when the time history of the boundary conditions is complex and in this case, step-by-step methods of the type subsequently discussed are preferred.

6.2.2 **Finite difference in time**

Let us assume that the time derivative in Equation 6.1 can be approximated in a finite difference form, for a sufficiently small time step Δt, as

$$\frac{\partial u}{\partial t} = \frac{u_{t+\Delta t} - u_t}{\Delta t} \tag{6.11}$$

Equation 6.1 can now be written as,

$$k\nabla^2 u_{t+\Delta t} - \frac{u_{t+\Delta t}}{\Delta t} = -\frac{u_t}{\Delta t} \tag{6.12}$$

This equation is similar in form to Equation 6.5 and its fundamental solution is then of the same type as Equation 6.8, i.e.

$$u^* = \frac{1}{2\pi k} K_0 \left[\frac{r}{(k\Delta t)^{1/2}} \right] \tag{6.13}$$

The boundary integral equation for this formulation is[6],

$$c_i u^i_{t+\Delta t} + k \int_\Gamma q^* u_{t+\Delta t} \, \mathrm{d}\Gamma = k \int_\Gamma u^* q_{t+\Delta t} \, \mathrm{d}\Gamma + \frac{1}{\Delta t} \int_\Omega u^* u_t \, \mathrm{d}\Omega \tag{6.14}$$

Starting from known initial values of u_t at $t = t_0$, we can advance the process in time by solving Equation 6.14 numerically. Values of $u_{t+\Delta t}$ at a sufficient number of internal points are then computed in order to be used as initial values for the next time step.

Notice that for each time step, a domain integration needs to be performed in Equation 6.14. Assuming a constant time step throughout the analysis, we can write recursive relations for the boundary nodes and internal points such that all the matrices involved will also be constant in time and thus can be computed only once and stored. However, very small time steps have to be employed if approximation Equation 6.11 is to produce good results and this can make the approach very time consuming.

6.2.3 Time dependent fundamental solution

Considering the time dependence of the problem directly in the integration by parts process, the boundary integral equation for this formulation becomes[10],

$$c_i u_i + k\int_{t_0}^{t_p}\int_{\Gamma} q^* u \, \mathrm{d}\Gamma \, \mathrm{d}t = k\int_{t_0}^{t_p}\int_{\Gamma} u^* q \, \mathrm{d}\Gamma \, \mathrm{d}t + \left[\int_{\Omega} u^* u \, \mathrm{d}\Omega\right]_{t=t_0} \tag{6.15}$$

in which the fundamental solution u^* is[11],

$$u^*(t_n,t) = \frac{1}{4\pi k(t_n - t)} \exp\left(-\frac{r^2}{4k(t_n - t)}\right) \tag{6.16}$$

and

$$q^*(t_n,t) = \frac{\partial u^*}{\partial n} = -\frac{r}{8\pi k^2(t_n - t)^2} \exp\left(-\frac{r^2}{4k(t_n - t)}\right)\frac{\partial r}{\partial n} \tag{6.17}$$

Fundamental solutions of Equation 6.1 for one- and three-dimensional problems can also be found in Carslaw and Jaeger[11] and for axisymmetric problems in Wrobel and Brebbia[12].

In principle, knowing the time variation of u and q, one can directly compute the time integrals in Equation 6.15 analytically and then solve the remaining equation numerically in order to find their spatial distribution for any value of t. However, as these time variations are part of the solution, a time stepping technique (not to be confused with the previous finite difference one) has to be introduced but as the fundamental solution itself is time dependent, large time steps can be adopted. This formulation is described in detail in the next section.

6.3 NUMERICAL FORMULATION

For the numerical solution of Equation 6.15, the boundary Γ is discretized into elements with prescribed space-time interpolation functions. Let us also assume that the domain Ω is divided into cells (although it is not always necessary).

Two different numerical procedures can then be employed. Assuming that functions u and q remain constant on time over each time step, Equation 6.15 can be written in matrix form, for the first process, as

$$\mathbf{H}\mathbf{U}_n = \mathbf{G}\mathbf{Q}_n + \mathbf{B}\mathbf{U}_{n-1} \tag{6.18}$$

where

$$H_{ij}=k\int_{\Gamma_j}\varphi^T\int_{t_{n-1}}^{t_n}q^*(t_n,t)\ \mathrm{d}t\ \mathrm{d}\Gamma+c_i\delta_{ij}$$

$$G_{ij}=k\int_{\Gamma_j}\psi^T\int_{t_{n-1}}^{t_n}u^*(t_n,t)\ \mathrm{d}t\ \mathrm{d}\Gamma \tag{6.19}$$

$$B_{il}=\int_{\Omega_l}u^*(t_n,t_{n-1})\ \mathrm{d}\Omega$$

in which δ_{ij} is the Kronecker delta, φ and ψ are spatial interpolation functions. Notice that at the end of each time step, potential values at a sufficient number of internal points need to be computed in order to be used as initial values for the next time step. Assuming that there are M of such internal points and N boundary nodes in the problem, the dimensions of matrices $\mathbf{G}$ and $\mathbf{H}$ are $N\times N$, $\mathbf{B}$ is a $N\times M$ matrix, $\mathbf{U}_n$ and $\mathbf{Q}_n$ are vectors of dimension N and $\bar{\mathbf{U}}_{n-1}$ of dimension M.

For the second process, the matrix form of Equation 6.15,

$$\sum_{s=1}^{n}\mathbf{H}_{sn}\mathbf{U}_s=\sum_{s=1}^{n}\mathbf{G}_{sn}\mathbf{Q}_s+\mathbf{B}_n\bar{\mathbf{U}}_0 \tag{6.20}$$

where

$$H_{snij}=k\int_{\Gamma_j}\varphi^T\int_{t_{s-1}}^{t_s}q^*(t_n,t)\ \mathrm{d}t\ \mathrm{d}\Gamma+c_i\delta_{sn}\delta_{ij}$$

$$G_{snij}=k\int_{\Gamma_j}\psi^T\int_{t_{s-1}}^{t_s}u^*(t_n,t)\ \mathrm{d}t\ \mathrm{d}\Gamma \tag{6.21}$$

$$B_{nil}=\int_{\Omega_l}u^*(t_n,t_0)\ \mathrm{d}\Omega$$

Since in this approach we always start the time integration process at time $t=t_0$, there is no need to compute internal potentials. Furthermore, if the initial conditions u_0 satisfy Laplace's equation, the domain integral of Equation 6.20 can be transformed into an equivalent boundary integral[8]. Notice that the values of $\mathbf{U}_s$ and $\mathbf{Q}_s$, for $s=1,\ldots,n-1$, are known from calculations at previous steps.

From Equations 6.19 and 6.21, it follows that the time integrals can be carried out analytically, giving

$$\int_{t_{s-1}}^{t_s} u^*(t_n,t)\ \mathrm{d}t=\frac{1}{4\pi k}[E_1(a_{s-1})-E_1(a_s)] \tag{6.22}$$

$$\int_{t_{s-1}}^{t_s} q^*(t_n,t)\ \mathrm{d}t=-\frac{1}{2\pi kr}\left(\frac{\partial r}{\partial n}\right)[\exp(-a_{s-1})-\exp(-a_s)] \tag{6.23}$$

where

$$a_s=\frac{r^2}{4k(t_n-t_s)} \tag{6.24}$$

and E_1 is the exponential-integral function. Notice that $E_1(a_n)=\exp(-a_n)=0$.

The previous assumption of constancy in time for u and q is not restrictive, and higher order time interpolation functions can also be used. Let us now assume a linear variation in time (over each time step) for u and q, according to the following interpolation function,

$$u=\frac{t-t_{s-1}}{t_s-t_{s-1}}u_s+\frac{t_s-t}{t_s-t_{s-1}}u_{s-1} \tag{6.25}$$

The discretized form of Equation 6.15 is now,

$$\mathbf{H}^I\mathbf{U}_{n-1}+\mathbf{H}^F\mathbf{U}_n=\mathbf{G}^I\mathbf{Q}_{n-1}+\mathbf{G}^F\mathbf{Q}_n+\mathbf{B}\bar{\mathbf{U}}_{n-1} \tag{6.26}$$

for the first approach, where

$$\begin{aligned}
H_{ij}^I&=\frac{k}{t_n-t_{n-1}}\int_{\Gamma_j}\varphi^T\int_{t_{n-1}}^{t_n}(t_n-t)q^*(t_n,t)\ \mathrm{d}t\ \mathrm{d}\Gamma\\
H_{ij}^F&=\frac{k}{t_n-t_{n-1}}\int_{\Gamma_j}\varphi^T\int_{t_{n-1}}^{t_n}(t-t_{n-1})q^*(t_n,t)\ \mathrm{d}t\ \mathrm{d}\Gamma+c_i\delta_{ij}\\
G_{ij}^I&=\frac{k}{t_n-t_{n-1}}\int_{\Gamma_j}\psi^T\int_{t_{n-1}}^{t_n}(t_n-t)u^*(t_n,t)\ \mathrm{d}t\ \mathrm{d}\Gamma\\
G_{ij}^F&=\frac{k}{t_n-t_{n-1}}\int_{\Gamma_j}\psi^T\int_{t_{n-1}}^{t_n}(t-t_{n-1})u^*(t_n,t)\ \mathrm{d}t\ \mathrm{d}\Gamma
\end{aligned} \tag{6.27}$$

with B_{il} being calculated as in Equation 6.19. For the second approach, we have the matrix equation

$$\sum_{s=1}^{n} (\mathbf{H}_{sn}^{I}\mathbf{U}_{s-1} + \mathbf{H}_{sn}^{F}\mathbf{U}_{s}) = \sum_{s=1}^{n} (\mathbf{G}_{sn}^{I}\mathbf{Q}_{s-1} + \mathbf{G}_{sn}^{F}\mathbf{Q}_{s}) + \mathbf{B}_{n}\bar{\mathbf{U}}_{0} \tag{6.28}$$

where

$$H_{snij}^{I} = \frac{k}{t_s - t_{s-1}} \int_{\Gamma_j} \varphi^T \int_{t_{s-1}}^{t_s} (t_s - t) q^*(t_n, t) \, dt \, d\Gamma$$

$$H_{snij}^{F} = \frac{k}{t_s - t_{s-1}} \int_{\Gamma_j} \varphi^T \int_{t_{s-1}}^{t_s} (t - t_{s-1}) q^*(t_n, t) \, dt \, d\Gamma + c_i \delta_{sn} \delta_{ij}$$

$$G_{snij}^{I} = \frac{k}{t_s - t_{s-1}} \int_{\Gamma_j} \psi^T \int_{t_{s-1}}^{t_s} (t_s - t) u^*(t_n, t) \, dt \, d\Gamma \tag{6.29}$$

$$G_{snij}^{F} = \frac{k}{t_s - t_{s-1}} \int_{\Gamma_j} \psi^T \int_{t_{s-1}}^{t_s} (t - t_{s-1}) u^*(t_n, t) \, dt \, d\Gamma$$

with B_{nil} computed as in Equation 6.21.

The time integrals can be carried out analytically as follows,

$$\int_{t_{s-1}}^{t_s} (t_s - t) u^*(t_n, t) \, dt = \frac{1}{4\pi k} \left\{ \left[(t_s - t_n) - \frac{r^2}{4k} \right] \left[E_1(a_{s-1}) - E_1(a_s) \right] + \frac{r^2}{4k} \left[\frac{1}{a_{s-1}} \exp(-a_{s-1}) - \frac{1}{a_s} \exp(-a_s) \right] \right\} \tag{6.30}$$

$$\int_{t_{s-1}}^{t_s} (t - t_{s-1}) u^*(t_n, t) \, dt = \frac{1}{4\pi k} \left\{ \left[(t_n - t_{s-1}) + \frac{r^2}{4k} \right] \times [E_1(a_{s-1}) - E_1(a_s)] - \frac{r^2}{4k} \left[\frac{1}{a_{s-1}} \exp(-a_{s-1}) - \frac{1}{a_s} \exp(-a_s) \right] \right\} \tag{6.31}$$

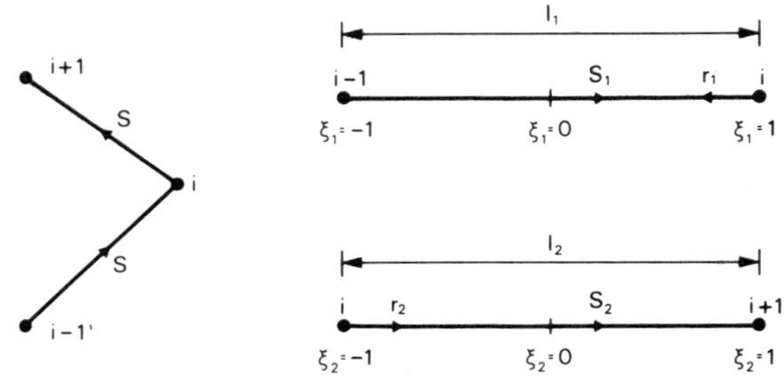

Figure 6.1 Definitions for integration purposes

$$\int_{t_{s-1}}^{t_s}(t_s-t)q^*(t_n,t)\,\mathrm{d}t=-\frac{r}{2\pi k}\frac{\partial r}{\partial n}\left\{\frac{t_s-t_n}{r^2}[\exp(-a_{s-1})-\exp(-a_s)]\right.$$
$$\left.+\frac{1}{4k}[E_1(a_{s-1})-E_1(a_s)]\right\} \tag{6.32}$$

$$\int_{t_{s-1}}^{t_s}(t-t_{s-1})q^*(t_n,t)\,\mathrm{d}t=-\frac{r}{2\pi k}\frac{\partial r}{\partial n}\left\{\frac{t_n-t_{s-1}}{r^2}\times\right.$$
$$\left.\times[\exp(-a_{s-1})-\exp(-a_s)]-\frac{1}{4k}[E_1(a_{s-1})-E_1(a_s)]\right\} \tag{6.33}$$

The remaining step in the calculation is the computation of the space integrals. In this work, linear spatial interpolation functions are employed for the geometry, potential and flux. A six-point Gauss quadrature rule is adopted for the numerical integration of the coefficients of matrices **G** and **H**. However, for the diagonal coefficients of some of these matrices, singular integrals occur and thus special care has to be taken in their calculation.

The terms H_{ii} in Equation 6.19 and H_{nnii} in Equation 6.21 (as well as H_{ii}^F in Equation 6.27 and H_{nnii}^F in Equation 6.29) contain integrals with a singularity of the $1/r$ type that must be evaluated in the Cauchy principal value sense. But since in this work we are discretizing the boundary Γ using straight segments, these integrals vanish due to the orthogonality of r and n, which makes $\frac{\partial r}{\partial n}=0$.

The terms G_{ii} in Equation 6.19 and G_{nnii} in Equation 6.21 (as well as G_{ii}^F in Equation 6.27 and G_{nnii}^F in Equation 6.29) have a logarithmic singularity, which is directly integrable. With references to Figure 6.1, we can write

$$G_{ii}=\frac{1}{4\pi}\left[\int_{-l_{1/2}}^{l_{1/2}}\frac{(1+\xi_1)}{2}E_1\left(\frac{r_1^2}{4k\Delta t}\right)\mathrm{d}S+\int_{-l_{2/2}}^{l_{2/2}}\frac{(1-\xi_2)}{2}E_1\left(\frac{r_2^2}{4k\Delta t}\right)\mathrm{d}S\right] \tag{6.34}$$

where

$$\xi_i = \frac{2S_i}{l_i} \tag{6.35}$$

$$r_1 = (1-\xi_1)\frac{l_1}{2}; \quad r_2 = (1+\xi_2)\frac{l_2}{2} \tag{6.36}$$

Expanding the exponential-integral in series[13],

$$E_1(x) = -C - \ln x + \sum_{n=1}^{\infty} (-1)^{n-1} \frac{x^n}{nn!} \tag{6.37}$$

in which C is Euler's constant, $C = 0.57721566\ldots$, the integrals in Equation 6.34 can be evaluated in closed form as,

$$G_{ii} = \frac{1}{8\pi}[l_1 F(\alpha_1) + l_2 F(\alpha_2)] \tag{6.38}$$

where $\alpha_i = \dfrac{l_i^2}{4k\Delta t}$ and

$$F(\alpha_i) = 3 - C - \ln(\alpha_i) + \sum_{n=1}^{\infty} (-1)^{n-1} \frac{\alpha_i^n}{n(2n+1)(n+1)!} \tag{6.39}$$

The series that appears in Equation 6.39 converges very quickly for small values of α but slowly as α increases. To overcome this problem, we can integrate analytically over a short segment near the singularity, thus ensuring that α is always less or equal to one, and numerically integrate over the rest of the elements using a standard Gaussian quadrature, as if these parts were separate elements.

Notice that when using linear or higher order time interpolation functions only additional non-singular integrals are introduced into the analysis, such that the singular terms can always by computed, as in Equation 6.38.

Computation of the **B** matrix coefficients in Equation 6.19 is carried out numerically. Triangular cells are employed to discretize the domain Ω and Hammer's quadrature rule is adopted, the potential being computed at each integration point. Alternatively, if we assume that the variation of the potential in each cell is also linear, a semi-analytical integration scheme using polar coordinates[10] can also be employed. Further discussions on the domain integration, regarding accuracy and computer efficiency, can be found in Wrobel[14].

A final remark on the computer efficiency of the numerical procedures developed in this section — Equation 6.18 as opposed to Equation 6.20, the same applying to Equations 6.26 and 6.28 — is that although in many problems the second process does not require a domain integration, thus effectively reducing by one the dimensionality of the problem, all matrices **G** and **H** in Equation 6.20 have to be recomputed for each time step, since they depend on the actual value of time. For the first process, however, they depend on the value of the time step itself and so, assuming that it is constant throughout the analysis, matrices **G**, **H** and **B** in Equation 6.18 can all be computed once only and stored. Thus the computation is reduced to solving a system of equations with the size of the number of boundary nodes and then computing potentials at internal points by simply performing products and sums of matrices.

From the above discussion and several different analyses carried out by the authors using both numerical procedures, we can conclude that in general, the first above-mentioned process is more efficient than the second, excluding the case of problems with regions extending to infinity.

6.4 SPECIAL BOUNDARY CONDITIONS

Up to now, we have only dealt with the solution of problems with Neumann, Dirichlet or Cauchy (mixed) boundary conditions. This means that on solving Equation 6.18, for instance, N boundary values of potentials or fluxes have to be prescribed in order for the remaining N boundary unknowns to be calculated.

Another common type of boundary condition that appears frequently in practical problems is the Robin condition, prescribing a linear combination of potential and flux,

$$au + bq = c \tag{6.40}$$

where a, b and c are constants. Notice that Equation 6.40 includes all the previous boundary conditions, since for $b = 0$ it becomes the Dirichlet condition while for $a = 0$ we have the Neumann condition.

Many different physical problems are governed by the diffusion equation 6.1. For example, transient heat conduction for which Equation 6.40 prescribes the so-called convection or 'radiation' condition, electromagnetic problems where Equation 6.40 represents the impedance boundary condition, etc. In matrix form, Equation 6.40 can be written as,

$$\mathbf{Q}_n = \mathbf{C} - \mathbf{A}\mathbf{U}_n \tag{6.41}$$

in which **C** is a vector of dimension N containing the values of c/b at each boundary node and **A** is a $N \times N$ diagonal matrix whose coefficients are the values of a/b at each boundary node.

Substituting Equation 6.41 into Equation 6.18 yields the following recursive equation,

$$(\mathbf{H} + \mathbf{G}\mathbf{A})\mathbf{U}_n = \mathbf{G}\mathbf{C} + \mathbf{B}\bar{\mathbf{U}}_{n-1} \tag{6.42}$$

where from the boundary (and internal) potential values can be advanced in time.

Let us now study more closely some problems related to transient heat conductionm which has been the main concern of the authors. The 'radiation' condition specifies that,

$$hu - Kp = hv \tag{6.43}$$

where u is temperature, q is the heat flux, K is the thermal conductivity, h the heat transfer coefficient and v the temperature of the medium that surrounds the domain Ω. It may occur, in many practical situations, that h and v are also time dependent. Assuming at first that u and v vary linearly in time over each time step according to Equation 6.25 while h remains constant, condition 6.43 gives,

$$q = \frac{h}{K}(u - v) = \frac{h}{K(t_s - t_{s-1})}[(t - t_{s-1})(u_s - v_s) + (t_s - t)(u_{s-1} - v_{s-1})] \tag{6.44}$$

Equation 6.28 — the same applies for Equation 6.26 — can then be rearranged for our numerical solution as,

$$\sum_{s=1}^{n} [(\mathbf{H}_{sn}^{I} + \mathbf{G}_{sn}^{I}\mathbf{A})U_{s-1} + (\mathbf{H}_{sn}^{F} + \mathbf{G}_{sn}^{F}\mathbf{A})\mathbf{U}_s] = \sum_{s=1}^{n} (\mathbf{G}_{sn}^{I}\mathbf{A}\mathbf{C}_{s-1} + \mathbf{G}_{sn}^{F}\mathbf{A}\mathbf{C}_s) + \mathbf{B}_n\bar{\mathbf{U}}_0 \tag{6.45}$$

where vector $\mathbf{C}$ contains the values of v/K and the diagonal matrix $\mathbf{A}$ the -h/K values, all of which are known.

Consider now that the heat transfer coefficient h also varies in time according to Equation 6.25. Applying condition 6.43 then gives,

$$\begin{aligned} q = \frac{1}{K(t_s - t_{s-1})^2}\{&(t - t_{s-1})^2 h_s(u_s - v_s) + (t - t_{s-1})(t_s - t) \times \\ &\times [h_s(u_{s-1} - v_{s-1}) + h_{s-1}(u_s - v_s)] + (t_s - t)^2 h_{s-1}(u_{s-1} - v_{s-1})\} \end{aligned} \tag{6.46}$$

The discretized form of the boundary integral Equation 6.15 for this kind of problem becomes,

$$\begin{aligned} &\sum_{s=1}^{n} [(\mathbf{H}_{sn}^{I} + \mathbf{G}_{sn}^{2}\mathbf{A}_s + \mathbf{G}_{sn}^{1}\mathbf{A}_{s-1})\mathbf{U}_{s-1} + (\mathbf{H}_{sn}^{F} + \mathbf{G}_{sn}^{3}\mathbf{A}_s + \mathbf{G}_{sn}^{2}\mathbf{A}_{s-1})\mathbf{U}_s] \\ &= \sum_{s=1}^{n} [(\mathbf{G}_{sn}^{2}\mathbf{A}_s + \mathbf{G}_{sn}^{1}\mathbf{A}_{s-1})\mathbf{C}_{s-1} + (\mathbf{G}_{sn}^{3}\mathbf{A}_s + \mathbf{G}_{sn}^{2}\mathbf{A}_{s-1})\mathbf{C}_s] + \mathbf{B}_n\bar{\mathbf{U}}_0 \end{aligned} \tag{6.47}$$

where

$$G_{snij}^{1} = \frac{k}{(t_s - t_{s-1})^2}\int_{\Gamma_j} \psi^T \int_{t_{s-1}}^{t_s} (t_s - t)^2 u^*(t_n, t)\, \mathrm{d}t\, \mathrm{d}\Gamma$$

$$G_{snij}^2 = \frac{k}{(t_s - t_{s-1})^2} \int_{\Gamma_j} \psi^T \int_{t_{s-1}}^{t_s} (t_s - t)(t - t_{s-1}) u^*(t_n, t) \, dt \, d\Gamma \tag{6.48}$$

$$G_{snij}^3 = \frac{k}{(t_s - t_{s-1})^2} \int_{\Gamma_j} \psi^T \int_{t_{s-1}}^{t_s} (t - t_{s-1})^2 u^*(t_n, t) \, dt \, d\Gamma$$

with H_{snij}^I, H_{snij}^F and B_{nil} being computed as in Equation 6.29.

The analytical evaluation of the time integrals in Equation 6.48 gives,

$$\int_{t_{s-1}}^{t_s} (t_s - t)^2 u^*(t_n, t) \, dt = \frac{1}{4\pi k} \left\{ \left[(t_s - t_n)^2 - \frac{r^2}{2k} \left((t_s - t_n) - \frac{r^2}{16k} \right) \right] \times \right.$$

$$\times [E_1(a_{s-1}) - E_1(a_s)] + \frac{r^2}{2k} \left[(t_s - t_n) - \frac{r^2}{16k} \right] \left[\frac{1}{a_{s-1}} \exp(-a_{s-1}) - \frac{1}{a_s} \exp(-a_s) \right]$$

$$\left. + \frac{r^4}{32k^2} \left[\frac{1}{a_{s-1}^2} \exp(-a_{s-1}) - \frac{1}{a_s^2} \exp(-a_s) \right] \right\} \tag{6.49}$$

$$\int_{t_{s-1}}^{t_s} (t_s - t)(t - t_{s-1}) u^*(t_n, t) \, dt = \frac{1}{4\pi k} \left\{ \left[(t_s - t_n)(t_n - t_{s-1}) + \frac{r^2}{4k} \left\{ (t_s - t_n) \right. \right. \right.$$

$$\left. \left. - (t_n - t_{s-1}) - \frac{r^2}{8k} \right\} \right] [E_1(a_{s-1}) - E_1(a_s)] - \frac{r^2}{4k} \left[(t_s - t_n) - (t_n - t_{s-1}) - \frac{r^2}{8k} \right]$$

$$\left. \times \left[\frac{1}{a_{s-1}} \exp(-a_{s-1}) - \frac{1}{a_s} \exp(-a_s) \right] - \frac{r^4}{32k^2} \left[\frac{1}{a_{s-1}^2} \exp(-a_{s-1}) - \frac{1}{a_s^2} \exp(-a_s) \right] \right\} \tag{6.50}$$

$$\int_{t_{s-1}}^{t_s} (t - t_{s-1})^2 u^*(t_n, t) \, dt = \frac{1}{4\pi k} \left\{ \left[(t_n - t_{s-1})^2 + \frac{r^2}{2k} \left((t_n - t_{s-1}) + \frac{r^2}{16k} \right) \right] \times \right.$$

$$\times [E_1(a_{s-1}) - E_1(a_s)] - \frac{r^2}{2k} \left[(t_n - t_{s-1}) + \frac{r^2}{16k} \right] \left[\frac{1}{a_{s-1}} \exp(-a_{s-1}) - \frac{1}{a_s} \exp(-a_s) \right]$$

$$\left. + \frac{r^4}{32k^2} \left[\frac{1}{a_{s-1}^2} \exp(-a_{s-1}) - \frac{1}{a_s^2} \exp(-a_s) \right] \right\} \tag{6.51}$$

6.5 APPLICATIONS

In order to show the applicability of the numerical procedures developed in this chapter, several transient heat conduction problems were analyzed. Examples 1 to 3 compare BEM and FEM results for simple problems with known analytical solutions[11]. Similar results were obtained employing both techniques described herein and as all these problems have zero initial temperature, no domain discretization was needed when using Equations 6.20 and 6.28.

The examples show that the FEM results present much more severe oscillations than the BEM ones. Furthermore, the assumption of constancy in time for functions u and q in the BEM 'damps' the solutions in such a way that no oscillation was detected in any of the problems studied.

Example 4 presents the study of a cooling hole in an infinite medium and demonstrates how easy it is to deal with such problems when using the BEM, as only the interface between hole and medium needs to be discretized. FEM analyses of the same problem require discretization of the infinite region, resulting in a much larger system of equations to be solved and an increase in the complexity of the input data.

Finally, Example 5 studies a practical problem with a complicated boundary condition history. The BEM results of this analysis compare favorably with a refined FEM solution employing parabolic isoparametric elements.

6.5.1 **Semi-infinite solid**

Example 1

As a first example, the one-dimensional problem of constant heat flux applied to a semi-infinite solid was chosen. The material properties of the solid are assumed to be unity, as well as the applied heat flux per unit time.

A FEM analysis of this problem was performed in Wilson and Nickell[15] using simple quadrilateral elements and assuming the temperature to vary linearly between time steps. Results for the surface temperature versus time are presented in Figure 6.2. BEM analyses employing constant and linear time interpolation functions produced the same results and showed no oscillation, as can be seen in the same figure.

6.5.2 **Plane plate**

Example 2

This example studies a plane plate, initially at 0°C, surrounded by a medium at 100°C. Its cross-section is 0.1×0.1 m and the values of the thermal conductivity, heat capacity and heat transfer coefficient are 18 kcal/(h m °C), 912 kcal/(m^3 °C) and 5000 kcal/(h m^2 °C), respectively.

The surface temperature of the plate is plotted in Figure 6.3. Due to symmetry, only one quarter of its cross section needs to be discretized. The FEM analyses employed 5 parabolic isoparametric elements in space and linear,

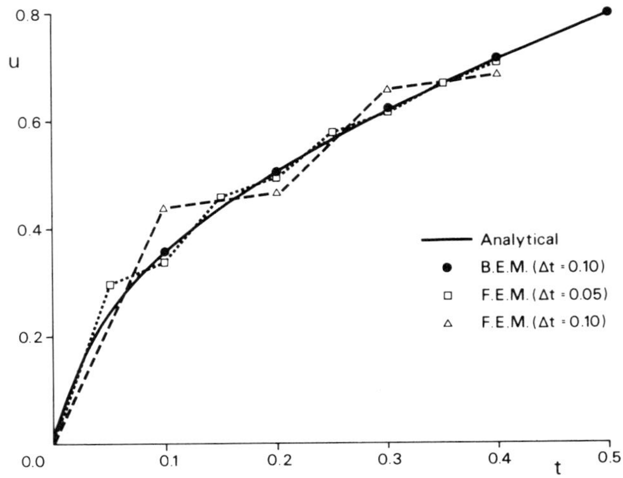

Figure 6.2 Surface temperature of semi-infinite solid

parabolic or cubic elements in time[16]. Results using a central finite difference scheme are also presented in the same reference.

The discretization for the BEM analyses consists of only 8 boundary elements (no elements are needed over the symmetry axes). As pointed out in reference 16, the time step adopted for this problem is extremely large and because of this, the BEM with linear time variation produced oscillations which are, however, much smaller than the FDM and FEM ones. No oscillation was detected for the BEM with constant time variation.

6.5.3 **One-dimensional bar**

Example 3

The subject of this investigation is a bar with unit material properties, at zero initial temperature. We assume that for any time $t>0$, a unit temperature is applied at $x=0$. The surface $x=4$ is insulated, as well as the surfaces $y=0$ and $y=1$. The problem was analysed with the FEM in reference 17, where the space domain was discretized into 4 linear, parabolic or cubic isoparametric elements and the variation with time approximated by the Crank-Nicholson scheme. For the BEM analysis, only one half of the cross-section was discretized into 12 boundary elements, due to symmetry.

Notice that in order to perform a BEM analysis with linear time variation the initial boundary fluxes (in addition to the initial temperature) have to be prescribed. These values, however, are unbounded at $x=0$. Although this problem can be overcome by simply applying the boundary conditions linearly over the first step[14], no attempt of this kind was made in this example and only results for constant time variation are presented in Figure 6.4.

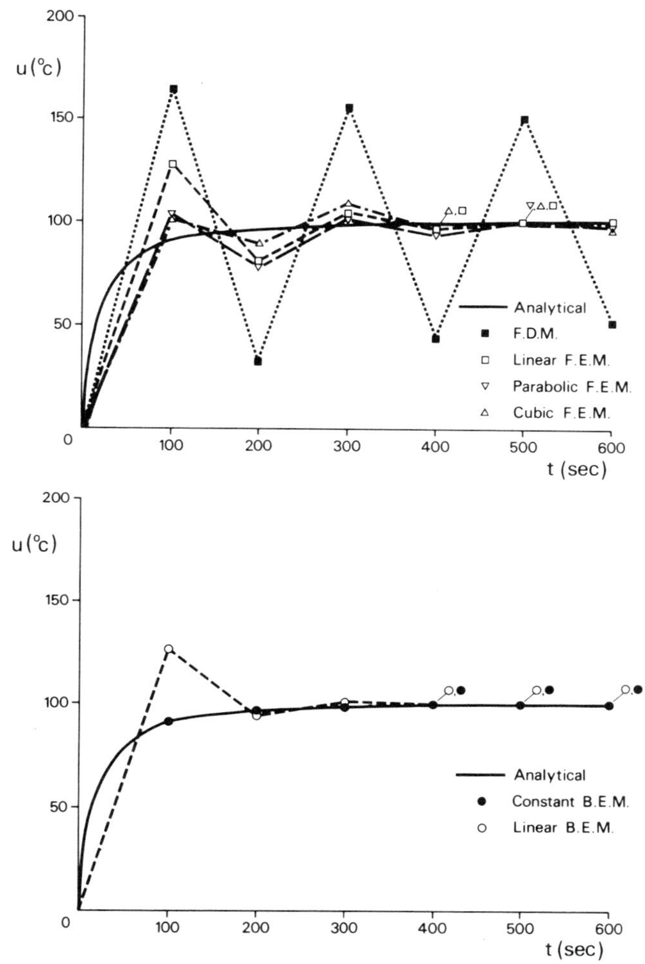

Figure 6.3 Surface temperature of plane plate

6.5.4 **Hole in an infinite medium**

Example 4

For problems with regions extending to infinity, BEM solutions are much more economical than FEM ones. In order to demonstrate this, we study a circular opening in an infinite plane region, initially at zero temperature. The radius of the hole is unity, its ambient temperature equals -10 and the material properties of the medium are also taken to be unity, for simplicity.

The variation of the surface temperature with time is presented in Figure 6.5, for various values of the heat transfer coefficient, and compared to an analytical solution[11]. The agreement between the two solutions is very good. A time step

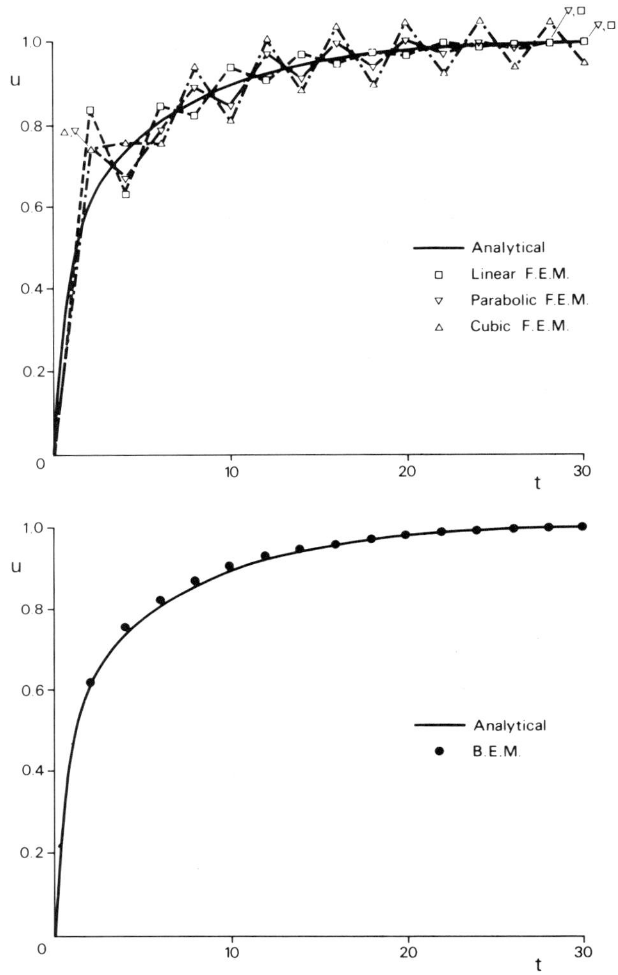

Figure 6.4 Temperature variation of bar at $x = 1.0$, for $\Delta t = 2.0$

of 0.5 was adopted and the analyses carried out until the surface temperature began to drop significantly. For obvious reasons, only the second numerical approach described in this chapter was employed in the BEM analyses and, due to symmetry, only one quarter of the interface between hole and medium was divided into 6 boundary elements.

This problem was also studied with the FEM[18] but since the FEM is a domain-type technique, the infinite region had to be limited by a finite, non-conducting boundary. In order to achieve the same level of accuracy, a time step ten times smaller ($\Delta t = 0.05$) was adopted and the domain discretized into 70 triangular elements or 3 cubic isoparametric elements.

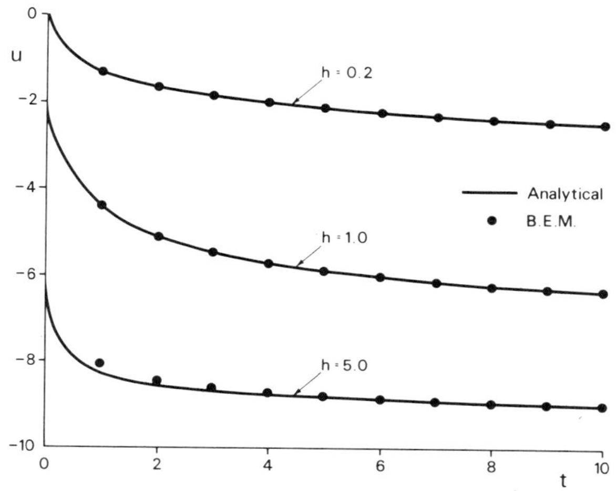

Figure 6.5 Surface temperature of cooling hole in an infinite medium

6.5.5 **Turbine disc**

Example 5

The final example is a more practical one, where an actual turbine disc temperature variation is studied. The initial temperature of the turbine disc is 295.1°K and the values of the thermal conductivity, density and specific heat of the material are 15 W/(m °K), 8221 kg/m^3 and 550 J/(kg °K), respectively. There are 18 different zones along the boundary, each of which with a different set of prescribed values for the heat transfer coefficient and the temperature of the surrounding gas. Their time variation at one of such boundary zones is shown in Figure 6.6.

The BEM analysis was performed by discretizing the boundary into 90 elements and applying Equation 6.47. In order to simplify the computation, a constant temperature was subtracted out so as to make the initial temperature equal to zero, thus avoiding the domain integration. This value was afterwards added to the solution.

Isotherms at various times are plotted in Figure 6.7 and compared to FEM results obtained with 85 parabolic isoparametric elements and 348 nodes. The agreement is, in general, excellent and the discrepancy between the results at $t = 2565$ s was attributed to the very large time step value ($\Delta t = 100$ s) employed for the BEM analysis after $t = 1065$ s.

6.6 CONCLUSIONS

This chapter has described the application of boundary element solutions to time-dependent problems. The advantages of employing fundamental solutions which are space and time dependent, in preference to Laplace

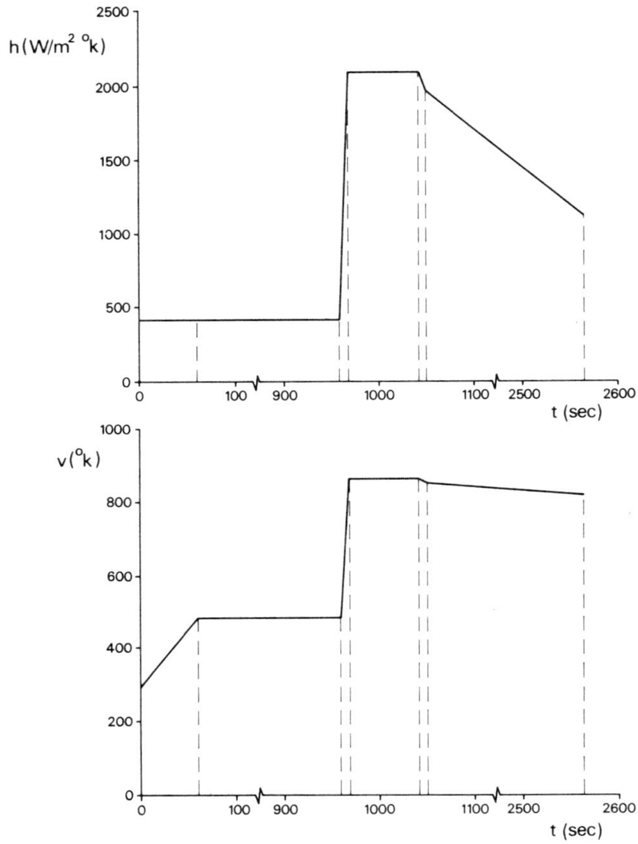

Figure 6.6 Time variation of heat transfer coefficient and temperature of surrounding gas for a typical boundary zone

transform and finite difference type formulations, are also pointed out. The Laplace transform approach requires a great deal of experience from the analyst in selecting the appropriate time function for the transform inversion process and may lead to poor results if the time variation of the boundary conditions is complex. Finite difference type schemes, on the other hand, are normally time consuming due to the fact that very small time steps have to be adopted.

The use of space-time fundamental solutions eliminates the need of integrating step-by-step in time in a finite difference fashion. But in order to compute the time integrals analytically, time stepping is also necessary although the steps can be comparatively large. Higher order time interpolation functions were introduced into the formulation. Their inclusion does not produce any additional singular integral and is essential in order to obtain accurate solutions for cases in which the boundary conditions vary in time, such as in the turbine disc example.

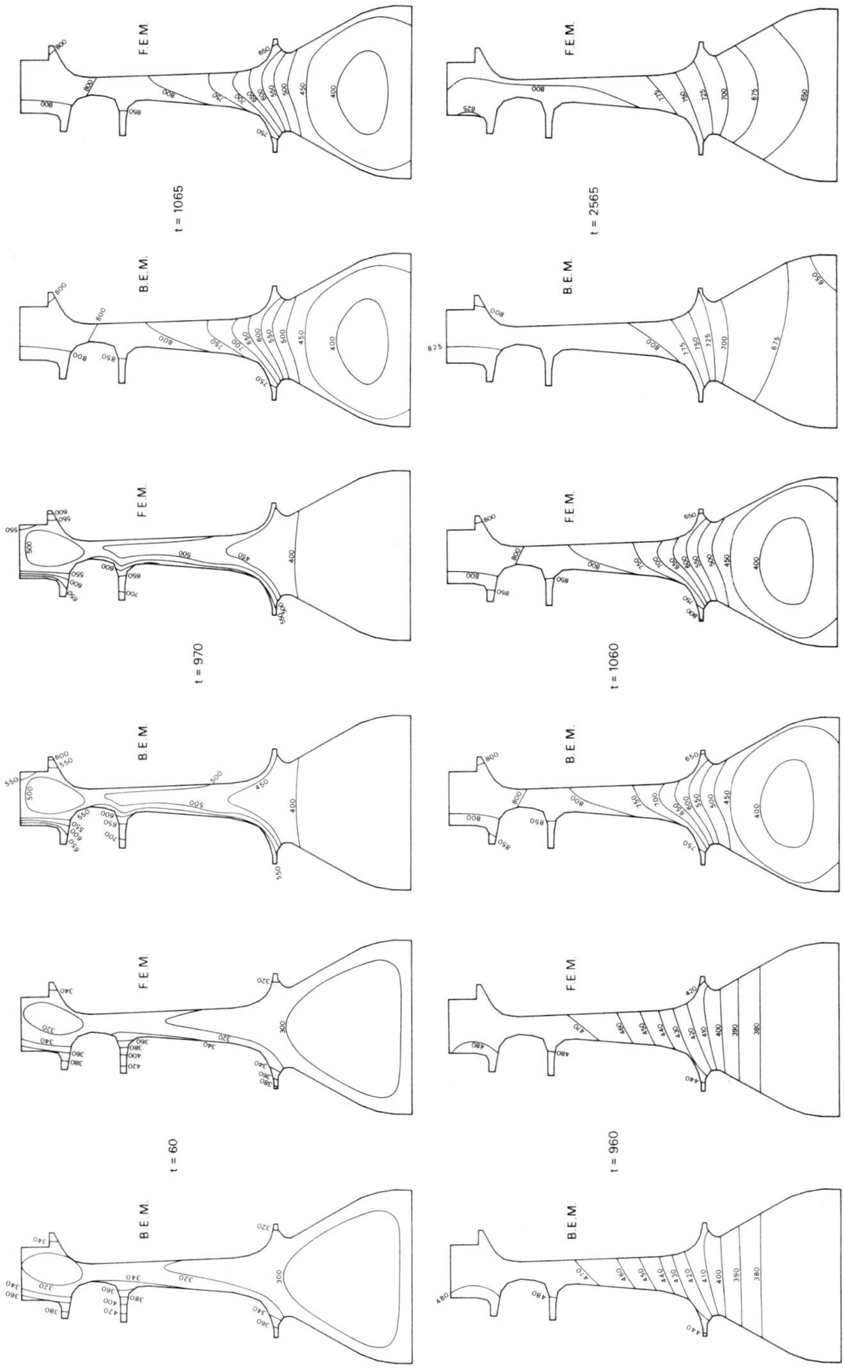

Figure 6.7 Isotherms at various times

Results of several different analyses are presented. Comparison with finite elements results showed that in general, larger time steps can be employed in conjunction with the BEM since less severe oscillations are expected. Furthermore, problems with regions extending to infinity are much more economically dealt with the BEM than using domain-type techniques. The method is also well suited for solving problems where high gradients occur.

REFERENCES

1 Rizzo, F. J. and Shippy, D. J., 'A method of solution for certain problems of transient heat conduction', AIAA J., **8**, No. 11, pp. 2004–2009 (1970)

2 Lachat, J. C. and Combescure, A., 'Laplace transform and boundary integral equations: application to transient heat conduction problems', *1st Int. Conf. on Innovative Numerical Analysis in Applied Engineering*, Versailles (1977)

3 Liggett, J. A. and Liu, P. L. F., 'Unsteady flow in confined aquifers: a comparison of two boundary integral methods', *Water Resources Res.*, **15**, No. 4, pp. 861–866 (1979)

4 Chang, Y. P., Kang, C. S. and Chen, D. J., 'The use of fundamental Green's functions for the solution of problems of heat conduction in anisotropic media', *Int. J. Heat Mass Transfer*, **16**, pp. 1905–1918 (1973)

5 Wrobel, L. C. and Brebbia, C. A., 'The boundary element method for steady-state and transient heat conduction', *1st Int. Conf. on Numerical Methods in Thermal Problems*, Swansea (1979)

6 Brebbia, C. A. and Walker, S., *Boundary Elements Techniques in Engineering*, Newnes-Butterworths, London (1980)

7 Brebbia, C. A., *The Boundary Element Method for Engineers*, 2nd revised edn., Pentech Press, London (1980)

8 Stippes, M. and Rizzo, F. J., 'A note on the body force integral of classical elastostatics', *ZAMP*, **28**, pp. 339–341 (1977)

9 Schapery, R. A., 'Approximate methods of transform inversion for visco-elastic stress analysis', *4th US National Congress on Applied Mechanics*, Vol. 2 (1962)

10 Wrobel, L. C. and Brebbia, C. A., 'Boundary elements in thermal problems', in *Numerical Methods in Heat Transfer*, R. W. Lewis (Ed.), Wiley, Chichester (1981)

11 Carslaw, H. S. and Jaeger, J. C., *Conduction of Heat in Solids*, 2nd edn, Clarendon Press, Oxford (1959)

12 Wrobel, L. C. and Brebbia, C. A., 'A formulation of the boundary element method for axisymmetric transient heat conduction', *Int. J. Heat Mass Transfer*, **24** (1981)

13 Abramowitz, M. and Stegun, I. A., *Handbook of Mathematical Functions*, Dover Publications, New York (1965)

14 Wrobel, L. C., Ph.D. Thesis, University of Southampton, Southampton, England (1981)

15 Wilson, E. L. and Nickell, R. E., 'Application of the finite element method to heat conduction analysis', *Nucl Engng Des.* 4, pp. 276–286 (1966)

16 Köhler, W. and Pittr, J., 'Calculation of transient temperature fields with finite elements in space and time dimensions', *Int. J. Num. Meth. Engng*, **8**, pp. 625–631 (1974)

17 Wood, W. L. and Lewis, R. W., 'A comparison of time marching schemes for the transient heat conduction equation', *Int. J. Num. Meth. Engng*, **9**, pp. 679–689 (1975)

18 Zienkiewicz, O. C. and Parekh, C. J., 'Transient field problems: two-dimensional and three-dimensional analysis by isoparametric finite elements', *Int. J. Num. Meth. Engng*, **2**, pp. 61–71 (1970)

7

Elastodynamics

by J. Dominguez and E. Alarcón

7.1 INTRODUCTION

In this chapter some applications of boundary element techniques to dynamic problems are presented. First, the basic theory is briefly reviewed in order to provide the necessary background to interpret the numerical results (for a fuller account of elastodynamic theory we recommend a study of the specialized literature, for instance, Graff[1], Achenbach[2] or Aki[26]).

The second part of the chapter is devoted to the numerical implementation of the BEM. The presentation is based on the steady-state solution because this is the area in which most experience exists. This is by no means a limitation of the BEM method, and the use of integral transformations to obtain transient solutions, see Cruse[3,4] and Chapter 6 of this book, is a well established procedure.

Finally, in the third part three examples are presented. The first example is the steady-state solution of a plate under cyclic forces with and without a crack. The second example relies on the determination of soil compliances necessary to study soil-structure interaction and the third example treats the problem of the influence of different incidence angles of incoming waves in foundations. The last two examples are relevant to earthquake engineering problems for which the BEM is very well suited.

7.2 GOVERNING EQUATIONS

7.2.1 Navier equations

The equilibrium equation of elasticity under dynamic forces is

$$(\lambda + G)\nabla(\nabla \cdot \vec{u}) + G\nabla^2\vec{u} + \vec{X} = \rho\ddot{\vec{u}} \tag{7.1}$$

or

$$(\lambda + 2G)\nabla(\nabla \cdot \vec{u}) - G\nabla\Lambda(\nabla\Lambda\vec{u}) + \vec{X} = \rho\ddot{\vec{u}} \tag{7.2}$$

where

λ and G are the elastic constants

$\vec{u}$ is the displacement vector
$\vec{X}$ is the body forces vector
ρ is the density of the medium

Equations 7.1 and 7.2 are the so-called Navier equations and have to be integrated in order to obtain the displacement vector $\vec{u}$ and the complete solution once the boundary conditions are imposed.

7.2.2 P and S waves

When $\vec{X}=0$ we have two different types of wave, i.e. *P* and *S* waves. As is well known from potential theory a gradient field is irrotational, so that taking the curl of both sides of Equation 7.1, one finds

$$G\nabla^2(\text{rot } \vec{u})=\rho\frac{\partial^2}{\partial t^2}(\text{rot } \vec{u}) \tag{7.3}$$

but

$$\text{rot } \vec{u}=2\vec{\omega}$$

where $\vec{\omega}$ is the rotation of the elementary particle and hence Equation 7.3 can be written

$$\frac{G}{\rho}\nabla^2\vec{\omega}=\frac{\partial^2\vec{\omega}}{\partial t^2} \tag{7.4}$$

Remembering also Equation 7.3 and that the curl is solenoidal, one can take the divergence of both sides of Equation 7.2 and obtain,

$$(\lambda+2G)\nabla^2\theta=\rho\ddot{\theta}$$

where $\theta=u_{k,k}=\text{div } \vec{u}=\nabla\cdot\vec{u}$, or

$$\frac{\lambda+2G}{\rho}\nabla^2\theta=\frac{\partial^2\theta}{\partial t^2} \tag{7.5}$$

Equations 7.4 and 7.5 both belong to the family of wave equations of the form

$$c^2\nabla^2\varphi=\ddot{\varphi} \tag{7.6}$$

where c is the phase velocity.

Because θ is the specific change of volume and $\vec{\omega}$ represents the distorsion without change of volume (div $\vec{\omega}=0$), Equation 7.4 is related to the so-called distortion waves or *S* waves, while 7.5 is the equation of compressional and dilational waves or *P* waves. Hence we define

$$c_S=\sqrt{\left(\frac{G}{\rho}\right)};\quad c_P=\sqrt{\left(\frac{\lambda+2G}{\rho}\right)} \tag{7.7}$$

Dividing both expressions

$$\frac{c_P}{c_S}=\sqrt{\left(\frac{\lambda+2G}{G}\right)}=\sqrt{\left(2+\frac{\lambda}{G}\right)} \tag{7.8}$$

Since λ and G are positive we have,

$$\frac{c_P}{c_S}>1 \tag{7.9}$$

which justifies the sub-index used: P stand for 'primae' and corresponds to the fastest wave, S for 'secundae' which arrives after the P wave.

7.2.3 **Wave potentials**

The partition of the Navier equations is sometimes obtained through the application of the Helmholtz decomposition of a vector. It is possible to write

$$\vec{u}=\text{grad } \varphi+\text{rot } \vec{\psi} \tag{7.10}$$

with the condition

$$\text{div } \vec{\psi}=0 \tag{7.11}$$

But we know that

$$\text{div } \vec{u}=\text{div grad } \varphi=\nabla^2\varphi \tag{7.12}$$

and

$$\text{rot } \vec{u}=\text{rot rot } \vec{\psi}$$

So that if $\vec{X}=0$, Equation 7.2 can be written as

$$(\lambda+2G) \text{ grad } (\nabla^2\varphi)-G \text{ rot } (\text{rot rot } \vec{\psi})=\rho(\text{grad } \ddot{\varphi}+ \text{ rot } \ddot{\vec{\psi}}) \tag{7.13}$$

Grouping together the operators we obtain

$$\text{grad } [(\lambda+2G)\nabla^2\varphi-\rho\ddot{\varphi}]=\text{rot } [G \text{ rot rot } \vec{\psi}+\rho\ddot{\vec{t}}] \tag{7.14}$$

Using $\nabla\Lambda\nabla\Lambda\vec{u}=\nabla(\nabla\cdot\vec{u})-\nabla^2\vec{u}$ it is clear that

$$\text{rot}[G \text{ rot rot } \vec{\psi}+\rho\ddot{\vec{\psi}}]=\text{rot}[-G\nabla^2\vec{\psi}+\rho\ddot{\vec{\psi}}] \tag{7.15}$$

A solution of Equation 7.14 is then

$$\frac{\lambda+2G}{\rho}\nabla^2\varphi=\ddot{\varphi} \tag{7.16}$$

$$\frac{G}{\rho}\nabla^2\vec{\psi}=\ddot{\vec{\psi}} \tag{7.17}$$

The phase velocities associated with φ and ψ are respectively, c_P and c_S, as expected because

$$\theta=\varphi_{,ii} \quad 2\omega_i=\psi_{i,kk} \tag{7.18}$$

7.3 PLANE WAVES

7.3.1 **Definitions and geometry**

Consider the one-dimensional equation

$$\frac{\partial^2 u}{\partial t^2}=c^2\frac{\partial^2 u}{\partial x^2} \tag{7.19}$$

and let us try the solution

$$u=e^{pt+qx} \tag{7.20}$$

Hence

$$\ddot{u}=p^2u=c^2\frac{\partial^2 u}{\partial x^2}=c^2q^2u \tag{7.21}$$

and

$$p=\pm cq$$

so that

$$u=Ae^{q(ct+x)}+Be^{-q(ct-x)} \tag{7.22}$$

If $q=+ik$ and we introduce the factor $e^{i\varepsilon}$, Equation 7.22 can be written as

$$u=Ae^{i[k(ct+x)+\varepsilon]}+Be^{i[k(-ct+x)+\varepsilon]} \tag{7.23}$$

Equation 7.23 is a solution of Equation 7.19 for every k and ε. The second term represents an outgoing plane wave and the first an incoming wave.

The obvious generalization is to write

$$u=A\ \mathrm{exp}i[k(\vec{r}\vec{v}-ct)+\varepsilon] \tag{7.24}$$

$\vec{v}$ being the normal to the wave front and $\vec{r}$ the direction of propagation. ε is called the phase angle and the following notation is used:

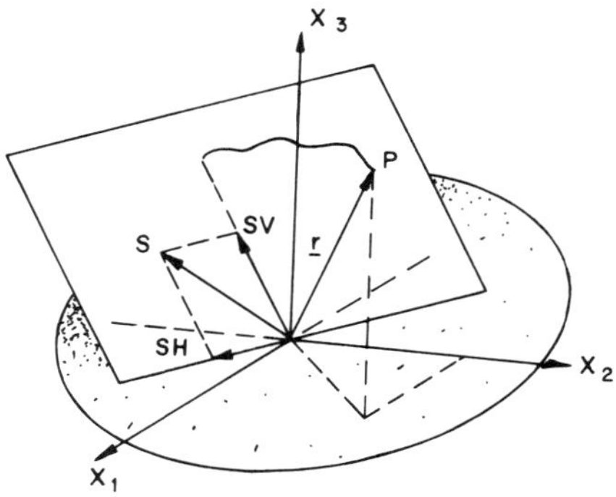

Figure 7.1

Frequency: $\omega = kc = \dfrac{2\pi}{T}$

Period: $T = 2\pi/kc$

Wave number: (or spatial frequency) $k = 2\pi/\lambda$

Wavelength: (or spatial period) $\lambda = 2\pi/k$ (7.25)

The frequency is

$$\omega = c\frac{2\pi}{\lambda} = \frac{2\pi}{T} \rightarrow c = \frac{\lambda}{T} \tag{7.26}$$

The *P* plane waves propagate in direction $\vec{r}$, the *S* waves in a plane normal to $\vec{r}$. It is usual to decompose them into *SV* and *SH* along the directions established in Figure 7.1, as follows;

(1) By the intersection of the plane normal to $\vec{r}$ with the vertical, defined by $\vec{r}$ and the x_3 axis.

(2) By the intersection of the plane normal to $\vec{r}$ with the x_1x_2 plane.

When one is studying, for example incoming *S* waves in the vertical plane, *SV* lies in the vertical plane while *SH* is normal to the plane. In the horizontal plane *SH* and *SV* are defined as in Figure 7.2.

7.3.2 Wave reflection

One of the more interesting topics in wave theory is the disturbance experienced by waves when meeting a discontinuity. Then reflection, refraction and difraction are produced and the resultant motion is a mixture of all these components.

A shortcoming of the finite element method when treating dynamic

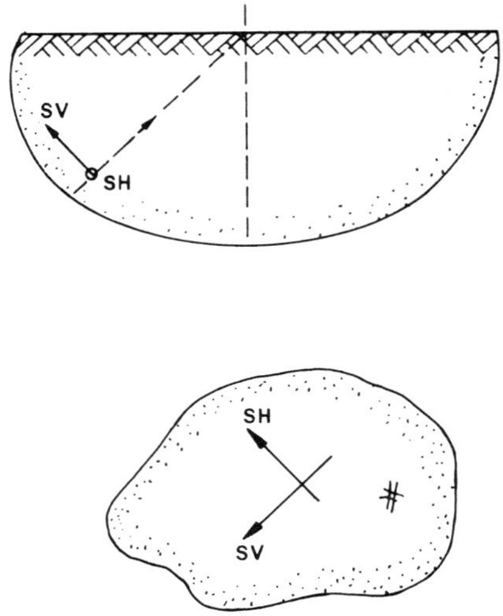

Figure 7.2

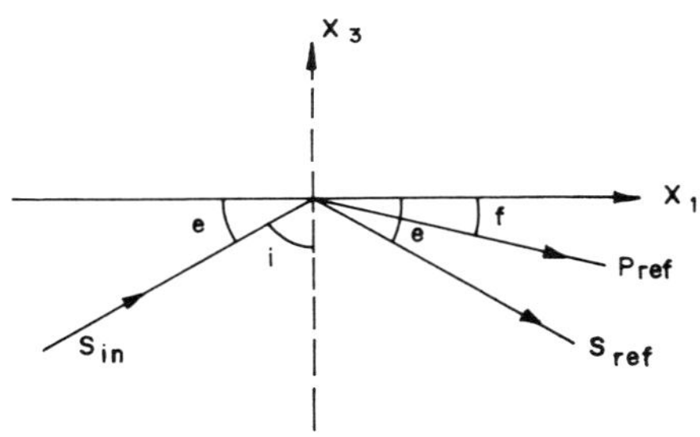

Figure 7.3

problems in unbounded media, is the necessity to truncate the media in order to establish a finite mesh. The artificial boundary so created produces waves that are reflected back from the boundaries. This reflection alters appreciably the results and this is why special devices, such as absorbing boundaries, have been developed in order to improve the accuracy of finite element solutions.

The field under study is very broad and the interested reader is recommended to study the classical work of Ewing, Jardtzeky and Press[5]. Here we shall study only the simple case of an SV wave meeting a horizontal boundary and its reflection giving a P as well as another SV wave.

In Figure 7.3, i is the incidence angle of the SVwave and according to Snell's law

$$\frac{\sin i}{c_S} = \frac{\cos e}{c_S} = \frac{\cos f}{c_P} \tag{7.27}$$

In order to represent the displacement, it is important to recognize that P and SV waves produce displacements with components in (x_1, x_3) while SH components act only along the x_2 axis. For this reason *three* potentials, φ, ψ and u_2 need to be defined so that

$$\begin{aligned}\ddot{\varphi} &= c_P^2 \nabla^2 \varphi \\ \ddot{\psi} &= c_s^2 \nabla^2 \psi \\ \ddot{u}_2 &= c_S^2 \nabla^2 u_2\end{aligned} \tag{7.28}$$

where $\psi = -\psi_2$ and the displacements are

$$\begin{aligned}u_1 &= \frac{\partial \varphi}{\partial x_1} + \frac{\partial \psi}{\partial x_3} \\ u_3 &= \frac{\partial \varphi}{\partial x_3} - \frac{\partial \psi}{\partial x_1} \\ u_2 &= u_2\end{aligned} \tag{7.29}$$

Calling

$$\begin{aligned}c &= c_S/\cos e \\ k &= k_\alpha \cos e \\ k' &= k \cos f \\ c' &= c_S/\cos f\end{aligned} \tag{7.30}$$

the potential of the incident wave

$$\psi = B_0 e^{ik_\alpha(x_1 \cos e + x_3 \sin e - c_s t)} \tag{7.31}$$

is transformed into

$$\psi = B_0 e^{ik(x_1 + x_3 \mathrm{tg} f - ct)} \tag{7.32}$$

The total potentials are then

$$\begin{aligned}\varphi &= A e^{ik'(x_1 - x_3 \mathrm{tg} e - ct)} \\ \psi &= B_0 e^{ik(x_1 + x_3 \mathrm{tg} f - ct)} + B e^{ik^1(x_1 - x_3 \mathrm{tg} f - c't)} \\ U_2 &= C e^{ik'(x_1 - x_3 \mathrm{tg} f - c't)}\end{aligned} \tag{7.33}$$

(By Snell's law, $k' = k$, $c' = c$).

To evaluate A, B, C we impose the boundary conditions ($x_3=0$),

$$\sigma_{33}=\sigma_{32}=\sigma_{31}=0 \tag{7.34}$$

From which one obtains $c=0$, no SH wave

$$\frac{A}{B_0}=\frac{4\text{tg}f(1+3\text{tg}^2 e)}{4\text{tg}e\text{tg}f+(1+3\text{tg}^2 e)^2}$$

$$\frac{B}{B_0}=\frac{4\text{tg}e\text{tg}f-(1+3\text{tg}^2 e)^2}{4\text{tg}e\text{tg}f+(1+3\text{tg}^2 e)} \tag{7.35}$$

As $c_S<c_P$, $e>f$, if $f=0$, it follows from Equation 7.27 that

$$e_c=\text{ arc cos}\frac{c_S}{c_P} \tag{7.36}$$

This is called the critical angle. The apparent velocity is

$$c=\frac{c_P}{\cos f}=\frac{c_S}{\cos e}$$

and

$$\text{tg}e=\left[\left(\frac{c}{c_S}\right)^2-1\right]^{1/2}, \quad \text{tg}f=\left[\left(\frac{c}{c_P}\right)^2-1\right]^{1/2}$$

If $f<f_c$ and $c_S<c<c_P$,

$$\text{tg}e=\left[\left(\frac{c}{c_S}\right)^2-1\right]^{1/2} \quad \text{tg}f=\left[1-\left(\frac{c}{c_P}\right)^2\right]^{1/2}$$

so that

$$B=B_0 e^{i2\delta}$$

$$\text{tg}\delta=\frac{4\left[1-\left(\frac{c}{c_P}\right)^2\right]^{1/2}\left[\left(\frac{c}{c_S}\right)^2-1\right]^{1/2}}{\left[2-\left(\frac{c}{c_S}\right)^2\right]^2}$$

that is the SV wave reflects as SV but *with a change of phase.*

If $\lambda=G$, we have a Poisson medium and $f_c=55°$. In this case the relationships in Equation 7.35 can be represented as shown in Figure 7.4.

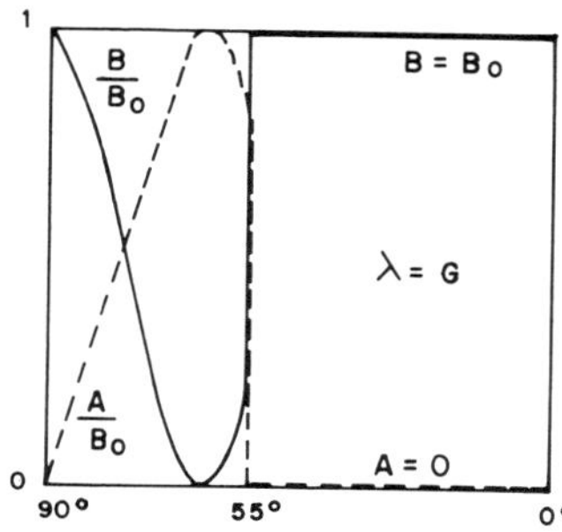

INCIDENT SV WAVE

Figure 7.4

7.3.3 **Surface waves**

Following the suggestions of seismologists, Rayleigh[6] tried to explain the existence of waves concentrated on a surface. Following the same scheme as before, he used equations 7.29 and the hypothesis

$$\begin{aligned} \varphi &= f(x_3)e^{ik(x_1 - ct)} \\ \psi &= g(x_3)e^{ik(x_1 - ct)} \\ u_2 &= h(x_3)e^{ik(x_1 - ct)} \end{aligned} \tag{7.37}$$

Substitution of (7.37) into (7.28) establishes conditions of the type

$$\frac{d^2 f}{dx_3^2} = -k^2\left(\frac{c^2}{\alpha^2} - 1\right)f$$

that is

$$f = Ae^{ikrx_3} + A'e^{ikrx_3}; \qquad r = \left[\left(\frac{c}{c_P}\right)^2 - 1\right]^{1/2}$$

and similarly for g and h.

When x_3 increases, f decreases, so that $A' = 0$ and

$$\begin{aligned} \varphi &= Ae^{ik(rx_3 + x_1 - ct)} \\ \psi &= Be^{ik(sx_3 + x_1 - ct)} \\ u_2 &= Ce^{ik(sx_3 + x_1 - ct)} \end{aligned}$$

$$r = \left[\left(\frac{c}{c_P}\right)^2 - 1\right]^{1/2}; \quad s = \left[\left(\frac{c}{c_S}\right)^2 - 1\right]^{1/2} \tag{7.38}$$

After establishing the boundary conditions, one obtains

$$
\begin{aligned}
C&=0\\
[c_P^2(1+r^2)-2c_S^2]A+2c_S^2 sB&=0\\
2rA+(1-s^2)B&=0
\end{aligned}
\tag{7.39}
$$

Hence there are no transversal components in the displacement and

$$
\begin{vmatrix} c_P^2(1+r^2)-2c_S^2 & 2sc_S^2 \\ 2r & 1-s^2 \end{vmatrix}=0 \tag{7.40}
$$

This is the Rayleigh equation which must be solved in order to obtain the value of c for a particular medium. If the medium is a Poisson one the result is

$$
c=0.919c_S \quad (\lambda=G) \tag{7.41}
$$

and for a surface point

$$
\begin{aligned}
u_1&=0.42Ak \sin k(x_1-ct)\\
u_3&=0.62Ak \cos k(x_1-ct)
\end{aligned}
\tag{7.42}
$$

The displacement is eliptic and counterclockwise. As seismologists discovered the existence of waves producing displacements orthogonal to the wave propagation direction, Love studied the possibility of obtaining them by studying wave propagation through a superficial layer. The problem is a bit more complicated than before and its development can be followed in the classical work of Love[7,8].

We can see how the dispersion phenomenon appears, as the phase velocity is a function of the wave number k. The dispersion equation is

$$
\frac{G\left[1-\left(\dfrac{c}{c_S}\right)^2\right]^{1/2}}{G'\left[\left(\dfrac{c}{c'_S}\right)^2-1\right]^{1/2}}=\operatorname{tg}kH\left[\left(\dfrac{c}{c'_S}\right)^2-1\right]^{1/2}
$$

where H is the layer depth, G' and c'_S are the layer characteristics and G, c_S are those of the underlying medium.

As can be seen, c is a function of k. There is then a complex path when a disturbance acts in a bounded medium. Figure 7.5 illustrates the waves generated by the disturbance acting at a hypocentre, of the type due to earthquakes.

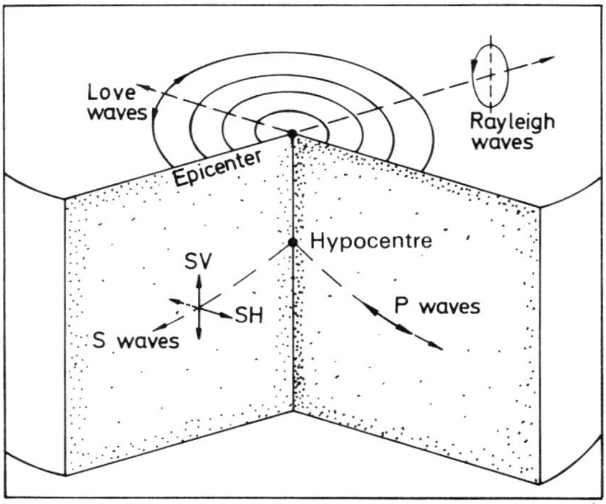

Figure 7.5

7.4 RETARDED POTENTIALS

It is useful, sometimes, to have a representation of wave potentials as a function of the so-called 'retarded potentials' which are the analogue to the Newtonian ones discussed in Chapter 3 on potential theory.

The wave equation in the case of spherical symmetry is

$$\frac{\partial^2 \varphi}{\partial t^2} = \frac{c^2}{r^2}\frac{\partial}{\partial r}\left(r^2 \frac{\partial \varphi}{\partial r}\right) \tag{7.43}$$

which admits the solution

$$\varphi = \frac{1}{r} f\left(t - \frac{r}{c}\right) \tag{7.44}$$

where f is an arbitrary function of the argument $\left(t - \frac{r}{c}\right)$. Corresponding to the volumetric Newtonian potential, it is possible to define

$$\varphi = \int_\Omega \frac{1}{r} f\left(t - \frac{r}{c}\right) \mathrm{d}\Omega \tag{7.45}$$

which can be interpreted as the Newtonian potential generated by a point of strength f but arriving after a time $\frac{r}{c}$, due to the propagation velocity c. Hence the name *retarded* potential.

Integrating round a sphere ε of radius ct it is possible to show that at an interior point

$$\varphi_P = \frac{1}{4\pi}\left[\frac{\partial}{\partial(ct)} ct \int_{\Gamma_\varepsilon} \varphi_{t=0}\, \mathrm{d}\bar{\omega} + t \int_{\Gamma_\varepsilon} \left(\frac{\partial \varphi}{\partial t}\right)_{t=0} \mathrm{d}\bar{\omega}\right], \quad \mathrm{d}\Gamma_\varepsilon = r^2\, \mathrm{d}\bar{\omega} \quad (7.46)$$

More information on this topic can be found, for example, in Carslaw and Jaeger[9].

7.5 TIME RESPONSE

The aim of this section is to represent the $\vec{u}$ solution in a similar fashiong to that in elastostatics. The representation is obtained in two separate ways: using the response to an impulse $P(t)$ acting in direction $\vec{a}$ at points

$$\vec{f}(\vec{x},t) = \vec{a}P(t)\delta|\vec{x} - \xi| \quad (7.47)$$

or using the response to a harmonic excitation $\vec{P}(\omega)$

$$\vec{f}(\vec{x},t) = \vec{P}(\omega)e^{i\omega t} \quad (7.48)$$

In both cases the solution is considered for the whole space and then applied to a particular surface.

The solution to Equation 7.48 can be used to obtain time responses by synthesizing the various frequency results using the Fourier transform. Both methods are based on the use of the Betti-Rayleigh dynamic reciprocal identity (proved, for instance, in Achenbach[2]) which establishes

$$\begin{aligned} &\int_\Gamma \overset{\vec{\nu}}{T} * \vec{u}'\, \mathrm{d}\Gamma + \rho \int_\Omega [\vec{f} * \vec{u}' + \vec{\mathring{u}}'\vec{u} + \vec{u}'\vec{\mathring{v}}]\, \mathrm{d}\Omega = \\ &= \int_\Gamma \overset{\vec{\nu}}{T}{}' * \vec{u}\, \mathrm{d}\Gamma + \rho \int_\Omega [\vec{f}' * \vec{u} + \vec{\mathring{u}}\vec{u}' + \vec{u}\vec{\mathring{v}}']\, \mathrm{d}\Omega \end{aligned} \quad (7.49)$$

where $\overset{\vec{\nu}}{T}, \vec{f}, \vec{u}$ are respectively the traction, body force and displacement vector of the dynamic state. The initial conditions are

$$\vec{u}(\vec{x},0) = \vec{\mathring{u}}; \quad \mathring{\vec{u}}(\vec{x},0^+) = \vec{\mathring{v}} \quad (7.50)$$

The primes $\overset{\vec{\nu}}{T}{}', \vec{f}', \vec{u}', \vec{\mathring{u}}', \vec{\mathring{v}}'$ are assigned to another state and the symbol $*$ represents the convolution product.

7.5.1 **Impulse response**

The response of an infinite medium to an impulse $P(t)$ located at the origin of the coordinate system, i.e.

$$\vec{f} = \vec{a}P(t)\delta(\vec{x}) \tag{7.51}$$

can be obtained using the Helmholtz representation of (7.51). If

$$\vec{f} = \text{grad } F + \text{rot } \vec{F}^* \tag{7.52}$$

Equation 7.14 can be written

$$\text{grad } [(\lambda + 2G)\nabla^2 \chi - \rho\ddot{\varphi} + F] = \text{rot } [-G\nabla^2 \vec{\psi} + \rho\ddot{\vec{\psi}} - \vec{F}^*] \tag{7.53}$$

and hence

$$\begin{aligned} \nabla^2 \varphi - \frac{1}{c_P^2}\ddot{\varphi} &= -\frac{1}{c_P^2}F \\ \nabla^2 \vec{\psi} - \frac{1}{c_s^2}\ddot{\vec{\psi}} &= -\frac{1}{c_s^2}\vec{F}^* \end{aligned} \tag{7.54}$$

It can be shown that

$$\begin{aligned} F &= -P(t) \text{ div} \frac{\vec{a}}{4\pi r} \\ \vec{F}^* &= P(t) \text{ rot} \frac{\vec{a}}{4\pi r} \end{aligned} \tag{7.55}$$

so that after the necessary substitutions we obtain

$$\begin{aligned} \varphi &= \text{div } \vec{\Phi}; \quad \vec{\Phi} = \varphi\vec{a} \\ \vec{\Psi} &= -\text{rot } \vec{\Psi}; \quad \vec{\Psi} = \Psi\vec{a} \end{aligned} \tag{7.56}$$

Equations (7.54) reduce to the pair

$$\begin{aligned} \nabla^2 \Phi - \frac{1}{c_P^2}\ddot{\Phi} &= \frac{P(t)}{c_P^2}\frac{1}{4\pi r} \\ \nabla^2 \Psi - \frac{1}{c_S^2}\ddot{\Psi} &= \frac{P(t)}{c_S^2}\frac{1}{4\pi r} \end{aligned} \tag{7.57}$$

which admit as solutions

$$4\pi\Phi(\vec{x},t)=\frac{1}{r}\int_{0}^{t-r/c_P}\tau P\left[t-\left(\frac{r}{c_P}+\tau\right)\right]\mathrm{d}\tau-\frac{1}{r}\int_{0}^{t}\tau P(t-\tau)\,\mathrm{d}\tau \tag{7.58}$$

$$4\pi\Psi(\vec{x},t)=\frac{1}{r}\int_{0}^{t-r/c_S}\tau P\left[t-\left(\frac{r}{c_S}+\tau\right)\right]\mathrm{d}\tau-\frac{1}{r}\int_{0}^{t}\tau P(t-\tau)\,\mathrm{d}\tau \tag{7.59}$$

Using Equation 7.10, i.e. $\vec{u}=\text{grad }\varphi+\text{rot }\vec{\psi}$ and Equation 7.56, we obtain

$$\vec{u}=\nabla^2\vec{\Phi}-\text{rot rot }\vec{\Psi}=\nabla^2(\vec{\varphi}-\vec{\Psi})+\nabla^2\Psi \tag{7.60}$$

Achenbach uses the notation,

$$4\pi u_i(\vec{x},t)=D_i^k[\vec{0},P(t)] \tag{7.61}$$

to express Equation 7.60 through 7.58 and 7.59 when the load is directed along the x_k axis ($\vec{a}=\vec{e}_k$). After the substitutions have been done we obtain,

$$D_i^k[\vec{0},P(t)]=\left(\frac{3x_ix_k}{r^3}-\frac{\delta_{ik}}{r}\right)\int_{1/c_p}^{1/c_s}\tau P(t-\tau r)\,\mathrm{d}\tau+$$
$$+\frac{x_ix_k}{r^3}\left[\frac{1}{c_P^2}P\left(t-\frac{r}{c_P}\right)-\frac{1}{c_S^2}P\left(t-\frac{r}{c_S}\right)\right]+\frac{\delta_{ik}1}{c_S^2r}P\left(t-\frac{r}{c_S}\right) \tag{7.62}$$

The corresponding stresses can be obtained with the use of Hooke's law, and are represented by

$$4\pi\sigma_{ij}(\vec{x},t)=S_{ij}^k[\vec{0},P(t)] \tag{7.63}$$

where

$$\rho^{-1}S_{ij}^k[\vec{0},P(t)]=-6c_S^2\left[\frac{5x_ix_jx_k}{r^5}-\frac{\delta_{ij}x_k+\delta_{ik}x_j+\delta_{jk}x_i}{r^3}\right]\int_{1/c_p}^{1/c_s}\tau P(t-r\tau)\,\mathrm{d}\tau+$$
$$+2\left[\frac{6x_ix_jx_k}{r^5}-\frac{\delta_{ij}x_k+\delta_{ik}x_j+\delta_{jk}x_i}{r^3}\right]\left[P\left(t-\frac{r}{c_S}\right)-\left(\frac{c_P}{c_S}\right)^2P\left(t-\frac{r}{c_P}\right)\right]+$$
$$+2\frac{x_ix_jx_k}{r^4c_S}\left[\dot{P}\left(t-\frac{r}{c_S}\right)-\left(\frac{c_S}{c_P}\right)^3\dot{P}\left(t-\frac{r}{c_P}\right)\right]-$$

$$-\frac{x_k\delta_{ij}}{r^3}\left[1-2\left(\frac{c_S}{c_P}\right)^2\right]\left[P\left(t-\frac{r}{c_P}\right)+\frac{r}{c_P}\dot{P}\left(t-\frac{r}{c_P}\right)\right]-$$

$$-\frac{\delta_{ik}x_j+\delta_{jk}x_i}{r^3}\left[P\left(t-\frac{r}{c_S}\right)+\frac{r}{c_S}\dot{P}\left(t-\frac{r}{c_S}\right)\right] \qquad (7.64)$$

If the point of application of the impulse is at $\vec{\xi}$, we have

$$\begin{aligned} 4\pi u_i &= D_i^k[\vec{\xi},\vec{P}(t)] \\ 4\pi\sigma_{ij} &= S_{ij}^k[\vec{\xi},\vec{P}(t)] \\ r &= |\vec{x}-\vec{\xi}| \\ \vec{f} &= \vec{a}P(t)\delta(\vec{x}-\vec{\xi}) \end{aligned} \qquad (7.65)$$

When these formulae are substituted in the reciprocal relationship, Equation 7.49, we obtain the Wheeler and Sternberg formula, that is

$$4\pi\rho u_k(\vec{r},t)=\sum_{i=1}^{3}\int_{\Gamma}\{D_i^k[\vec{\xi},\overset{v}{T}_i(\vec{x},t)]-v_jS_{ij}^k[\vec{\xi},u_i(\vec{x},t)]\}\ \mathrm{d}\Gamma+$$

$$+\rho\sum_{i=1}^{3}\int_{\Omega}\{D_i^k[\vec{\xi},\vec{f}_i(x,t)]\}\ \mathrm{d}\Omega \qquad (7.66)$$

This formula is valid for a body which starts at rest. Otherwise it is necessary to add another integral containing the influence of the initial values. In vector form Equation 7.66 now becomes,

$$4\pi\rho\vec{u}(\vec{r},t)=\int_{t_0=0}^{\infty}\int_{\Gamma}[\vec{D}\overset{\vec{v}}{T}(\xi,t_0)-\vec{u}(\xi,t_0)\vec{S}\vec{v}]\ \mathrm{d}\Gamma_0\ \mathrm{d}t_0+$$

$$+\int_{t_0=0}^{\infty}\int_{\Omega}\vec{D}\vec{f}(\vec{\xi},t_0)\ \mathrm{d}\Omega_0\ \mathrm{d}t_0+ \qquad (7.67)$$

$$+\int_{\Omega}\rho\left[\vec{D}\frac{\partial}{\partial t_0}\vec{u}(\vec{\xi},t_0)-\vec{u}\frac{\partial\vec{D}}{\partial t_0}\right]\Bigg|_{t_0=0}\mathrm{d}\Omega_0$$

The last integral represents the effect of the initial values.

It is important to notice the following properties of $\vec{D}$ (Cole *et al.*[10]):

(1) Causality

$$D_{ij}(\vec{x},t,\vec{\xi},t_0)=0, \text{ if } c_s(t-t_0)<r \tag{7.68}$$

(2) Time translation

$$D_{ij}(\vec{x},t+t_1,\vec{\xi},t_0+t_1)=D_{ij}(\vec{x},t,\vec{\xi},t_0) \tag{7.69}$$

At the limit Equation 7.67 can be used to represent the displacements on the boundary and the equation is then a dynamic counterpart of the static Somigliana identity.

The formula is then

$$4\pi\rho\vec{c}\vec{u}(\vec{r},t) = \int_{t_0=0}^{\infty}\int_{\Gamma}[\vec{D}\overset{\vec{v}}{T}-\vec{u}\vec{s}\vec{v}]\mathrm{d}\Gamma_0\,\mathrm{d}t_0+\vec{R}(\vec{r},t), \quad \vec{r}\in\Gamma \tag{7.70}$$

where $\vec{c}$ is a matrix depending on the local characteristics of Γ at $\vec{x}$, and $\vec{R}$ represents the two last integrals of Equation 7.67. The limit procedure is well documented in Cole *et al.*[10].

7.5.2 Harmonic response

It is possible to transform the equations of the problem by using the Fourier integral transformation, for example. As is well known, if

$$\varphi^*(\omega)=\mathscr{F}\varphi=\int_{-\infty}^{\infty}\varphi(t)e^{-i\omega t}\,\mathrm{d}t$$

and

$$\varphi(t)=\mathscr{F}^{-1}\varphi^*=\frac{1}{2\pi}\int_{-\infty}^{\infty}\varphi^*(\omega)e^{i\omega t}\,\mathrm{d}\omega \tag{7.71}$$

integrating by parts it is possible to show that

$$\mathscr{F}\frac{\mathrm{d}^2\varphi}{\mathrm{d}t^2}=(i\omega)^2\varphi^*(\omega) \tag{7.72}$$

When this procedure is applied to Equation 7.1, one obtains

$$(\lambda+G)u^*_{i,ij}+Gu^*_{j,ii}+X^*_j=-\rho\omega^2u^*_j \tag{7.73}$$

or

$$(c_P^2 - c_S^2)u^*_{i,ij} + c_2^2 u^*_{j,ii} + \frac{X_j^*}{\rho} + \omega^2 u_j^* \qquad (7.74)$$

The definition of the tractions becomes

$$\overset{\vec{v}}{T_i^*} = \sigma_{ij}^* v_j \qquad (7.75)$$

and the dynamic reciprocity relationship, Equation 7.49, is reduced to a form completely analogous to the static form, because the convolution products in time-space are transformed to normal ones in the frequency domain. Thus, if initial conditions are zero,

$$\int_\Gamma \overset{\vec{v}}{T}{}^* \vec{u}'^* \, d\Gamma + \rho \int_\Omega \vec{f}^* \vec{u}'^* \, d\Omega = \int_\Gamma \overset{\vec{v}}{T}{}'^* \vec{u}^* \, d\Gamma + \rho \int_\Omega \vec{f}'^* \vec{u}^* \, d\Omega \qquad (7.76)$$

By the use of a special solution designated by a prime (′) it is possible to establish the Somigliana identity. In order to do this we can use the singular Kelvin type solution[11]. Writing the components along the direction of the unit vectors $\vec{e}_i$, it is possible to find the response to a unit load, i.e.

$$U'_{ij} = \frac{1}{\alpha\pi\rho c_s^2}[\psi\delta_{ij} - \chi v_{,i} v_{ij}]$$

$$T'_{ij} = \frac{1}{\alpha\pi}\left\{\left(\frac{\partial\psi}{\partial r} - \frac{1}{r}\chi\right)\left(\delta_{ij}\frac{\partial r}{\partial v} + r_{ij}v_i\right) - \right.$$

$$-\frac{2}{r}\chi\left(v_j r_{ij} - 2r_{,i}r_{,j}\frac{\partial r}{\partial v}\right) - 2\frac{d\chi}{dr}r_{,i}r_{,j}\frac{\partial r}{\partial v} \qquad (7.77)$$

$$\left. + \left[\left(\frac{c_P}{c_S}\right)^2 - 2\right]\left(\frac{d\psi}{dr} - \frac{d\chi}{dr} - \frac{\alpha}{2r}\chi\right) r_{,i} v_j\right\}$$

where for 2-dimensional problems, $\alpha = 2$ and the functions ψ and χ are defined as

$$\psi = K_0\left(\frac{i\omega r}{c_S}\right) + \frac{c_S}{i\omega r}\left[K_1\left(\frac{i\omega r}{c_S}\right) + \frac{c_S}{c_P}K_1\left(\frac{i\omega r}{c_P}\right)\right]$$

$$\chi = K_2\left(\frac{i\omega r}{c_S}\right) - \left(\frac{c_S}{c_P}\right)^2 K_2\left(\frac{i\omega r}{c_1}\right) \qquad (7.78)$$

and K_i is the modified Bessel function of the second kind and order i.

For 3-dimensional problems, $\alpha=4$ and the functions ψ and χ are

$$\psi=\left(1-\frac{c_S^2}{\omega^2r^2}+\frac{c_S}{i\omega r}\right)\frac{e^{-\frac{i\omega r}{c_S}}}{r}-\left(\frac{c_S^2}{c_P^2}\right)\left(-\frac{c_P^2}{\omega^2r^2}+\frac{c_P}{i\omega r}\right)\frac{e^{-\frac{i\omega r}{c_P}}}{r} \tag{7.79}$$

$$\chi=\left(-\frac{3c_S^2}{\omega^2r^2}+\frac{3c_S}{i\omega r}+1\right)\frac{e^{-\frac{i\omega r}{c_P}}}{r}-\left(\frac{c_S^2}{c_P^2}\right)\left(-\frac{3c_P^2}{\omega^2r^2}+\frac{3c_P}{i\omega r}+1\right)\frac{e^{-\frac{i\omega r}{c_P}}}{r} \tag{7.80}$$

Using these solutions, the Somigliana identity is, as usual

$$u_j^*(P)=-\int_\Gamma u_i^*T'_{ji}\,d\Gamma+\int_\Gamma \overset{v}{T}{}_i^*U'_{ij}\,d\Gamma+\int_\Gamma X_i^*U'_{ji}\,d\Omega \tag{7.81}$$

where p is the point where the Kelvin load is applied.

Referring to Equation 7.71 the solution can be interpreted as the response to a harmonic component of the forcing function, so that using Equation 7.71 we can find the total response. In general this can only be done in a numerical way. Hence, the emergence of the Fast Fourier Transformation (FFT) algorithm, has been of great help in practical applications.

7.6 DISCRETIZATION

In what follows, we shall try to present the discretization of Equations 7.70 and 7.81 using the BEM. In the case of harmonic response the procedure is the same as that in elastostatics but for an arbitrary time-domain excitation, the solution is generally obtained using a time-stepping technique.

7.6.1 Time response

The discretization of Equation 7.70 can be done in different ways. The simplest way is to assume the spatial values of the field variables are constants on selected parts of the boundary (the elements), just as it was done in potential theory or elastostatics. Simultaneously, these values are considered as constants along a conveniently chosen time step.

An interesting example has been presented by Cole *et al.*[10]. They take J nodes on the boundary and a set $\{t_n=n\Delta t,\ n=1,2,\ldots N\}$ of equally spaced time steps and, as an approximation,

$$\begin{aligned} u(\vec{r},t)&=\sum_j\sum_n U_j^n(\vec{r},t)u_j^n \\ \overset{v}{T}{}_j^n(\vec{r},t)&=\sum_j\sum_n T_j^n(\vec{r},t)\overset{v}{T}{}_j^n \end{aligned} \tag{7.82}$$

with the usual assumptions

$$u(\vec{r}_i,t_m)=u_i^m$$
$$\overset{v}{T}(\vec{r}_i,t_m)=\overset{v}{T}{}_i^m \tag{7.83}$$

Substitution in Equation 7.70 produces the following system:

$$4\pi\rho c_j u_j^m=\sum_{m=1}^{N}\sum_{j=1}^{J}\{G_{ji}^{nm}\overset{v}{T}{}_i^m+\mathbf{K}_{ji}^{nm}u_i^m\}+F(r_j,t_n)$$
$$j=1,2,\ldots J; \quad n=1,2,\ldots N \tag{7.84}$$

where

$$G_{ji}^{nm}=\int_0^\infty\int_\Gamma \vec{D}(\vec{r}_j,t_n\vec{\xi}t_0)\overset{v}{T}_i(\vec{\xi},t_0)\ \mathrm{d}\Gamma_0\ \mathrm{d}t_0$$
$$\mathrm{K}_{ji}^{nm}=\int_0^\infty\int_\Gamma \vec{n}\nabla_0\vec{D}(\vec{r}_j t_n\vec{\xi}t_0)U_i^m(\vec{\xi},t_0)\ \mathrm{d}\Gamma_0\ \mathrm{d}t_0 \tag{7.85}$$

Equation 7.84 is solved for the unknown coefficients by setting $n=1$ then we put $n=2$ and repeat the solution, progressing in this manner along the time axis. As Cole *et al.*[10] point out there are only NJ^2 kernels to be calculated, due to the properties of Equations 7.68 and 7.69. More information and an interesting example can be found in that reference.

7.6.2 Steady-state solution

In the previous sections, the integral equation that relates displacements and tractions over the boundary of a region has been obtained. The Equation 7.81 (Somigliana's identity) may be written in the form

$$u_j(p)=\int_\Gamma U_{ji}t_i\ \mathrm{d}\Gamma-\int_\Gamma T_{ji}u_i\ \mathrm{d}\Gamma$$

where all the * super-indexes have been suppressed for simplicity and the symbol $\overset{v}{T}{}_i^*$ for the boundary traction has been changed for the more common t_i since the variable t for time is not going to be used further.

When the point p is taken to the boundary, a singularity appears in the kernels of the integrals. Removing the singularities, the previous equation transforms into

$$c_p u_j(p) = \int_\Gamma U_{ji} t_i \, \mathrm{d}\Gamma - \int_\Gamma T_{ij} u_i \, \mathrm{d}\Gamma \tag{7.86}$$

where c_p is a constant $0 \leqslant c_p \leqslant 1$ ($c_p = 0.5$ for smooth boundary at p). Equation (7.86) may be written in the form:

$$c_p u_p = \int_\Gamma U t \, \mathrm{d}\Gamma - \int_\Gamma T u \, \mathrm{d}\Gamma \tag{7.87}$$

Application of boundary conditions to this equation will produce an integral equation of the first kind in the case of motion prescribed over the whole boundary, an integral equation of the second kind in case of tractions prescribed and a mixed type of equation when motions are prescribed on certain parts of the boundary and tractions over the rest.

Solution of an integral equation like the one above in a systematic way is only possible by numerical procedures. To do so, the boundary will be assumed to be formed by a number of elements (Figure 7.6). The geometry of these elements will be linear (a straight line for two dimensions and a plane for three), quadratic, cubic, etc. For each direction, at any point of the boundary, either the motion u or the traction t is unknown. In order to reduce the numer of unknowns to a finite number, the motions and tractions over the boundary will be assumed to be represented by their values at a number of 'nodes'. The motions and tractions at points different from the nodes will be obtained by polynomial interpolation among the nodal values. According to the order of the interpolation function are the number of nodes that are needed in each element. Thus, in two-dimensional problems where the boundary elements are lines, when one node per element is used the tractions and motions will be constant along the elements. A linear variation will be obtained, with two nodes; with three, the variation will be quadratic, and so on. For three-dimensional problems, the possible shapes of the elements are more complex. For instance, if elements with a triangular shape are chosne, a constant value of the variables will be obtained with one node per element, a linear variation with three nodes, a quadratic variation with six nodes, etc. In Figure 7.7., a few types of elements are represented.

The motions and tractions over the elements are written in terms of their nodal values using polynomial shape-functions. Thus,

$$\begin{aligned} u_i &= N^1 U_i^1 + N^2 U_i^2 + \ldots + N^M U_i^M = N^J U_i^J \\ t_i &= N^1 t_i^1 + N^2 t_i^2 + \ldots + N^M t_i^M = N^J U_i^J \end{aligned} \tag{7.88}$$

where U_i and t_i are motion and traction in the 'i' direction for a point within

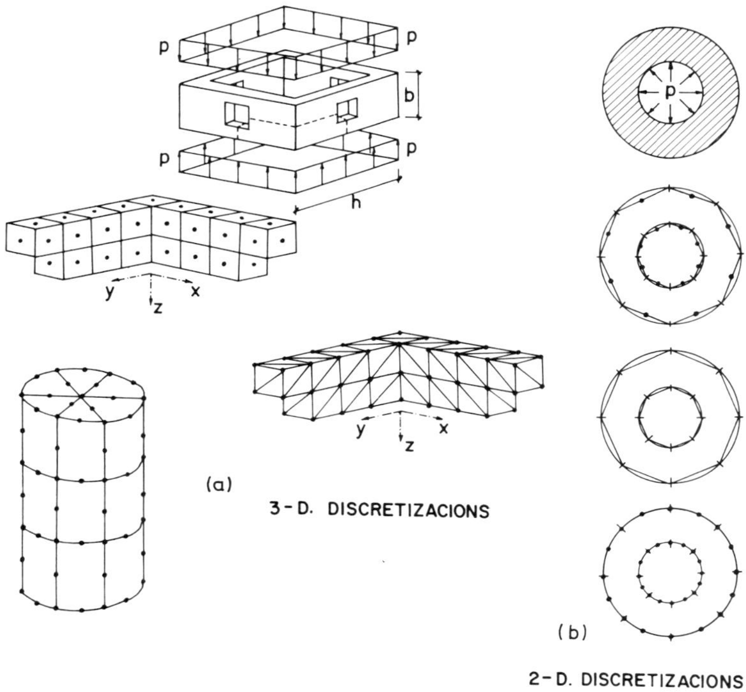

Figure 7.6 Boundary discretization: (a) 3-dimensional, (b) 2-dimensional

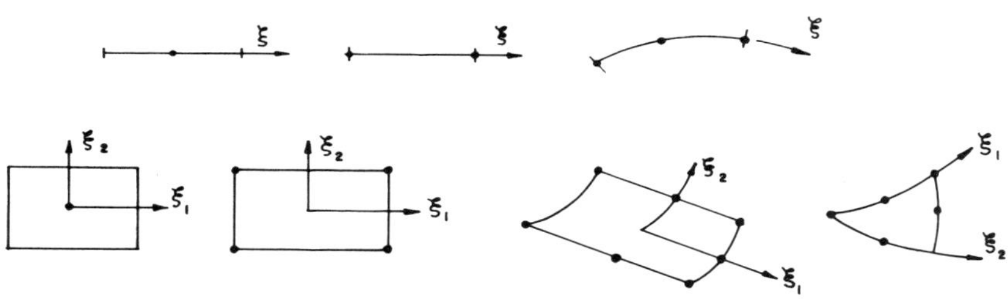

Figure 7.7 Boundary elements

the element. N^J is the shape function corresponding to the J node and U_i^J and t_i^J are motion and traction in the 'i' direction for the J node (in what follows a capital index denotes mode and a small inkex denotes direction).

The same polynomial functions are normally used to generate the geometry of the elements from the coordinates of the nodal points.

$$x_i = N^J x_i^J \tag{7.89}$$

Taking into account the boundary discretization and functions representation, the Somigliana's identity can be written for a node k:

$$\vec{c}_k\vec{u}_k+\sum_{\substack{n=1\\ \text{elements}}}^{N}\left[\int_{\Gamma}\vec{T}\vec{N}^n\,\mathrm{d}\Gamma_n\right]\vec{u}^n=\sum_{\substack{n=1\\ \text{elements}}}^{N}\left[\int_{\Gamma}\vec{U}\vec{N}^n\,\mathrm{d}\Gamma_n\right]\vec{t}^n \tag{7.90}$$

where, for the 3-D case

$$\vec{N}^n=\left[\begin{array}{ccc|ccc|c|ccc} N^1 & 0 & 0 & N^2 & 0 & 0 & & N^M & 0 & 0\\ 0 & N^1 & 0 & 0 & N^2 & 0 & \cdots & 0 & N^M & 0\\ 0 & 0 & N^1 & 0 & 0 & N^2 & \cdots & 0 & 0 & N^M \end{array}\right]$$

where M = number of nodes per element, and

$$\begin{aligned}\vec{u}^n&=(u_1^1u_2^1u_3^1u_1^2u_2^2u_3^2\ldots u_1^Mu_2^Mu_3^M)^T\\ \vec{t}^n&=(t_1^1t_2^1t_3^1t_1^2t_2^2t_3^2\ldots t_1^Mt_2^Mt_3^M)^T\end{aligned} \tag{7.91}$$

In intrinsic coordinates

$$\mathrm{d}\Gamma_n=|J|\,\mathrm{d}\xi \tag{7.92}$$

where $|J|$ is the Jacobian. Thus, the integral equation can be written:

$$\vec{c}_K\vec{u}_K+\sum_{\substack{n=1\\ \text{elements}}}^{N}\left[\int_{\Gamma_n}\vec{T}\vec{N}^m|J|\,\mathrm{d}\xi\right]\vec{u}^n=\sum_{\substack{n=1\\ \text{elements}}}^{N}\left[\int_{\Gamma}\vec{U}\vec{N}^m|J|\,\mathrm{d}\xi\right]\vec{t}^n \tag{7.93}$$

The integrals can be carried out numerically or analytically over the surface Γ_n of the elements, being a function not only of the surface on which the integration takes place but also of the point k under consideration.

The last equation can be written as a sum for all the boundary nodes.

$$\vec{c}_K\vec{u}_K+\sum_{\text{nodes}}\vec{H}^*_{KJ}\vec{u}_J=\sum_{\text{nodes}}\vec{G}_{KJ}\vec{t}_J \tag{7.94}$$

where $\vec{u}_J$ and $\vec{t}_J$ are motion and traction vectors for node J. $\vec{H}^*_{KJ}$ is the sum of all the elements to which the node belongs, for the integrals

$$\int\vec{T}\vec{N}^L|J|\,\mathrm{d}\xi$$

where $\vec{N}^L=\vec{I}N^L$ and L is the position occupied by J within the element n. The same can be said for $\vec{G}_{KJ}$ and the integrals

$$\int_{\Gamma_n} \vec{U}\vec{N}^L|J|\ \mathrm{d}\xi \tag{7.95}$$

Making

$$\begin{aligned} \vec{H}_{KJ} &= \vec{H}^*_{KJ}, && \text{if } K \neq J \\ \vec{H}_{KJ} &= \vec{H}^*_{KJ} + \vec{c}, && \text{if } K = J \end{aligned} \tag{7.96}$$

the equation for node k will be:

$$\sum_{\text{nodes}} \vec{H}_{KJ}\vec{u}_J = \sum_{\text{nodes}} \vec{G}_{KJ}\vec{t}_J \tag{7.97}$$

And writing the equations for all nodes in a compact form:

$$\vec{H}\vec{u} = \vec{G}\vec{t} \tag{7.98}$$

where $\vec{u}$ and $\vec{t}$ are vectors that include the displacements and tractions of all the boundary nodes. $\vec{H}$ and $\vec{G}$ are square matrices formed from the terms $\vec{H}$ and $\vec{G}$, their total dimension being three times (two in the case of 2-D problems) the number of nodes. For any boundary node and for each direction, either t or u is known; consequently there are as many equations as unknowns.

7.7 SOME COMPUTATIONAL PROBLEMS

7.7.1 **Two-dimensional constant elements**

If a constant value of $\vec{u}$ and $\vec{t}$ is considered over each element, the terms $\vec{H}_{KJ}$ and $\vec{G}_{KJ}$ that relate the nodes K and J will be:

$$\vec{H}_{KJ} = \int_{\Gamma_J} \vec{T}\,\mathrm{d}s \quad \text{and} \quad \vec{G}_{KJ} = \int_{\Gamma_J} \vec{U}\ \mathrm{d}s \tag{7.99}$$

where $\mathrm{d}s = \mathrm{d}\Gamma_J$. Each one of these terms has four components which can be written in the form:

$$H_{ij}(K,J) = \int_{\Gamma_J} T_{ij}(K,J)\ \mathrm{d}s \quad \text{and} \quad G_{ij}(K,J) = \int_{\Gamma_J} U_{ij}(K,J)\ \mathrm{d}s \tag{7.100}$$

The terms T_{ij} and U_{ij} depend on the modified Bessel functions of complex argument $K_0(z)$, $K_1(z)$ and $K_2(z)$ and their derivatives with respect to r. The variable z will be either $z=i\omega r/c_S$ or $z=i\omega r/c_P$, in all cases r being the distance from node K to the points along the integration segment J.

The modified Bessel functions $K_0(z)$, $K_1(z)$ and $K_2(z)$ are functions that have a singularity and consequently their integration presents some problems when $r \to 0$. They also have certain recurrence relations that are useful for differentiation and integration, namely:

$$
\begin{aligned}
&K_{v-1}(z)-K_{v+1}(z)=\frac{2v}{z}K_v(z)\\
&K_v^1(z)=K_{v-1}(z)-\frac{v}{z}K_v(z)\\
&K_{v-1}(z)+K_{v+1}(z)=2K_v^1(z)\\
&K_v^1(z)=K_{v+1}(z)+\frac{v}{z}K_v(z)
\end{aligned}
\tag{7.101}
$$

These relations taken from Olver[12] permit us to obtain all $K_v(z)$ and $K_v'(z)$ in terms of $K_0(z)$ and $K_1(z)$, these two being the only ones that have to be integrated. When the element J is different from K, the functions $K_0(z)$ and $K_1(z)$ will not reach the vicinity of their singular point and a standard numerical integration scheme will be adequate. The values of $K_0(z)$ and $K_1(z)$ at the integration point will be obtained by the series expansions:

$$
\begin{aligned}
K_0(z)=&-\left(\ln\frac{z}{2}+\gamma\right)+\frac{\left(\frac{1}{4}z^2\right)}{(1!)^2}\left(1-\ln\frac{z}{2}-v\right)+\frac{\left(\frac{1}{4}z^2\right)^2}{(2!)^2}\left(1+\frac{1}{2}-\ln\frac{z}{2}-\gamma\right)+\\
&+\ldots+\frac{\left(\frac{1}{4}z^2\right)^n}{(n!)^2}\left(1+\frac{1}{2}+\ldots+\frac{1}{n}-\ln\frac{z}{2}-\gamma\right)+\ldots
\end{aligned}
\tag{7.102}
$$

$$
\begin{aligned}
K_1(z)=&\frac{1}{z}+\frac{1}{2}z\left\{\left(\ln\frac{z}{2}+\gamma-\frac{1}{2}\right)+\left(\ln\frac{z}{2}+\gamma-1-\frac{1}{2\times 2}\right)\frac{\left(\frac{1}{4}z^2\right)}{1!\,2!}+\right.\\
&\left.+\ldots+\left(\ln\frac{z}{2}+\gamma-1-\frac{1}{2}-\ldots-\frac{1}{n}-\frac{1}{2(n+1)}\right)\frac{\left(\frac{1}{4}z^2\right)^n}{n!(n+1)!}+\ldots\right\}
\end{aligned}
\tag{7.103}
$$

γ being a constant.

When T_{ij} and U_{ij} have to be integrated over the element where the

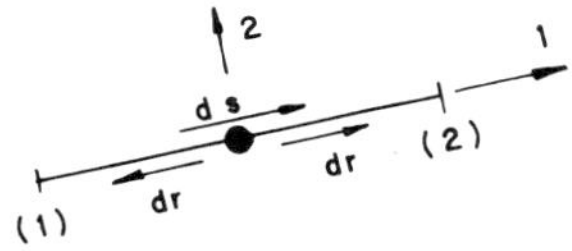

$r_{,2} = 0$

$\eta_{,1} = 0$

Figure 7.8

concentrated load is applied, an integration of $K_0(z)$ and $K_1(z)$ in the vicinity of the singular point will be necessary in order to obtain the terms H_{JJ} and G_{JJ}. For simplicity H_{JJ} and G_{JJ} are obtained in a system of local axes (Figure 7.8). It can be seen that all $H_{ij}=0$; either because all the terms of the kernel are zero (when $i=j$) or because the integrals on both sides of the node compensate.

It can be shown very easily that $G_{12}=G_{21}=0$ and consequently, the only terms that require treatment are G_{11} and G_{22}. Using the relation

$$K_\nu(z)=\frac{z}{\nu}[K_{\nu-1}(z)-K_\nu^1(z)] \tag{7.104}$$

the integrals of all the modified Bessel functions will be reduced to values of those functions at $r=L/2$ and to the integrals

$$\int_{(1)}^{(2)} K_0\left(\frac{i\omega r}{c_S}\right) \mathrm{d}S \text{ and } \int_{(1)}^{(2)} K_0\left(\frac{i\omega r}{c_P}\right) \mathrm{d}S \tag{7.105}$$

transformed as follows:

$$\int_{(1)}^{(2)} K_0\left(\frac{i\omega r}{c}\right) \mathrm{d}S = -2\int_0^R K_0\left(\frac{i\omega r}{c}\right) \mathrm{d}r = -2\frac{c}{i\omega}\int_0^Z K_0(z)\, \mathrm{d}z \tag{7.106}$$

with $z=\omega/c$ and $Z=\omega(L/2)/c$.

The value of the last integral can be calculated from its series expansion which is obtained by integration of the K_0 series expansion.

$$\int_0^Z K_0(z)\,\mathrm{d}z = Z\left\{\left(1-\gamma-\ln\frac{Z}{2}\right)+\frac{\frac{1}{4}Z^2}{(1!)^2}\frac{1}{3}\left(1+\frac{1}{3}-\gamma-\ln\frac{Z}{2}\right)+\right.$$
$$+\frac{\left(\frac{1}{4}Z^2\right)^2}{(2!)^2}\frac{1}{5}\left(1+\frac{1}{2}+\frac{1}{5}-\gamma-\ln\frac{Z}{2}\right)+\ldots \tag{7.107}$$
$$\left.+\frac{\left(\frac{1}{4}Z^2\right)^n}{(n!)^2}\frac{1}{2n+1}\left(1+\frac{1}{2}+\ldots+\frac{1}{n}+\frac{1}{2n+1}-\gamma-\ln\frac{Z}{2}\right)+\ldots\right\}$$

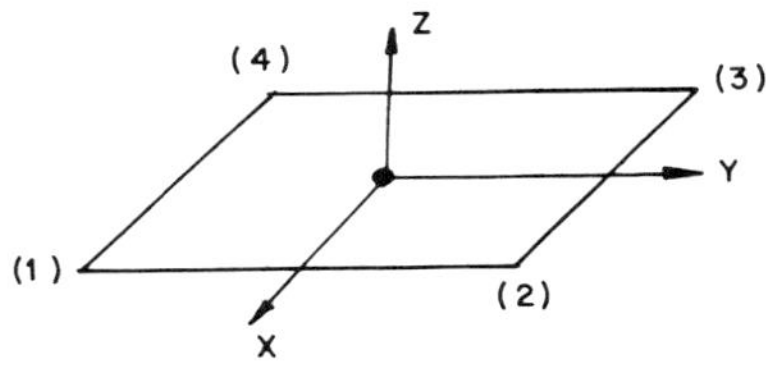

Figure 7.9 3-dimensional constant elements

Once the $\vec{H}$ and $\vec{G}$ matrices are known, the boundary problem is solved and any integral point can be computed doing the corresponding numerical integrations over the boundary elements.

7.7.2 Three-dimensional constant elements

One type of element for three-dimensional problems that do not present many difficulties is the rectangular constant element (Figure 7.9). It has only one node and the tractions and displacements are constant throughout the element. Because of the increase in one dimension, the computation of the $\vec{H}$ and $\vec{G}$ matrices is greater but on the other hand T_{ij} and U_{ij} are now formed by exponential functions that are easier to treat than the modified Bessel functions.

As usual, the fundamental solution can be integrated over a certain element J using a standard two-dimensional numerical quadrature; except when the concentrated load is applied at the node J. When the diagonal terms $\vec{H}_{JJ}$ and $\vec{G}_{JJ}$ are computed, the integration includes the vicinity of the singularity and some manipulation will be needed to obtain a reliable value for the integrals.

The integration over the element will be done using the local axes of Figure 7.9. Considering the rectangular shape of the elements, it can be shown that all terms H_{ij} are zero; either because T_{ij} is zero over the element or because it takes positive and negative values. The same happens to all terms G_{ij} when $i \neq j$. The rest of the terms are:

$$G_{11}=\frac{1}{4\pi\rho c_S^2}\left[\int\int\psi \, dx \, dy-\int\int\chi r_{,1}r_{,1} \, dx \, dy\right]$$

$$G_{22}=\frac{1}{4\pi\rho c_S^2}\left[\int\int\psi \, dxdy-\int\int\chi r_{,2}r_{,2} \, dx \, dy\right] \tag{7.108}$$

$$G_{33}=\frac{1}{4\pi\rho c_S^2}\int\int\psi \, dx \, dy$$

The second integral of G_{11} and G_{22} can be computed analytically, at least in one direction. Consequently, those reduce to a line integral along two opposite sides of the rectangle. Because the singularity is not close to the sides, a numerical integration formula is used for the integrals along the sides.

The integral $\int\int\psi \, dx \, dy$ can be computed by means of a series expansion of the exponential terms. By integrating the terms of the series analytically, the series corresponding to the integral is obtained.

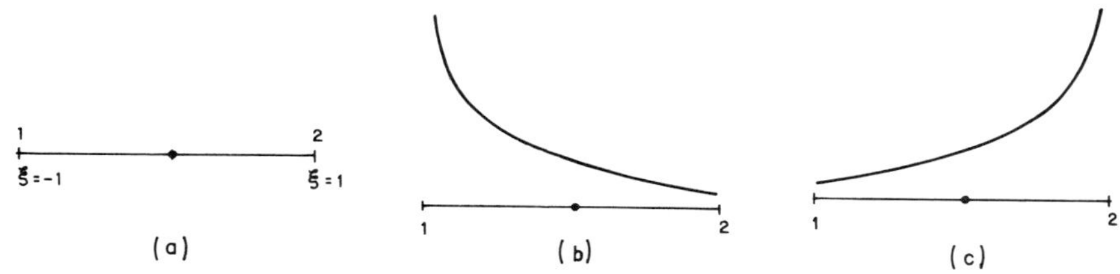

Figure 7.10 Singular element

7.7.3 **Two-dimensional singular elements**

The boundary element method is suitable for problems that include a singularity. In order to get a good representation of the behaviour near the singularity with a small number of elements a special type of element may be used. The element includes a singularity at one of the extremes using a logarithmic shape function for the tractions.

In Figure 7.10(a) an element with a single node and intrinsic coordinate ξ is represented. The tractions along the elements are given by:

$$t_i = \left[1 + \ln\left(\frac{1}{1+\xi}\right)\right] t_i^1 \tag{7.109}$$

when the singularity is located at extreme 1 (Figure 7.10(b)) and by

$$t_i = \left[1 + \ln\left(\frac{1}{1-\xi}\right)\right] t_i^1 \tag{7.110}$$

when the singularity is located at extreme 2 (Figure 7.10(c)), t_i^1 being constant along the element and associated with the node. In a general form:

$$t_i = N t_i^1 \tag{7.111}$$

The terms of the matrix $\vec{G}$ corresponding to integrations over the singular elements will be of the form:

$$G_{ij}(K,J) = \int_{\Gamma_J} U_{ij}(K,J) N \, \mathrm{d}S \tag{7.112}$$

The problem is now how to obtain the values of these integrals that in the general case have a singularity at one of the extreme points. In addition, when the concentrated load is applied to the node J, there is not only the singularity of the extreme point but also that at the middle.

In the case that $K \neq J$, the integral can be easily transformed into the sum of two integrals. One with no singularity and the other which corresponds to the type:

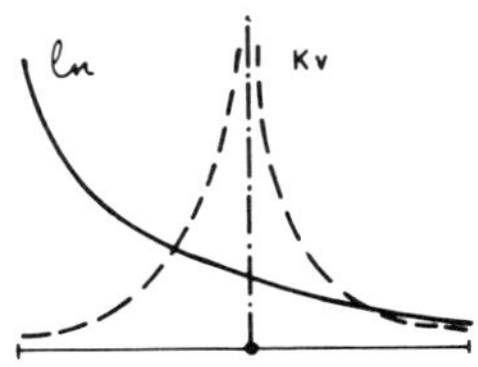

Figure 7.11

Figure 7.12 Strip foundations

$$\int_0^1 \ln\left(\frac{1}{x}\right) U_{ij}\,\mathrm{d}x \tag{7.113}$$

The first integral can be evaluated by means of a normal Gauss quadrature formula mean, while the second can be computed using a special type of numerical integration formula, given for this kind of function by Stroud and Secrest[13], and also shown in Brebbia[14].

When $K = J$, the problem becomes somewhat more complicated because the function to be integrated is the product of two functions like the ones in Figure 7.11. However, with some manipulation, it is always possible to express the product as the sum of certain functions that have either a close form integral or can be numerically computed.

7.8 TWO-DIMENSIONAL APPLICATIONS

In the following, two applications of the method for dynamic problems are presented. The first is related to the dynamic soil-structure interaction problem. This field has acquired a considerable importance in civil engineering due to the increasing demands for reliability in modern structures, in particular power plants. Part of the process consists of computing the dynamic stiffness of foundations, that is, the forces that are needed to produce a unit harmonic motion in a massless foundation on the surface or embedded in a halfspace (Figure 7.12). The halfspace will be considered to be linear elastic but it is very easy to stimulate a viscoelastic Voigt-type media by introducing a complex value of the shear modulus. This is equivalent to assuming a hysteretic damping model.

Due to the fact that the fundamental solution corresponds to the complete space, the boundary to the discretized will be the soil-foundation interface and

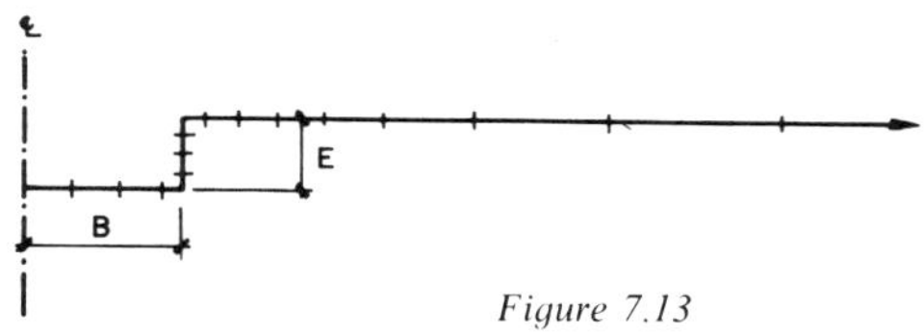

Figure 7.13

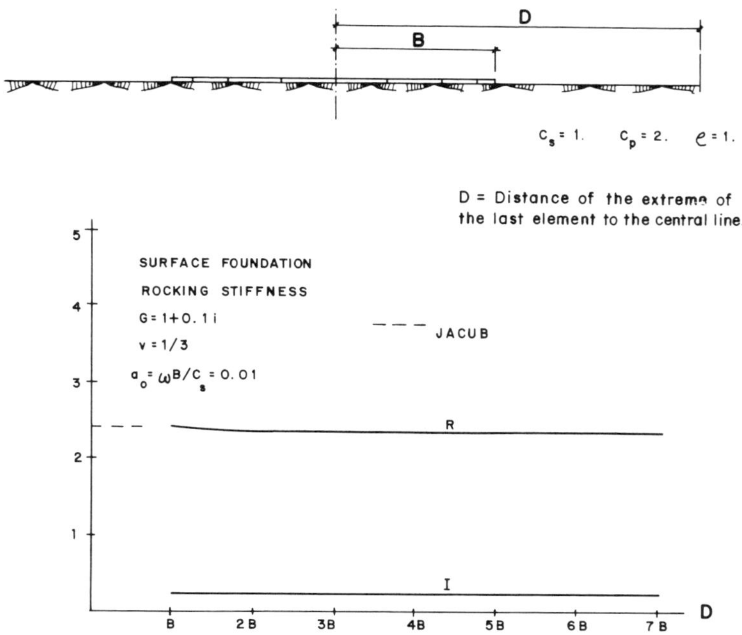

Figure 7.14

the free surface (Figure 7.13). However, it can be shown that only a small amount of free-field around the foundation has to be discretized in order to obtain accurate results.

In the case that the foundation is resting on the surface, a good solution can be obtained without any free field element. It can be shown that in this case, all terms that represent the influence of the elements outside the interface are zero except those responsible for the influence of the vertical motion on the horizontal, and viceversa. However, this influence is small and is usually neglected in soil structure interaction when 'smooth footing' conditions are assumed.

In Figure 7.14, the value of the rocking stiffness for a surface foundation is shown in terms of the amount of free field discretized for a very low dimensionless frequency.

The soil-foundation interface was discretized into eight elements and as expected, the influence of the extent of the free field discretized is negligible.

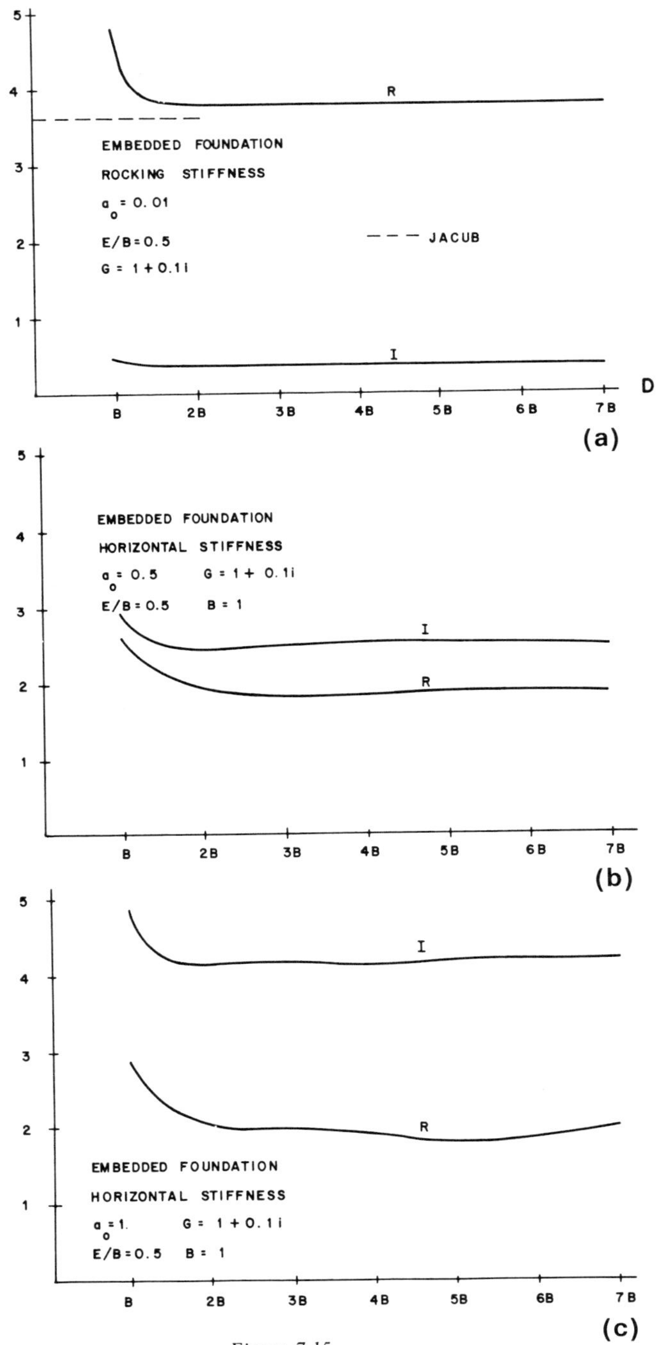

R
EMBEDDED FOUNDATION
ROCKING STIFFNESS
a_o = 0.01
E/B = 0.5
G = 1 + 0.1i
– – – JACUB
I
D
B 2B 3B 4B 5B 6B 7B
(a)
EMBEDDED FOUNDATION
HORIZONTAL STIFFNESS
a_o = 0.5 G = 1 + 0.1i
E/B = 0.5 B = 1
I
R
(b)
I
R
EMBEDDED FOUNDATION
HORIZONTAL STIFFNESS
a_o = 1. G = 1 + 0.1i
E/B = 0.5 B = 1
(c)

Figure 7.15

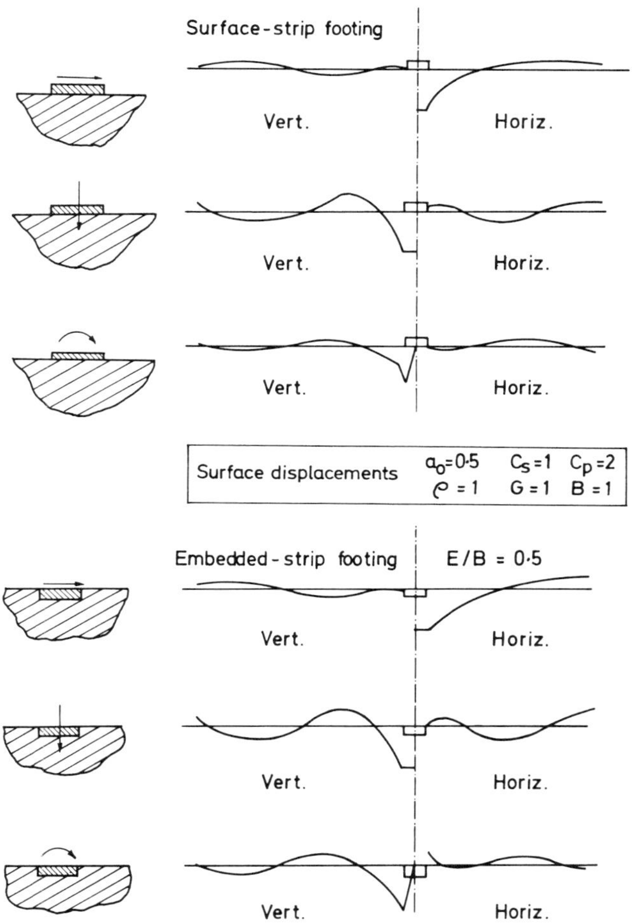

Figure 7.16

The same is true for higher frequencies. In Figure 7.14 the result is compared with that of Jakub[15].

In Figure 7.15, the influence of the amount of discretized free-field is shown for an embedded foundation. First, the rocking stiffness is represented for the same case as Figure 7.14 but with an embedment ratio E/B = 0.5. Here, the extent of free field discretized has an important influence until it reaches $D = 2B$. In Figures 7.15(b) and 7.15(c) the same curves are plotted for the horizontal stiffness and two values of the dimensionless frequency a_0. The amount of free field that has to be discretized to obtain good results increases with the embedment ratio.

Figure 7.16 shows the amplitude of the free-surface motion when a unit harmonic motion of frequency $a_0 = 0.5$ is applied to a surface or embedded

Table 7.1

8 Elements	*4 Elements*	
2.189 + 0.453*i*	1.965 + 0.387*i*	No singular elements
2.27 + 0.478*i*	2.249 + 0.472*i*	Two singular elements

$\omega = 0.5$ rad/s, $G = 1 + 0.1i$

massless foundation. In the work by Dominguez and Alarcon[16] the reader can find plots of the variation of the stiffness components with frequency for foundations on the surface or embedded in an elastic halfspace with no damping.

In a previous section the logarithmic singular elements were presented. For the cases under consideration, where a motion is prescribed to the rigid foundation, stress singularities will appear at the extremes of the footing. Two singular elements at those extremes will permit not only a better representation of the stress distribution under the footing but also good results for the stiffness components with a small number of boundary elements: Table 7.1 presents some of the values obtained using singular elements compared to the case when normal constant elements were used. It can be seen that with eight elements both results are close but when the number is reduced to four, the value of the vertical stiffness changes more if no singular elements are used. The variation of the normal stress under the foundation is presented in Figure 7.17 for the case of applied unit harmonic vertical motion.

The second application of dynamic boundary elements that will be shown corresponds to a finite body. A square plate under in-plane dynamic loads and different boundary conditions. The dimensions and material properties of the plate are as shown in Figure 7.18. Three problems were treated:

(1) The plate under a uniform harmonic traction over two opposite sides and no possible transverse motion (Figure 7.19(a)).

(2) The plate under a uniform harmonic traction over two opposite sides and free on the other sides (Figure 7.19(b)).

(3) The plate with the same conditions as (2) but with a crack transverse to the loading as shown in Figure 7.19(c).

The discretization was the same in all cases in order to use the same $\vec{G}$ and $\vec{H}$ matrices. Symmetry was taken into account and only a quarter of the plate was discretized with 35 constant boundary elements (Figure 7.19(d)).

Due to the boundary conditions of problem (1), only *P* waves were expected to appear. All points with a horizontal line should have the same motion and the same stresses. The results obtained agree very well with the expected. In Figure 7.20(a) the motion of the lower and upper boundaries are represented versus frequency. Figure 7.20(b) shows the normal stresses at points of the horizontal central line. Both Figures 7.20(a) and (b) show the typical behaviour of a single degree of freedom system with a natural frequency at $\omega \simeq 56500$ rad/s. According to the properties of the material, the *P* wave velocity was computed to be $c_P = 5397.17$ m/s. Resonance should appear when the plate length equals half a wave length, i.e.

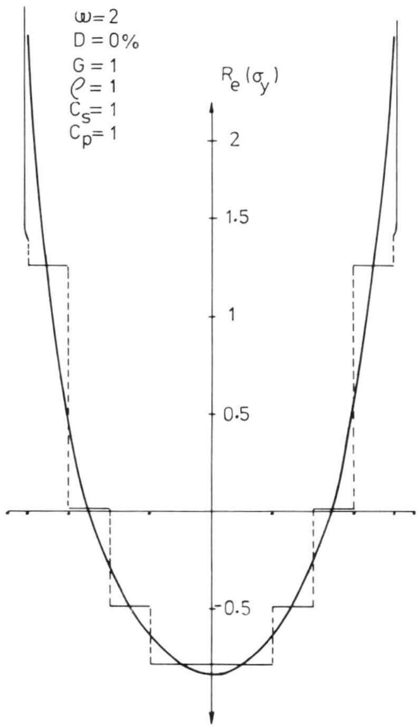

Figure 7.17 Normal stress under footing

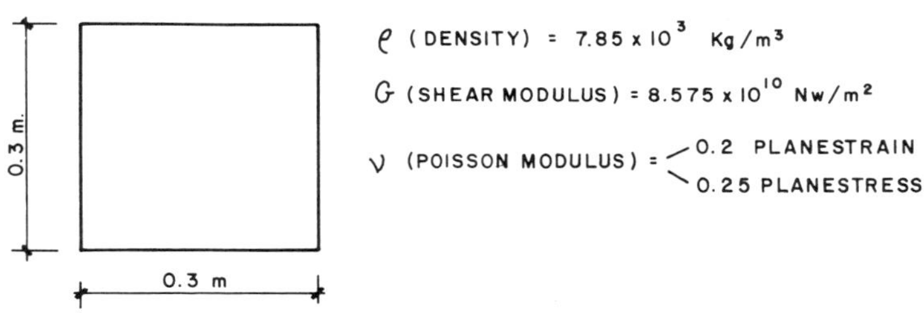

Figure 7.18 Plate properties

$$0.3 = \lambda/2 = \frac{\pi}{\omega} c_P$$

$$\omega = \frac{c_P \pi}{0.3} = 56519 \text{ rad/s} \tag{7.114}$$

which agrees very well with the value in Figure 7.20.

Problem (2) presents a more complicated behaviour due to the possibility of transverse motion and consequently the appearance of other kinds of waves. Figure 7.21(a) shows the modulus and argument of the vertical motion of point

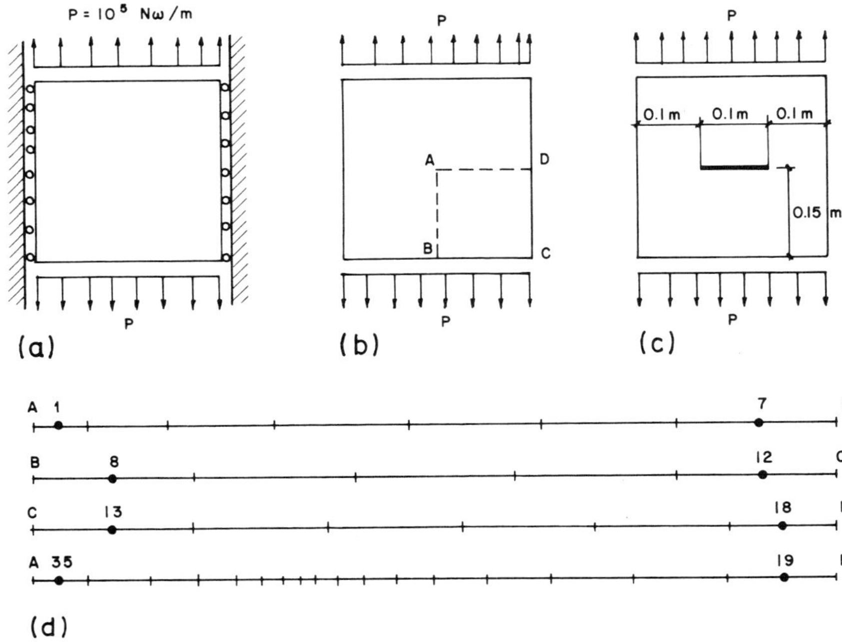

Figure 7.19 Problem definition and discretization

$P1$, which corresponds to node 7 of the discretization, versus frequency ω. In Figure 7.21(b), the vertical traction over the central point (node 35) is represented versus ω.

Problem (3) in addition to the conditions of (2) presents a transverse crack. Figure 7.22 shows the motion of point $P2$ (node 35) versus frequency. Modulus and argument are shown. Figure 7.23 shows the motion of the lower and upper boundaries. Again a regular behaviour is observed for low and high frequencies and large differences among points for frequencies between 35 000 and 65 000 rad/s. In Figure 7.24(a), the motion of the crack lips are shown for a number of frequencies, while in Figure 7.24(b) the normal vertical stress for the mid-horizontal line is shown; again both figures show irregular behaviour for the mid-range of frequencies.

7.9 THREE-DIMENSIONAL APPLICATIONS

In the same way that the boundary element method is applied to two-dimensional soil-structure interaction problems, it can be extended to three-dimensional problems. However, it is in this kind of problem where the advantages of the boundary element method over domain methods really come through. The cost of a finite element analysis for a three-dimensional problem, including a halfspace, can be very high. Because of this, some other numerical methods that work only with the soil-foundation interface have

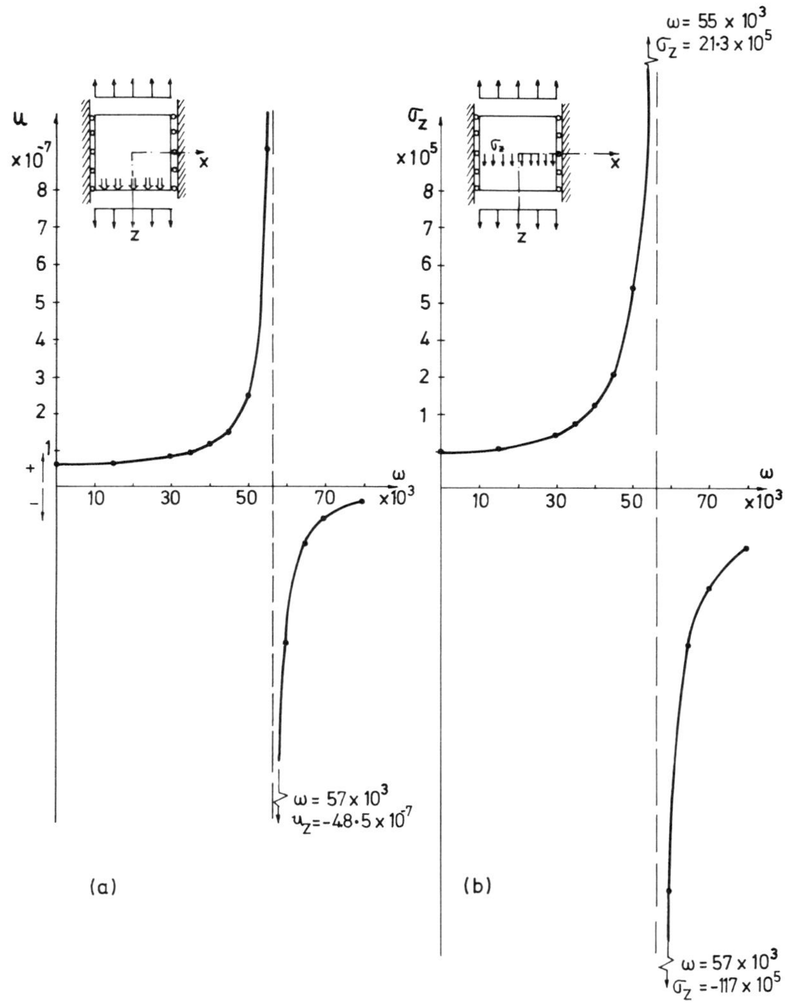

Figure 7.20

been developed. These methods (Wong and Luco[17]) usually work with Lamb's solution[18] and are only suitable for surface foundations. In the following, some results for surface and embedded foundations will be shown.

Dynamic foundation stiffnesses were computed using constand rectangular boundary elements. In Figure 7.25(a) and (b) two discretizations are shown, one with sixty-four boundary elements for a surface square foundation and the other with thrity-seven elements, some of them on the soil-foundation interface and the others on the free field. The first of these two discretizations is more than enough for this problem unless the frequency a_0 is very high.

The soil is considered to be linear elastic and again, there is a possible

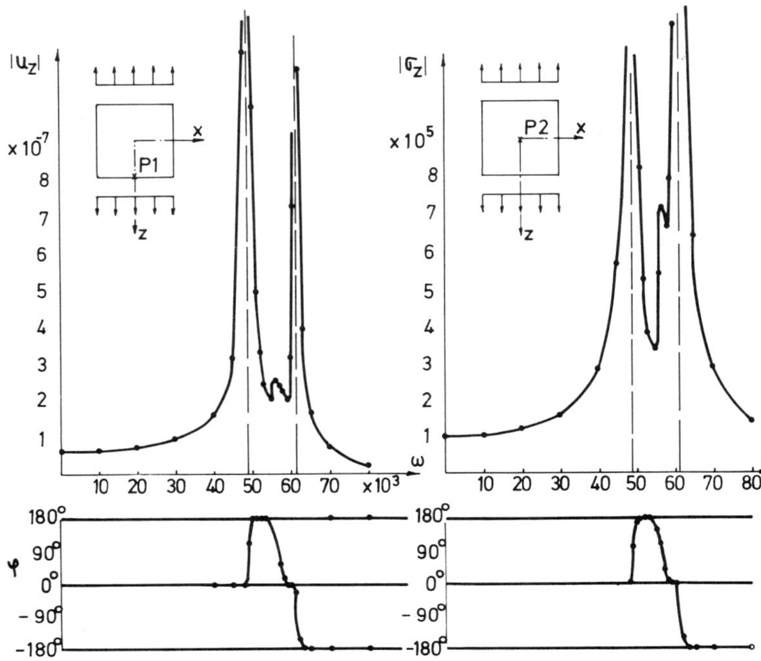

Figure 7.21

hypothesis that simplifies the computation. If the two horizontal motions are considered to be coupled but independent of the vertical one, it can be easily shown that for the case of surface foundations, no free field elements are needed because these elements do not influence the result. In the case of embedded foundation, free field elements are needed but good results can be obtained with not many of them.

To study the variation of the stiffness components with frequency, they are expressed in the form recommended by Kausel[20].

$$K_{ij} = K_{ij}^{0}(k_{ij} + i a_0 c_{ij}) \qquad (7.115)$$

where

K_{ij} is the static value

k_{ij} and c_{ij} are frequency-dependent coefficients

$a_0 = \omega\ B/C_s$ is a dimensionless frequency

B is the half-side for square foundations and half of the smallest side for rectangular footings

C_S is the shear wave velocity of the soil.

In Figures 7.26 and 7.27, the horizontal and rocking dynamic stiffness coefficients are shown versus a_0. Results obtained with 'relaxed' (x and y independent of z) and 'non-relaxed' boundary conditions are compared and also compared with the equivalent square foundation obtained from the

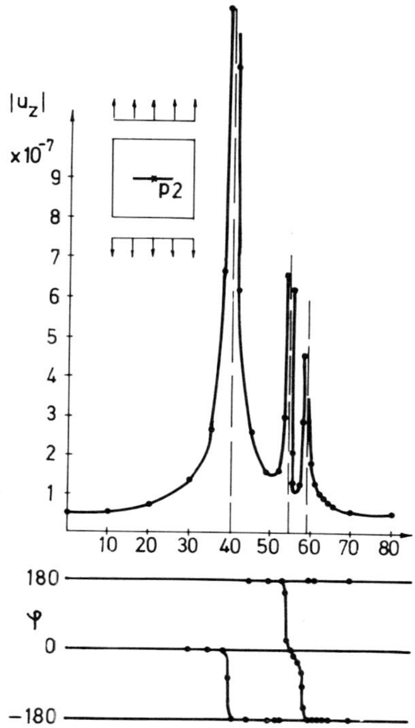

Figure 7.22

results for circular footing given by Veletsos and Wei[21] the static values obtained for these two stiffness components are:

$$K^0_{xx}/GB = 5.61$$
$$K^0_{\phi\phi}/GB^3 = 5.69 \qquad (7.116)$$

Dynamic stiffness for embedded square and rectangular foundations were also computed assuming relaxed and non-relaxed boundary conditions. Table 7.2 shows the values of the dynamic stiffness components obtained for an embedded square foundation with an embedment ratio $E/B = 2/3$ and a dimensionless frequency $a_0 = 0.9$. Results are compared when relaxed and non-relaxed boundary conditions are used and also for two different free-field discretizations, one as shown in Figure 7.25(b) and the other with 40 free field elements covering a square surface the side of which is seven times the side of the foundation. The soil-foundation interface was discretized in all cases, as shown in Figure 7.25(b). Taking into account the amount of uncertainty that is involved in soil behaviour, any of the models in Table 7.2 can be considered to be useful for normal practice and consequently, the relaxed boundary conditions with a line of free-field elements around the foundation may be used.

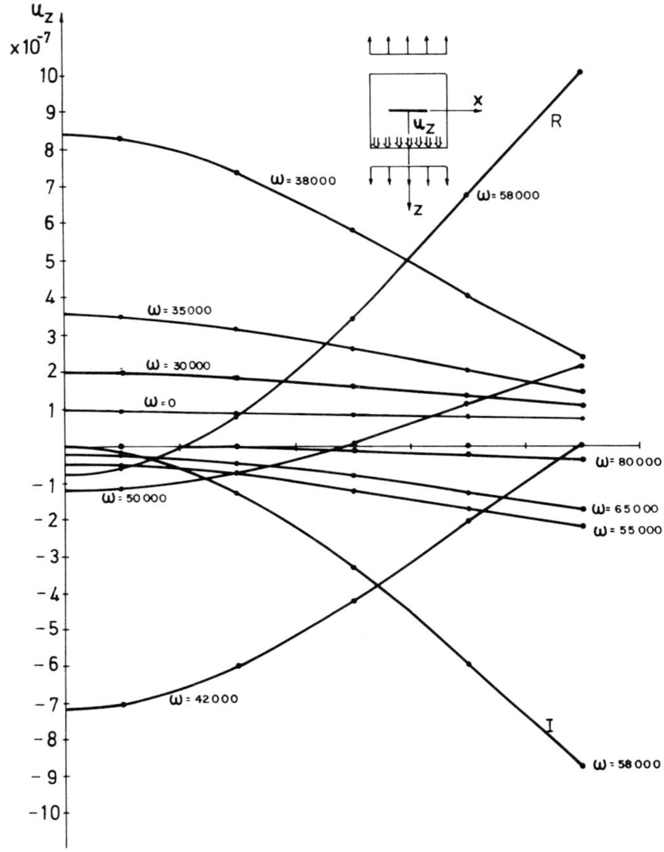

Figure 7.23

When relaxed boundary conditions are prescribed (coordinates x and y are considered to be independent of z), the rocking stiffness for embedded foundations must be obtained in two steps; one considering the vertical component of the motion and the other doing the same for the horizontal component of the motion of the side walls.

Figure 7.28 shows the static stiffness components versus the embedment ratio, taking the value for the surface foundations equal to one.

In Figure 7.29, the horizontal stiffness coefficients for several embedment ratios are shown. The results were obtained using relaxed boundary conditions.

The reader can find a parametric study of dynamic stiffness components for square and rectangular embedded foundations in the work by Dominguez[22]. Results obtained using boundary elements are presented there for foundations with different aspect and embedment ratios.

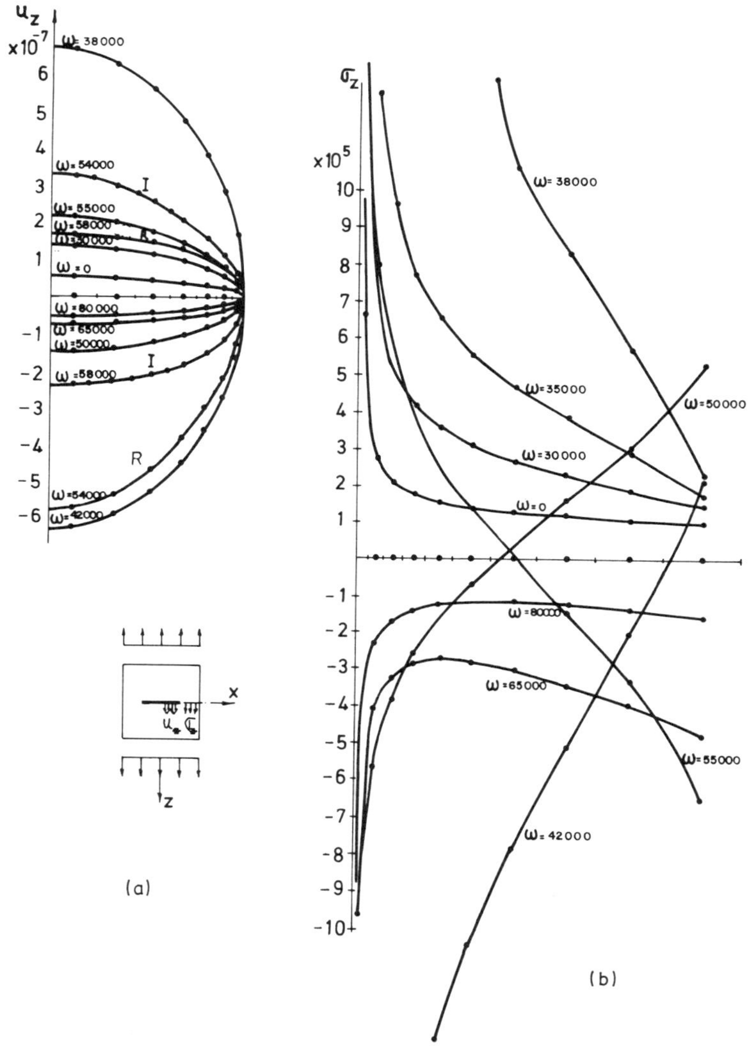

Figure 7.24

Another application of the method in the soil structure interaction field will be briefly discussed in what follows.

It is often assumed in seismic analysis of structures that the base motion is applied uniformly at all points of the foundation. This assumption would be correct only for shear or dilational waves propagating vertically through the soil and in the case of a surface foundation. The effect of inclined (or travelling) waves on the foundation motion gives rise to a difference in phase between the motions of the points under the foundation. As a result torsional and rocking components of the motion appear on surface and embedded foundations.

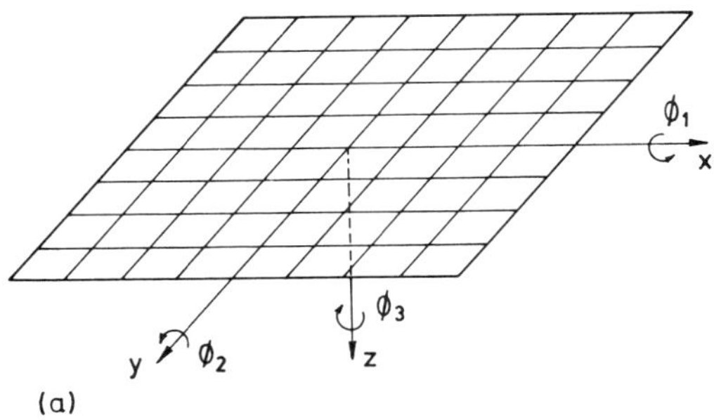

Figure 7.25(a) 8 × 8 element discretization and system of coordinates

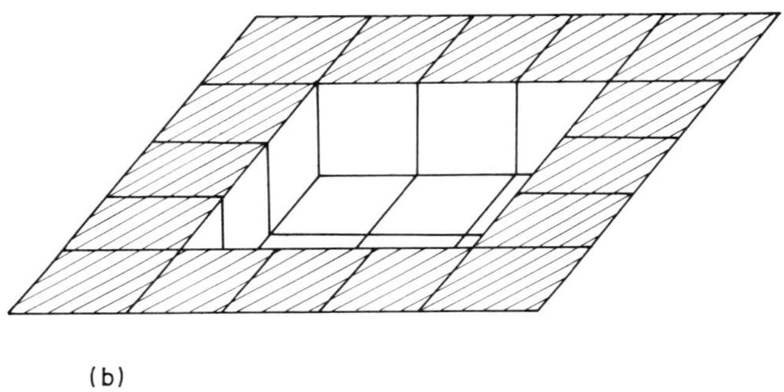

Figure 7.25(b) Model for embedded foundation

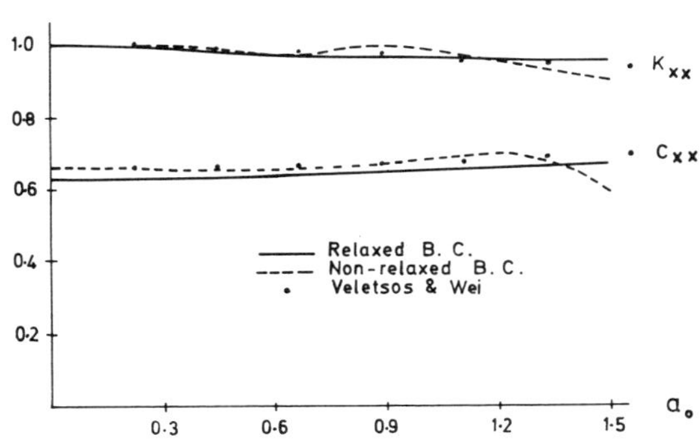

Figure 7.26

For embedded foundations, when the waves propagate vertically, there is not only horizontal and vertical motion of the foundations but also a rocking motion which is a consequence of the difference of motion that the points along the vertical axis of the foundation would have in free field conditions.

The problem may be treated using the boundary element method. The soil-structure interface and the free-field will be discretized in the way it is done

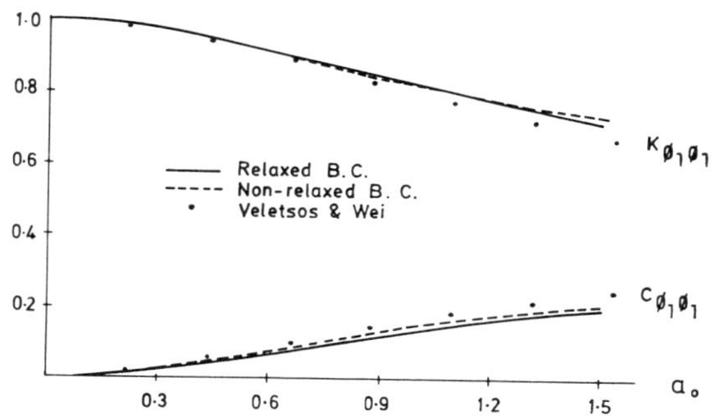

Figure 7.27

Table 7.2 EMBEDDED SQUARE FOUNDATION; E/B = 2/3, $a_0 = 0.9$

	Relaxed b.c. 16 free-field elements	*Non-relaxed b.c. 16 free-field elements*	*Non-relaxed b.c. 40 free-field elements*
$Re[K_{xx}]/GB$	9.772	9.841	9.627
$Im[K_{xx}]/GB$	8.879	8.922	9.606
$Re[K_{zz}]/GB$	8.650	7.930	8.367
$Im[K_{zz}]/GB$	10.023	9.740	10.573
$Re[K_{\varphi_1\varphi_1}]/GB^3$	8.987	9.146	9.378
$Im[K_{\varphi_1\varphi_1}]/GB^3$	2.811	3.105	3.428
$Re[K_{\varphi_3\varphi_3}]/GB^3$	17.47	17.452	17.289
$Im[K_{\varphi_3\varphi_3}]/GB^3$	3.895	3.882	4.325

when foundation stiffnesses are computed. The system of equations will be written for the relative motions and tractions with respect to the free-field values:

$$\vec{G}(\vec{t}-\vec{t}_{FF})=\vec{H}(\vec{u}-\vec{u}_{FF}) \tag{7.117}$$

where $\vec{t}_{FF}$ and $\vec{u}_{FF}$ are the free-field values of the variables at the nodal points.

Since the foundation is rigid, all nodal motions can be written in terms of six degrees-of-freedom.

$$\underset{(n\times 1)}{\vec{u}} = \underset{(n\times 6)}{\mathbf{L}}\underset{(6\times 1)}{\vec{u}^*} \tag{7.118}$$

On the other hand, the resultant of tractions under the foundation may be written as

$$\underset{(6\times 1)}{\vec{F}^*} = \underset{(6\times n)}{\mathbf{M}}\underset{(n\times 1)}{\vec{t}} \tag{7.119}$$

Introducing the last two equations into Equation 7.17 an equilibrium

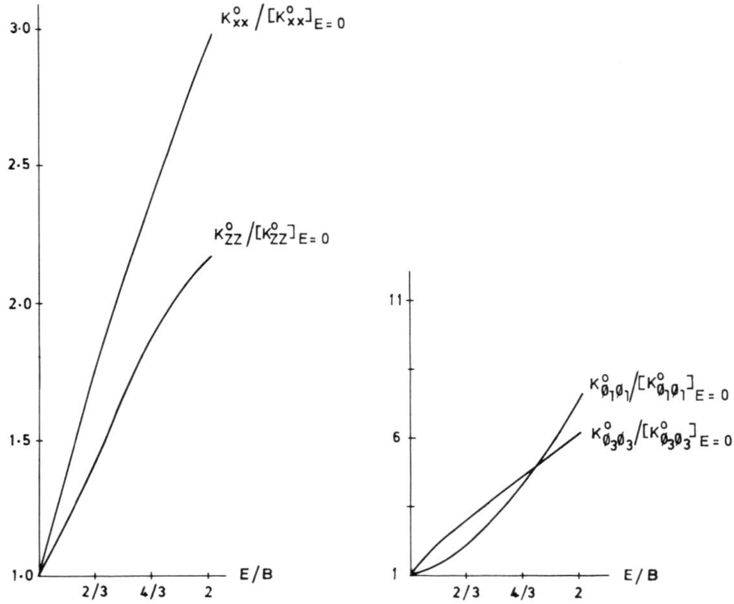

Figure 7.28 Static stiffness for embedded square foundation

equation relating $\vec{F}^*$ and $\vec{u}^*$ is obtained. From this equation $\vec{u}^*$ can be computed since $\vec{F}^*=0$. (The formulation may be found in the work by Dominguez[22,23].)

A computer program was implemented to obtain the response of foundations to travelling waves. Figure 7.30 shows the amplitude of the horizontal motion obtained for a square surface foundation when *SH* waves propagate with angles $\varphi=0.45$ and 90° with the x-axis. Results are compared with those reported by Wong and Luco[24] for the same angles using a more time-consuming numerical procedure. The agreement is satisfactory.

The torsional motion of the massless square foundation is shown in Figure 7.31 for the same angles of incidence and the same range of frequencies. The results also show a good agreement with Wong and Luco's[24].

Finally, in Figure 7.32, the response of a square foundation for two different levels of embedment is shown for the case of vertically propagating *S* waves. The foundation had levels of embedment $E/B=4/3$ and $E/B=2$. The model had 16 rectangular elements in the free field and 33 or 45 equal square elements on the soil foundation interface, 9 on the bottom and two or three rows of 12 elements around the side walls. The motions are plotted versus the natural frequency of the layer of soil corresponding to the embedment, $f_0=c_S/4E$. The values of the horizontal and rocking motions are compared with the approximate formulas of Elsabee and Morray[25] for circular foundations embedded in a soil stratum. An equivalent radius was taken for comparison.

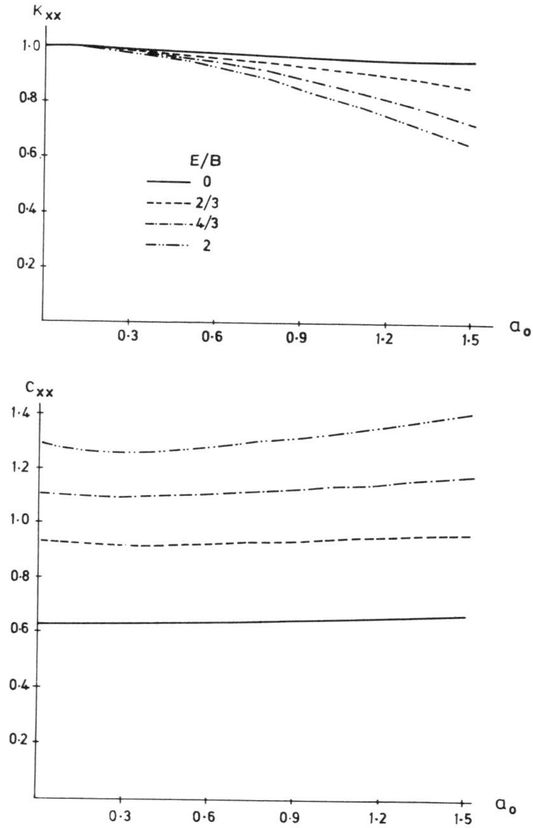

Figure 7.29 Horizontal stiffness coefficients for square embedded foundation

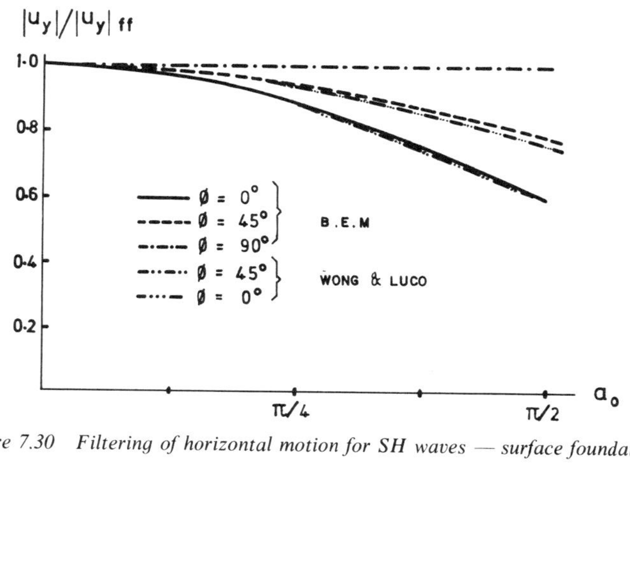

Figure 7.30 Filtering of horizontal motion for SH waves — surface foundation

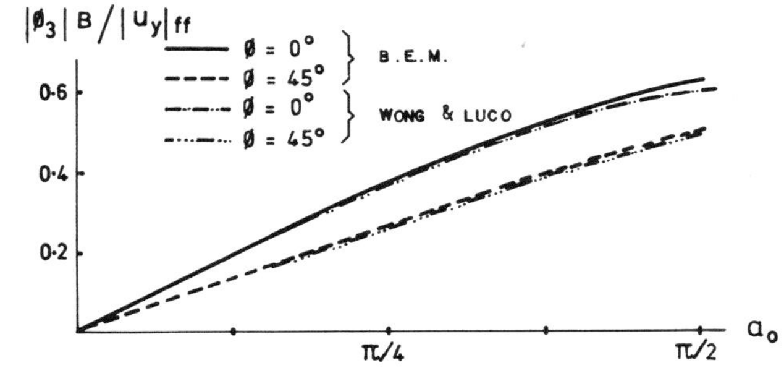

Figure 7.31 Torsional motion for SH waves — surface foundation

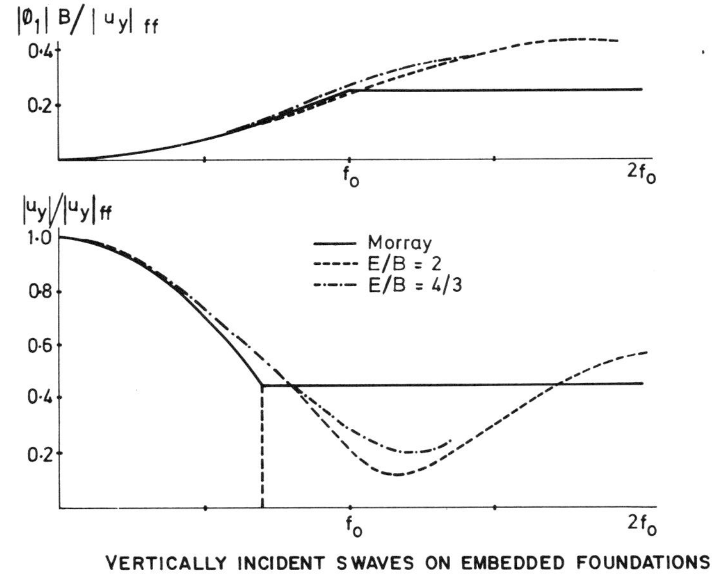

Figure 7.32 Vertically incident S waves on embedded foundations

The oscillation of the values around the approximate solution was also obtained by Elsabee and Morray[25].

REFERENCES

1 Graff, K. F., *Wave Motion in Elastic Solids*, Ohio State University (1975)

2 Achenbach, J. D., *Wave Propagation in Elastic Solids*, NorthHolland (1973)

3 Cruse, T. A. and Rizzo, F. J., 'A Direct Formulation and Numerical Solution of the General Transient Elastodynamic Problem I', *J. Math. Analysis and Applications*, **22** (1968)

4 Cruse, T. A. and Rizzo, F. J., 'A Direct Formulation and Numerical Solution of the General Transient Elastodynamic Problem II', *J. Math. Analysis and Applications*, **22** (1968)

5 Ewing, W. M., Jardetzley, W. S. and Press, F., *Elastic Waves in Layered Media*, McGraw-Hill (1957)

6 Rayleigh, J. W. S., *The Theory of Sound*, Dover (1945)

7 Love, A. E. H., *A Treatise on the Mathematical Theory of Elasticity*, Dover (1944)

8 Love, A. E. H., *Some Problems of Geodynamics*, Dover (1967)

9 Carslaw, H. S. and Jaeger, J. C., *Conduction of Heat in Solids*, Clarendon Press, Oxford (1959)

10 Cole, D. M., Kosloff, D. D. and Minster, J. B., *A Numerical Boundary Integral Equation Method for Elastodynamics I*, Seismological Laboratory Pasadena, Calif.

11 Doyle, J. M., 'Integration of the Laplace Transformed Equations of Classical Elastokinetics', *J. Math. Analysis and Applications*, **13** (1966)

12 Olver, F. W. J., 'Bessel Functions of Integer Order', in *Handbook of Mathematical Functions*, M. Abramowitz and I. A. Stegun (Ed.) U.S. National Bureau of Standards (1964)

13 Stroud, A. M. and Secrest, D., *Gaussian Quadrature Formulas*, Prentice Hall (1966)

14 Brebbia, C. A., *The Boundary Element Method for Engineers*, Pentech Press, London (1978)

15 Jakub, M., 'Dynamic Stiffness of Foundations: 2-D vs 3-D Solutions', M.I.T. Research Report R77–36, Dept. of Civil Engineering, October (1977)

16 Dominguez, J. and Alarcon, E., 'Dynamic Stiffness of Foundations', *2nd Int. Conf. on Innovative Numerical Analysis*, Ecole Polytechnique, Montreal (1980)

17 Wong, H. L. and Luco, J. E., 'Dynamic response of Rectangular Foundations to Obliquely Incident Seismic Waves', *Earthq. Engng and Struct. Dynamics*, **6** (1978)

18 Lamb, H., *The Dynamical Theory of Sound*, Dover (1960)

19 Lamb, H., 'On the Propagation of Tremors over the Surface of an Elastic Solid', *Philosophical Trans. Roy. Soc. London*, Series A, **203** (1904)

20 Kausel, E., 'Forced Vibrations of Circular Foundations on Layered Media', Research Report R74–11, Dept. of Civil Engineering, M.I.T., January (1974)

21 Beletsos, A. S. and Wei, Y. T., 'Lateral and rocking vibration of footings', Report No. 8, Structural research at Rice, Department of Civil Engineering, Rice University (1971)

22 Dominguez, J., 'Response of embedded foundations to travelling waves', M.I.T. Research Report R–78–24, Civil Engineering Department, August (1978)

23 Dominguez, J., 'Dynamic Stiffness of Rectangular Foundations', M.I.T. Research Report R–78–20, Civil Engineering Department, August (1978)

24 Wong, H. L. and Luco, J. E., 'Dynamics Response of Rigid Foundations of Arbitrary Shape', *Earthq. Engng and Struct. Dynamics*, **4** (1976)

25 Elsabee, F. and Morray, J. P., 'Dynamic Behaviour of Embedded Foundations', M.I.T. Research Report R–77–33, September (1977)

26 Aki, K. and Richards, P. G., *Quantitative Seismology*, Freeman (1980)

8

Combination of the boundary element and finite element methods

by M. Margulies

8.1 INTRODUCTION

The Finite Element Method (FEM) is, by now, a well known procedure for solving many of the equations and problems encountered in the day-to-day practice of various engineering disciplines and physics. A wide range of analyses can be tackled by the FEM; in fact, extensive theoretical investigations in recent years have made it clear that the FEM is a very general approximation process, capable of solving most properly defined continuum problems. In addition to its widespread use in structural mechanics, the method has been applied in such diverse areas as fluid mechanics[1-7], flow in porous media[8-10], electromagnetic field theory[11-13], diffusion studies in nuclear engineering, geophysical modelling[14-19]. Yet, despite its enormous field of application and its flexibility (or is it because it must be all things to every user?), the *standard* Finite Element process (i.e. one implemented with locally based polynomial expansions) has certain drawbacks.

The first and foremost difficulty is manifested when modelling a physical system of infinite extent. Examples are found in geophysical systems[15,19], ocean engineering[20-24], mining applications[25] and electromagnetic steady-state problems[11,12] to name just a few. In cases such as these, an appropriate approximation to reality by means of a finite number of truly *finite* elements is clearly impossible. On the other hand, approximate domain truncations are often questionable, due to the slow decay of the solutions and the absence of a natural length parameter by which to gauge the distance at which the relevant physical fields become truly negligible.

Another situation in which the FEM almost invariably encounters difficulties is when stress singularities are present in the modelled system (see, e.g. Irwin[26]). These are of great practical importance as they can initiate crack progression[27], starting at a re-entrant corner of the irregularities or cavities common in most solids. Whether crack propagation will indeed take place depends on how stresses decay at some distance from a crack tip — hence the need to obtain a reasonably accurate solution to the modelled problem in that region. But at stress singularities it is well known that theoretically infinite stresses occur and they cannot, of course, be adequately represented by locally based polynomial expansions. Under these circumstnaces, then, a standard finite element procedure may result in a completely meaningless analysis of crack propagation. Similarly, difficulties also crop up in the course of modelling concentrated loads, (for example when 'point sources' are present in potential problems) since the solution then becomes infinite at the location of the load. If we happen to need the fields in the immediate vicinity of the 'point source' (as in modelling certain geophysical problems[15,16]), it is generally impossible to adequately account for their rapid variation in that region by means of standard FEM polynomial expansions, no matter how small the elements.

Finally, we often need to solve a problem to a high degree of accuracy in successively different parts of the modelled domain (for example, the response to different load locations may be sought). Under these conditions, the FEM can become very uneconomical in terms of the required amounts of computer time and/or main storage owing to the generality of the gridding necessary to deal with the various loading conditions. Unlike the drawbacks described above, this does not represent a theoretical shortcoming, as such, of the FEM but rather one of a practical nature. Consequently, this aspect has not been dealt with extensively in the past, although it may severely limit the general applicability of the method to realistic models.

Various means have been devised to circumvent the aforementioned difficulties. The use of so-called 'infinite finite elements' has been advocated[21–23,25,28,29] to remedy the situation when the modelled domain includes regions infinite in extent. Such elements are defined as radiating strips (in two or three dimensions) in the exterior (infinite) region. On the basis of the known or suspected physical behaviour of the solution at large distances, a particular shape function form is adopted, and numerical integration is carried out with appropriate weights, taking advantage of the fact that the newly chosen shape functions must decay to zero at infinity, somehow or other. This technique overcomes in a pragmatic manner the difficulty resulting from the existence of an infinite domain without destroying the narrow banding of the final system of equations. Yet, from a theoretical viewpoint there are still several shortcomings:

(1) Unless extra shape functions are introduced (thus greatly complicating the formalism) or special care is taken, one runs the risk of nonconformity at element boundaries.

(2) In this formulation, the meaning of convergence and the existence of convergence proofs becomes rather shaky.

(3) It is usually not possible to incorporate correctly the behaviour of the solution at large distances, since it generally can only be guessed; rather, one must be content to include the 'leading order' term(s) of some asymptotic expansion; this kind of approximation may in turn generate errors on the boundaries which will propagate into the interior region, with a resulting loss of accuracy.

Difficulties generated by stress singularities such as occur at crack tips, can be obviated by the use of singular basis functions which incorporate the form of the predominant singularity near the crack tip[30,31]. The major drawbacks of this approach are similar to those described above for 'infinite finite elements'. Without careful precautions (such as the introduction of hybrid stress-displacement elements), nonconformity at element boundaries will invariable occur; because of the singular behaviour of the basis functions, convergence proofs become almost meaningless and finally, integration over the elements increases substantially in complexity. Several methods have been proposed to overcome some of these remaining shortcomings, none entirely successful so far.

The last objection mentioned to the universal applicability of the FEM can be easily eliminated by repetitive application of the technique to the same physical configuration, but with varying griddings and loading conditions. While rather straightforward, this procedure is of course quite costly in terms of both the computer resources needed and the manpower-time required to carry out the analysis.

As we will see, all of these shortcomings can be overcome by marrying the BEM to the FEM (or, sometimes, by using the BEM as a stand-alone). However, in all fairness, this recipe is not always applicable and it does have drawbacks of its own, as it destroys the banded nature of the resulting final system of equations.

Basically, the difficulties can be removed because, unlike the FEM, the BEM essentially frees itself from the constraints imposed by the necessary existence of a mesh over which the solution must be sought. In other words, at the most elementary level, the BEM can be viewed as a technique which has implicit in it as coarse or as fine a mesh as is dictated by the pecularities of the problem at hand. That this is indeed possible is a consequence of the following: knowledge of certain classes of solutions to problems belonging to the same general family as the one under study (the so-called fundamental solutions), together with computation of quantities which can be viewed as the strengths of elementary excitations (and which are usually located at safe distance from the trouble-some spots) allows us in principle to evaluate the solution at arbitrary points, be they very remote or within arbitrarily small neighbourhoods of possible singularities.

Coupling of the boundary element solution procedure to the finite element solution procedure can be described in very general terms as follows: let D be a domain in which a given problem is to be solved, and define S as a surface which partitions D into two parts, D_1 and D_2, i.e.

$$D = D_1 \cup D_2$$

$$S = D_1 \cap D_2$$

In D_1, the usual finite element solution procedure is used, i.e. an approximate parametric representation for the unknown solution is imposed *a priori*. Such a representation typically uses locally based shape functions and the parameters to be solved for are the values of the unknown function at the grid points of the chosen mesh. D_2, on the other hand, is regarded as a single 'finite element'. A representation for the unknown function in this domain is derived such that its functional form identically satisfies the governing equations (and may include a singular behaviour and/or an appropriate decay at large distances). By a judicious choice of representation, one can in addition arrange things so that the unknown parameters in that representation are identical with the values of the unknown function at the finite element nodes located on S. With these choices it thus becomes possible to reformulate the original problem in D by means of a variational technique or weighted residuals method, with the unknown parameters being the values of the solution at the finite element grid points in D_1 only. Thus by combining the FEM with BEM we can include *all* the points of D_2 as possible target points and yet ensure, wherever necessary, the correct behaviour of the unknown solution. A further advantage of this approach, particularly when D_2 is infinite, lies in the fact that the contribution from that region is now completely accounted for without having to increase the number of degrees of freedom in the modelling. In fact, with the BEM it is usually possible to *decrease* the total number of degrees of freedom, since large domains can be fully taken into account by means of a limited number of parameters defined on S (a set of points with dimension one less than that of D).

Were that the only side of the coin, there would be little or no reason to ever use the FEM. Unfortunately, the BEM does have drawbacks of its own which preclude universal application. Most important is the necessity of defining a correct representation for the solution in D_2. This, in practice, is equivalent to finding fundamental solutions under certain simplified conditions — a difficult task, but often manageable provided that the material properties in D_2 are homogeneous (so as to insure that the governing equations reduce to partial differential equations with *constant* coefficients). There are general techniques to derive such solutions when uniformity is present, such as Hörmander's method[32]. The hypothesis of uniformity is, for problems of practical use, a rather severe restriction, although it is possible in many cases to repartition D so as to include all inhomogeneities in D_1, leaving D_2 indeed homogeneous.

A second drawback results from the fact that all the boundary grid points now belong to a single 'finite element' (which we identify with D_2). Hence, an inescapable consequence is that the final stiffness matrix will contain non zero entries connecting these surface nodes. That matrix therefore loses, possibly to a large extent, the narrow banding that characterizes standard FEM stiffness matrices. Depending on the size of the problem being investigated, this may or may not represent an insurmountable difficulty. In all cases, though, it should be possible to minimize the degradation in bandedness by judicious node ordering.

8.2 COUPLING OF THE BEM AND THE FEM

At the heart of the BEM lies Green's second identity, for potential problems, or its equivalent (e.g. Somigliana's identity), for non-potential problems. For ease of exposition we will restrict our attention in this chapter to potential theory. The formulae necessary to apply to other types of problems the ideas developed below appear in an appendix at the end of the chapter.

Let D be a regular region assumed to have homogeneous material properties and φ a function which satisfies Poisson's equation in D,

$$k\nabla^2\varphi = F \tag{8.1}$$

subject to boundary conditions on the surface ∂D bounding D

$$\varphi = f, \text{ on } \partial D_d \tag{8.2a}$$

$$\hat{n}\vec{\nabla}\varphi = g, \text{ on } \partial D_n \tag{8.2b}$$

where $\hat{n}$ represents a unit vector directed along the exterior normal to ∂D and $\partial D = \partial D_d \cup \partial D_n$. In Equation 8.1, k denotes the *constant* value of some physical parameter relevant to the problem being studied, for example the medium's electrical conductivity, dielectric constant, or thermal conductivity. Henceforth, unless explicitly stated to the contrary, D will stand for a two- or three-dimensional region interchangeably. Accordingly, the terminology 'surface ∂D' will denote a line or surface, and all symbols $\int_{\partial D} (\ldots)\, da$ are to be understood, respectively, as line or surface integrals.

Without loss of generality we can assume that $F = 0$, so that Equation 8.1 reduces to Laplace's equation. Indeed, it is always possible to write $\varphi = \varphi_0 + \varphi'$, where

$$k\nabla^2\varphi_0 = F \tag{8.3}$$

in all of space, subject to regularity and smoothness conditions at large distance, φ' then satisfies the following boundary-value system:

$$k\nabla^2\varphi' = 0, \qquad \text{in } D$$

$$\varphi' = f - \varphi_0 \equiv f', \qquad \text{on } \partial D_d$$

$$\hat{n}\vec{\nabla}\varphi' = g - \hat{n}\vec{\nabla}\varphi_0 \equiv g', \qquad \text{on } \partial D_n$$

φ_0, on the other hand, can easily be found by using the principle of linear superposition. If we use $\psi^*(\mathbf{r};\mathbf{r}')$ to represent the solution to Equation 8.3 when $F = \delta(\mathbf{r} - \mathbf{r}')$, it immediately follows that

$$\varphi_0 = \int F(\mathbf{r}')\psi^*(\mathbf{r};\mathbf{r}')\, dv \tag{8.4}$$

the integration being carried out over all space. ψ^* is often referred to as a 'fundamental solution' to Laplace's equation. More rigorously stated, ψ^* is in fact the Green's function of free space with material properties identical with those of D. Explicitly,

$$\psi^* \equiv -\frac{1}{4\pi k|\mathbf{r} - \mathbf{r}'|}, \text{ in 3-D} \tag{8.5}$$

or

$$\psi^* \equiv -\frac{1}{2\pi k} \ln |\mathbf{r}-\mathbf{r}'|, \text{ in 2-D} \tag{8.6}$$

Since Equation 8.4 gives an explicit representation for φ_0, our initial inhomogeneous partial differential equation has effectively been reduced to a homogeneous one, and from here on, we will implicitly assume that only homogeneous equations are to be dealt with.

Green's second identity (sometimes called Green's theorem) is well known (see, e.g. Kellog[33]): let u and v be two functions continuously differentiable in D and let them also have continuous partial derivatives of the second order in D. Green's theorem can then be written in the form

$$\int_D [u\nabla^2 v - v\nabla^2 u]\, dv = \oint_{\partial D} [u\hat{n}\vec{\nabla} v - v\hat{n}\vec{\nabla} u]\, da \tag{8.7}$$

This equation remains valid when D is infinite in extent (for instance that part of space exterior to a closed surface $\partial D'$); in that case, ∂D stands for the union $\partial D' \cup S_\infty$, where S_∞ denotes the 'surface at infinity'.

We now apply Equation 8.7 taking $u=\varphi(\mathbf{r})$ and $v=\psi^*(\mathbf{r};\mathbf{r}')$, where $\mathbf{r}$ is the coordinate vector of the observation point P and $\mathbf{r}'$ is the integration variable. Assume that P belongs to D; in that case, Equation 8.7 cannot be applied directly since ψ^* does not satisfy the necessary continuity conditions. This difficulty can be circumvented by the usual limiting procedure, leading finally to the result

$$\varphi(\mathbf{r}) = k \oint_{\partial D} [\varphi(\mathbf{r}')\hat{n}\vec{\nabla}_{r'}\psi^*(\mathbf{r};\mathbf{r}') - \psi^*(\mathbf{r};\mathbf{r}')\hat{n}\vec{\nabla}_{r'}\varphi(\mathbf{r}')]\, da \tag{8.8}$$

where use has been made of the fact that φ is harmonic (i.e. satisfies Laplace's equation) in D.

Equation 8.8 forms the basis for the application of the BEM. Note that it is *not* a solution to the given boundary value problem. Indeed, for such a problem to be well-posed, φ and $\partial\varphi/\partial n$ cannot be specified simultaneously on the same parts of D. Rather, on letting the observation point P tend to the surface D, we are led to an integral equation. This equation must be solved for the unknowns φ and $\partial\varphi/\partial n$ in terms of the given boundary data on ∂D_d and ∂D_n. The integral equation can be written explicitly as

$$c(\mathbf{r})\varphi(\mathbf{r}) = k \,⨍_{\partial D} [\varphi(\mathbf{r}')\hat{n}\vec{\nabla}_{r'}\psi^*(\mathbf{r};\mathbf{r}') - \psi^*(\mathbf{r};\mathbf{r}')\hat{n}\vec{\nabla}_{r'}\varphi(\mathbf{r}')]\, da$$

$$\mathbf{r} \in \partial D \tag{8.9}$$

where $⨍$ denotes the principal value (in the Cauchy sense) of the integral and $c(\mathbf{r})$ is a factor which depends on the smoothness of the boundary ∂D at the point $\mathbf{r}$ ($c(\mathbf{r})=1/2$ for a smooth surface; as we will see later on, however, actual knowledge of $c(\mathbf{r})$ is unnecessary). Equation 8.9 is entirely equivalent to the original boundary value problem. Upon solving it, φ can be computed at any point of D by means of Equation 8.8 this requires only the evaluation of 'straightforward' quadratures. The BEM, at the most elementary level of interpretation, consists simply in the resolution of Equation 8.9 by some suitable discretization-collocation method.

Under some conditions, we may wish to combine the BEM with the FEM. This could occur, as mentioned earlier, when finite parts of D no longer have uniform material properties (in which case the governing equation becomes $\vec{\nabla}[k(\mathbf{r})\vec{\nabla}\varphi]=0$), or when D extends to infinty, or when singularities may exist as isolated points of D, etc. Let S denote a surface that separates the original domain D into two subdomains D_1 and D_2 and define

$$\partial D_{d2}=\partial D_d \cap D_2$$

$$\partial D_{n2}=\partial D_n \cap D_2$$

$$\partial D_2=\partial D_{d2} \cap \partial D_{n2} \cap S$$

In D_1 the solution is to be obtained by means of the FEM (D_1 is thus finite in size, may contain material inhomogeneities but no singularity points), whereas in D_2 the BEM will be used (hence D_2 may be infinite in extent, may contain singular points and necessarily has homogeneous material properties*).

Two similar approaches can be used to combine the BEM with the FEM. They differ in one respect: where one uses free space Green's functions, the other uses bounded Green's functions; this latter approach[34] was originally introduced to remedy some of the difficulties inherent in the former. A description of both of these methods now follows.

8.2.1 First method[35–37]

Step one

Our first task is to choose a trial form for the unknown φ on the boundary S (denote it by φ_S). Since S belongs to D_1 (as well as to D_2), we usually already have such a form at our disposal. Indeed, by the very nature of the FEM we know that φ is approximated in D_1 by a function of the general type

$$\varphi \simeq \sum_{i=1}^{M} \varphi_i u_i(\mathbf{r}) \tag{8.10}$$

* This restriction may be relaxed in some special cases. For clarity of exposition we will assume it to hold here; on page 277, however, we shall briefly indicate how to generalize the methods presented below to these special non-uniform situations.

In this equation, M is the total number of FEM grid points in D_1. For ease of notation, we assume that the nodes are numbered so that nodes $1,2,\ldots,N<M$ are located on S. The u_i represent the 'shape functions' used in D_1; they may be such as to result in a piecewise constant, piecewise linear, piecewise parabolic or in any other type of approximation for φ in D_1. The φ_i are the unknown parameters to be solved for; customarily, they are the values of the unknown solutions at the nodal points. With Equation 8.10 in mind, it is then natural to write

$$\varphi_S(\mathbf{r}) \simeq \sum_{i=1}^{N} \varphi_i v_i(\mathbf{r}) \tag{8.11}$$

where

$$v_i(\mathbf{r}) \equiv u_i(\mathbf{r};\mathbf{r}\in S) \tag{8.12}$$

It is not imperative that a trial function of the form given by Equation 8.11 be used. This requirement can in fact be relaxed with the proper use of Lagrange multipliers; in the following, however, we will take it for granted that the unknown φ_S is approximated as shown in Equation 8.11 and 8.12.

Step two

We now must solve the following boundary value problem:

$$k\nabla^2\varphi = 0, \text{ in } D_2 \tag{8.13a}$$

$$\varphi = f, \text{ on } \partial D_{d2} \tag{8.13b}$$

$$\hat{n}\vec{\nabla}\varphi = g, \text{ on } \partial D_{n2} \tag{8.13c}$$

$$\varphi = \varphi_S, \text{ on } S \tag{8.13d}$$

D_2 may extend to infinity; in that case, ∂D_{d2} and ∂D_{n2}, as appropriate, include the 'surface at infinity'. Note that φ_S is as yet unknown; the solution to the boundary value problem is therefore to be viewed as a functional of φ_S:

$$\varphi = \varphi[\varphi_S(\mathbf{r})] \tag{8.14}$$

Equations 8.13(a)–8.13(d) constitute a well-posed problem; its solution can be found by a straightforward application of the usual BEM. To this effect, we adopt representations for φ and $\partial\varphi/\partial n$ on the boundary of D_2, i.e. we let

$$\varphi \simeq \sum_{i=1}^{n} \varphi_i v_i(\mathbf{r}), \quad \mathbf{r}\in\partial D_2 \tag{8.15a}$$

$$\frac{\partial\varphi}{\partial n} \simeq \sum_{i=1}^{n} \psi_i v_i(\mathbf{r}), \quad \mathbf{r}\in\partial D_2 \tag{8.15b}$$

It is obvious that some of the ψ_i are known *a priori* (for $\mathbf{r}\in\partial D_{n2}$) and that some of the φ_i are either fixed (for $\mathbf{r}\in\partial D_{d2}$) or identical with those introduced in Equation 8.11 (for $\mathbf{r}\in S$).

We can now insert the trial functions 8.15(a) and 8.15(b) into Equation 8.9 and express that it must be satisfied at a certain number of boundary points*. This allows us then to determine the unknowns φ_i and ψ_i. Explicitly, we find

$$c_j\varphi_j = -\sum_i \varphi_i \hat{A}_{ij} + \sum_i \psi_i B_{ij} \quad i,j=1,2,\ldots \tag{8.16}$$

where $c_j = c(\mathbf{r}_j)$ and

$$\hat{A}_{ij} = -k \oint_{\partial D_2} \hat{n}\vec{\nabla}\psi^*(\mathbf{r}_j;\mathbf{r}')v_i(\mathbf{r}')\,\mathrm{d}a \tag{8.17a}$$

$$B_{ij} = -k \oint_{\partial D_2} \psi^*(\mathbf{r}_j;\mathbf{r}')v_i(\mathbf{r}')\,\mathrm{d}a \tag{8.17b}$$

Let

$$A_{ij} = \hat{A}_{ij} + c_j\delta_{ij}$$

with δ_{ij} the usual Kronecker symbol. Note, then, that one can in fact easily avoid having to compute the diagonal elements of A (hence, knowledge of $c(\mathbf{r})$ is not required): indeed, if D_2 is a closed domain the solution to Equations 8.13(a)–8.13(d) when φ is constant everywhere on ∂D_2 is trivial and such that $\partial\varphi/\partial n = 0$ at all points of that boundary. In such a case, it immediately follows from Equation 8.16 that

$$\sum_i A_{ij} = 0, \quad j=1,2,\ldots \tag{8.18}$$

which leads to a trivial determination of the diagonal elements A_{ii} as soon as the off-diagonal ones are known. On the other hand, when D_2 is an open domain (exterior to some closed surface), we simply use

$$c_j^{int} + c_j^{ext} = 1, \quad j=1,2,\ldots$$

$$\hat{A}_{ij}^{int} = -\hat{A}_{ij}^{ext}, \quad j\neq i,j=1,2,\ldots$$

* This number is left purposely vague. When the boundary ∂D_2 is smooth (i.e. contains no sharp corners), the number of points is obviously equal to the number of grid points on that boundary. When there are sharp corners, on the other hand, $\partial\varphi/\partial n$ suffers a discontinuity at such points. It then becomes necessary to write the collocation equation twice at such points, it being understood that $\partial\varphi/\partial n$ takes on *different* values (whether known or not) at each one of them. To further complicate matters, the reader may disregard this footnote when piecewise constant trial functions are used: in that case the number of collocation points simply equals the number of constant value boundary elements.

together with Equation 8.18 to find

$$\sum_i A_{ij} = 1, \quad j = 1,2,\ldots \tag{8.19}$$

Equation 8.19 can then be used to evaluate very simply the diagonal elements from the off-diagonal ones. Calculation of the latter follows well established numerical techniques. Typically, Gaussian quadratures are used; care must be exercised, however, when the observation point $\mathbf{r}_j$ belongs to the integration domain. For a piecewise constant representation of φ, this happens when $j = i$; for a piecewise linear trial function, it will occur for $j = i$ or $j = i \pm 1$. In these special cases, it is necessary to integrate out the singularity analytically or to use special weights in the numerical quadrature (remember, though, that there is never a need to evaluate A_{ii}). Further details on the integrations can be found in References 38–42.

Equation 8.16 can be written in matrix form as

$$\mathbf{A}\boldsymbol{\varphi} = \mathbf{B}\boldsymbol{\psi} \tag{8.20}$$

Let us now partition $\boldsymbol{\varphi}$ and $\boldsymbol{\psi}$ into three parts:

$$\boldsymbol{\varphi} \equiv \begin{bmatrix} \boldsymbol{\varphi}_d \\ \boldsymbol{\varphi}_n \\ \boldsymbol{\varphi}_s \end{bmatrix}; \quad \boldsymbol{\psi} \equiv \begin{bmatrix} \boldsymbol{\psi}_d \\ \boldsymbol{\psi}_n \\ \boldsymbol{\psi}_s \end{bmatrix}$$

where $\boldsymbol{\varphi}_d$ and $\boldsymbol{\psi}_n$ are known vectors; their elements are essentially the values of the given functions f and g at the boundary element nodes that belong to ∂D_d and ∂D_n respectively. We can thus recast Equation 8.20 as

$$\begin{bmatrix} A_{11} & A_{12} & A_{13} \\ A_{21} & A_{22} & A_{23} \\ A_{31} & A_{32} & A_{33} \end{bmatrix} \begin{bmatrix} \boldsymbol{\varphi}_d \\ \boldsymbol{\varphi}_n \\ \boldsymbol{\varphi}_s \end{bmatrix} = \begin{bmatrix} B_{11} & B_{12} & B_{13} \\ B_{21} & B_{22} & B_{23} \\ B_{31} & B_{32} & B_{33} \end{bmatrix} \begin{bmatrix} \boldsymbol{\psi}_d \\ \boldsymbol{\psi}_n \\ \boldsymbol{\psi}_s \end{bmatrix}$$

or, collecting the unknowns on one side:

$$\begin{bmatrix} B_{11} & A_{12} & B_{13} \\ B_{21} & A_{22} & B_{23} \\ B_{31} & A_{32} & B_{33} \end{bmatrix} \begin{bmatrix} \boldsymbol{\psi}_d \\ \boldsymbol{\varphi}_n \\ \boldsymbol{\psi}_s \end{bmatrix} = \begin{bmatrix} A_{11} & B_{12} & A_{13} \\ A_{21} & B_{22} & A_{23} \\ A_{31} & B_{32} & A_{33} \end{bmatrix} \begin{bmatrix} \boldsymbol{\varphi}_d \\ \boldsymbol{\psi}_n \\ \boldsymbol{\varphi}_s \end{bmatrix} \tag{8.21}$$

(note that we treat φ_S as if it were known). This system of equations can now be solved, at least in principle, and we finally obtain

$$\begin{bmatrix} \boldsymbol{\psi}_d \\ \boldsymbol{\varphi}_n \\ \boldsymbol{\psi}_s \end{bmatrix} = \begin{bmatrix} P_{11} & P_{12} & P_{13} \\ P_{21} & P_{22} & P_{23} \\ P_{31} & P_{32} & P_{33} \end{bmatrix} \begin{bmatrix} \boldsymbol{\varphi}_d \\ \boldsymbol{\psi}_n \\ \boldsymbol{\varphi}_s \end{bmatrix} \tag{8.22}$$

With this result, it is obvious that we have found the representation we sought for the solution to the boundary value problems, Equations 8.13(a)–8.13(d). This representation can be expressed solely in terms of the given functions f and g and the values of φ_S, as yet unknown, since we can apply Green's theorem (Equation 8.8) to write at *any* point of D_2,

$$\begin{aligned}\varphi(\mathbf{r}) = \sum_i \{ &\varphi_{d,i}\tau_{d,i}(\mathbf{r}) + \varphi_{n,i}\tau_{n,i}(\mathbf{r}) + \varphi_{s,i}\tau_{s,i}(\mathbf{r}) \\ &+ \psi_{d,i}\sigma_{d,i}(\mathbf{r}) + \psi_{n,i}\sigma_{n,i}(\mathbf{r}) + \psi_{s,i}\sigma_{s,i}(\mathbf{r})\}\end{aligned} \tag{8.23}$$

where

$$\tau_{d,i}(\mathbf{r}) \equiv k \int_{\partial D_{d2}} v_i(\mathbf{r}')\hat{n}\vec{\nabla}\psi^*(\mathbf{r};\mathbf{r}')\,\mathrm{d}a$$

$$\sigma_{d,i}(\mathbf{r}) \equiv k \int_{\partial D_{d2}} \psi^*(\mathbf{r};\mathbf{r}')\hat{n}\vec{\nabla}v_i(\mathbf{r}')\,\mathrm{d}a$$

$$\tau_{n,i}(\mathbf{r}) \equiv k \int_{\partial D_{d2}} v_i(\mathbf{r}')\hat{n}\vec{\nabla}\psi^*(\mathbf{r};\mathbf{r}')\,\mathrm{d}a, \text{ etc...}$$

and

$$\varphi_{n,i} \equiv [\mathbf{P}_{21}\boldsymbol{\varphi}_d + \mathbf{P}_{22}\boldsymbol{\psi}_n + \mathbf{P}_{23}\boldsymbol{\varphi}_s]_i$$

$$\psi_{d,i} \equiv [\mathbf{P}_{11}\boldsymbol{\varphi}_d + \mathbf{P}_{12}\boldsymbol{\psi}_n + \mathbf{P}_{13}\boldsymbol{\varphi}_s]_i$$

etc.

Step three

Having obtained a representation for φ in D_2, which from here on can be regarded as a single element in the finite element method spirit, it is now a simple matter to carry out the remainder of the solution procedure. For the sake of definiteness, we assume that the original problem was formulated as a variational principle; entirely similar results, however, are found when a Galerkin weighted-residuals formulation is used. The next step is to compute the 'energy' of the field associated with the domain D_2. This is nothing but the quantity which must be minimized to express the equivalence of the variational principle with the original partial differential equation. Using the known form of the variational functional

$$E_2[\varphi] \equiv \int_{D_2} \frac{k}{2}[\vec{\nabla}\varphi]^2 \, dv$$

one obtains, as in the usual FEM,

$$E_2(\boldsymbol{\varphi}_s) \equiv \frac{1}{2}\boldsymbol{\varphi}_s^T \mathbf{K}_2 \boldsymbol{\varphi}_s + \mathbf{Q}_2^T \boldsymbol{\varphi}_s \tag{8.24}$$

In the above equation, $\mathbf{K}_2$ is the stiffness matrix of domain D_2; it is easily shown to be symmetric, a result not restricted to the variational formalism and also obtained with Galerkin's approach. $\mathbf{Q}_2$ is a vector whose elements are easily calculable functions of $\boldsymbol{\varphi}_d$ and $\boldsymbol{\psi}_n$.

Step four

A similar 'energy' calculation can be performed for domain D_1. This procedure is simply equivalent to the standard FEM stiffness matrix assembly in D_1. It leads to

$$E_1(\boldsymbol{\varphi}) \equiv \frac{1}{2}\boldsymbol{\varphi}^T \mathbf{K}_1 \boldsymbol{\varphi} + \mathbf{Q}_1^T \boldsymbol{\varphi} \tag{8.25}$$

where the elements of $\mathbf{Q}_1$ are found from the *imposed* flux conditions at nodes located on the boundary of D_1.

Step five

In this last step, we add $E_1[\boldsymbol{\varphi}]$ and $E_2[\boldsymbol{\varphi}_s]$ (recall that the elements of $\boldsymbol{\varphi}_s$ form a subset of those of $\boldsymbol{\varphi}$) and minimize the resulting expression with respect to the φ_i $(i=1,2,\ldots,M)$. We then impose appropriate Dirichlet conditions on the boundary of D_1 and finally obtain a system of linear equations in the φ_i. Upon solving it, we can use the newly found values of φ_i $(i=1,2,\ldots,N<M)$ together with Equation 8.23 to evaluate $\varphi(\mathbf{r})$ at any point of D_2.

This completes the description of the first technique to combine the BEM with the FEM.

At this juncture, a few comments are in order. Because the nodes $i=1,2,\ldots,N$ all belong to the single element D_2 (though they may belong to different elements in D_1), non-zero matrix entries will connect the unknowns at these nodes in the final system of equations. Consequently, the narrow banding that is usually a dominant feature of the finite element formulation is lost to some extent, although a judicious choice of node numbering should prevent too severe an increase in matrix bandwith.

More serious than the degradation in bandwith (unavoidable in any case), is the fact that the free space BEM (FBEM), as Method One will be referred to, leads to a representation for φ in D_2 which is inconsistent with that used in D_1. Indeed, the boundary value problem 8.13(a)–8.13(d) (or the equivalent integral equation) usually cannot be solved exactly, owing to the complicated shape of the surface S that separates D_1 and D_2 (S is commonly chosen on physical grounds). Customarily, then, as in Step Two above, the integral equation is solved approximately by a discretization-collocation technique. As a result of this approximate solution, when we evaluate $\varphi(\mathbf{r})$ in D_2 by Green's theorem (see Equation 8.23, the result does *not* tend to the original $\varphi_s(\mathbf{r})$ the observation point tends to an arbitrary points of S. Rather, $\varphi(\mathbf{r}) \rightarrow \varphi_S(\mathbf{r})$ as $\mathbf{r} \rightarrow \mathbf{r} \in S$ only at those points where the integral equation was collocated. In other words, the approximate solution procedure has led to the creation of a 'gap' between the solutions in D_1 and D_2. This point is generally overlooked; it involves a question of mathematical rigour, although it has also been observed that the imperfect matching may lead to detectable errors in the FEM solution. When the latter is needed to only a few percent relative accuracy, whether a gap exists may not matter. When high accuracy is required, however, it becomes imperative that this imperfect matching be avoided.

In potential problems, the existence of a gap is physically equivalent to the creation of a spurious double layer on S (since φ now has a discontinuity upon crossing that surface). If D_1 is a small interior region (often the reason for implementing a combined BEM–FEM in the first place), the extraneous potential due to this double layer may then result in unacceptable errors in the solution in that domain.

The possible occurence of nonconformity between solutions in different regions is a difficulty often encountered in the application of the FEM. In that context, Lagrange multipliers are commonly introduced in order to ensure matching between the various solutions. A similar approach[35] has been proposed when combining the FBEM and the FEM*. Even with this added constraint, however, one can only achieve, at best, a matching of the solutions in D_1 and D_2 in some 'average sense'. Indeed, in the variational technique,

$$\int_S \lambda(\mathbf{r})(\varphi_1 - \varphi_2)\, da = 0$$

guarantees $\varphi_1 = \varphi_2$ at every point of S only for a very particular λ, one which is unknown *a priori* and cannot in general be found exactly; hence, the most we can hope for is to minimize the integral appearing in the above equation, rather than rendering it identically zero. Similarly, in the weighted residuals formulation, conditions of the kind

$$\int_S w_j(\varphi_1 - \varphi_2)\, da = 0$$

where the w_j form a *finite* set of weighting functions clearly do not imply $\varphi_1 = \varphi_2$ at all points of S.

* The subsequent remarks also apply when we use a representation for φ_S which is different from that given by Equation 8.11, i.e. when φ_S is *a priori* incompatible with the solution in D_1, a situation which inherently calls for the use of Lagrange multipliers.

As mentioned previously, the imperfect matching arises from having to solve the integral Equation 8.9 by some approximate means. (In the collocation method, for instance, the equation is satisfied only at a discrete number of points.) This in turn, usually stems from the fact that the surface S is chosen on physical grounds and often is of complicated shape, precluding the possibility of exact solution. Now recall that our goal is to combine the BEM with the FEM. There is, then, nothing to prevent us from replacing the physically relevant surface S by a slightly different one of much simpler shape (since in one of the two regions the problem must be solved by the FEM anyway). Choosing a simpler surface makes it possible to solve the integral equation in such a way that the solution will satisfy it *identically* at any point of S, though not necessarily on ∂D_2 if the latter is at finite distance*. In other words, we are now guaranteed that the potential φ computed in region D_2 tends rigorously to that assumed on S at *any* point of that surface. This is achieved in practice by using a bounded Green's function, rather than the free space one. Therefore, we refer to this method as the Green's function boundary element method (GBEM). The procedure is based on the following observation: let $\psi_G(\mathbf{r};\mathbf{r}')$ be a solution of

$$k\nabla^2\psi_G(\mathbf{r};\mathbf{r}')=\delta(\mathbf{r}-\mathbf{r}'), \quad \mathbf{r},\mathbf{r}'\in D_2 \tag{8.26}$$

subject to the boundary condition

$$\psi_G(\mathbf{r};\mathbf{r}')=0, \text{ for } \mathbf{r} \text{ or } \mathbf{r}'\in S \tag{8.27}$$

(If S is an open finite surface, as may happen when D is closed domain, it is permissible for ψ_G to satisfy additional boundary conditions outside of D — to render its determination simpler, for instance — as long as Equations 8.26 and 8.27 are verified.)

If we now use $\varphi(\mathbf{r})$ and $\psi_G(\mathbf{r};\mathbf{r}')$ in Green's second identity (rather than $\varphi(\mathbf{r})$ and $\psi^*(\mathbf{r};\mathbf{r}')$), it follows immediately that

$$\begin{aligned}\varphi(\mathbf{r};\mathbf{r}\in D_2)&=k\int_S \varphi_S(\mathbf{r}')\hat{n}\vec{\nabla}_{r'}\psi_G(\mathbf{r};\mathbf{r}')\,\mathrm{d}a\\&\quad+k\int_{\overline{\partial D}}[\varphi(\mathbf{r}')\hat{n}\vec{\nabla}_{r'}\psi_G(\mathbf{r};\mathbf{r}')-\psi_G(\mathbf{r};\mathbf{r}')\hat{n}\vec{\nabla}_{r'}\varphi(\mathbf{r}')]\,\mathrm{d}a\end{aligned} \tag{8.28}$$

instead of Equation 8.8 previously obtained with the free space Green's function ($\overline{\partial D}=\partial D_{d2}\cup\partial D_{n2}$). Equation 8.28 forms the basis for the application

* For simplicity, we still talk about S, D_1, D_2 although it should be clear that for a given problem these are now slightly different from the ones introduced in the preceding discussion.

of the GBEM. The dependence on $\varphi_S(\mathbf{r})$ is now explicit in contrast to the earlier implicit dependence (via $\partial\varphi/\partial n(\mathbf{r};\mathbf{r}\in S)$). Upon letting the observation point $\mathbf{r}$ tend to a point of $\overline{\partial D}$ in Equation 8.28 we still get an integral equation (for φ in terms of $\partial\varphi/\partial n$ or vice-versa, depending on the given boundary data). Its solution may not be found exactly on $\overline{\partial D}$, but it will be necessarily such that $\varphi(\mathbf{r})\rightarrow\varphi_S(\mathbf{r})$ as $\mathbf{r}$ tends to *any* point of S, thus eliminating the mismatch on that surface. We see therefore that provided ψ_G can indeed by found, a consistent formulation of the problem becomes possible. Using the available freedom in the choice of S, it turns out that appropriate Green's functions can often be readily obtained. Here, then, is a more detailed description of this second approach.

8.2.2 Second method[34]

Step one

This step is identical with the one given in the First Method. In essence it consists in the choice of a trial form for the unknown on the surface separating D_1 and D_2. We thus write as in Equation 8.11:

$$\varphi_S(\mathbf{r})\simeq\sum_{i=1}^{N}\varphi_i v_i(\mathbf{r})$$

Step two

Next, we solve the integral equation resulting from Equation 8.28 on letting $\mathbf{r}\rightarrow\mathbf{r}\in\overline{\partial D}$. To this effect we set

$$\varphi\simeq\sum_{i=N+1}^{n}\varphi_i v_i(\mathbf{r}),\quad \mathbf{r}\in\overline{\partial D}$$

$$\frac{\partial\varphi}{\partial n}\simeq\sum_{i=N+1}^{n}\psi_i v_i(\mathbf{r}),\quad \mathbf{r}\in\overline{\partial D}$$

Inserting the above expressions into the integral equation and performing the collocation at points $\mathbf{r}_j\in\overline{\partial D}$, $j=N+1,N+2,\ldots,n$ we find:

$$c_j\varphi_j=-\sum_{i=1}^{N}\varphi_i\hat{A}_{ij}-\sum_{i=N+1}^{n}\varphi_i\hat{A}_{ij}+\sum_{i=N+1}^{n}\psi_i B_{ij},\quad j=N+1,\ldots \tag{8.29}$$

where now

$$\hat{A}_{ij}=-k\int_{\partial D}\hat{n}\vec{\nabla}\psi_G(\mathbf{r}_j;\mathbf{r}')v_i(\mathbf{r}')\,\mathrm{d}a \tag{8.30a}$$

$$B_{ij} = -k \int_{\partial D} \psi_G(\mathbf{r}_j;\mathbf{r}')v_i(\mathbf{r}')\,da \tag{8.30b}$$

Define again $A_{ij} = c_j\delta_{ij} + \hat{A}_{ij}$; the diagonal matrix elements A_{ii} $(i = N+1,\ldots,n)$ can be evaluated by the same method as described earlier. (Note that the A_{ii}, $i = 1,2,\ldots,N$, do not appear in Equation 8.29 and hence are unneeded.) As in Method One, we collect all the unknowns on one side of the equations, invert the resulting matrix and finally obtain

$$\begin{bmatrix} \boldsymbol{\psi}_d \\ \boldsymbol{\varphi}_n \end{bmatrix} = \begin{bmatrix} P_{11} & P_{12} \\ P_{21} & P_{22} \end{bmatrix} \begin{bmatrix} \boldsymbol{\varphi}_d \\ \boldsymbol{\psi}_n \end{bmatrix} + \mathbf{Q}\boldsymbol{\varphi}_S \tag{8.31}$$

where $\mathbf{Q}$ is a rectangular matrix with $n-N$ rows and N columns.

Equations 8.28 and 8.31 define a representation for φ in D_2 in terms of the assumed values of φ on S and of the given boundary data only. The remaining steps in the solution procedure, then, are identical with those outlined earlier. Thus:

Step three: As in Method One
Step four: As in Method One
Step five: As in Method One

8.2.3 **Illustrative example**

To illustrate this method, let us describe the various steps assuming that D_2 is a two-dimensional domain, infinite in extent and exterior to a circle of radius R. (The case of a three-dimensional infinite domain is treated in detail in Margulies[34].) We also assume that the original problem (Laplace's equation) was cast as a variational principle, although identical results are obtained when using Galerkin's method. The bounded Green's function for the exterior of a cricle of radius R is well known. It is explicitly given by

$$\begin{aligned} \psi_G(\mathbf{r};\mathbf{r}') &\equiv \psi_G(r,\theta;r',\theta') \\ &= \frac{-1}{2\pi k}\left\{\ln\frac{r_<}{R} + \sum_{n=1}^{\infty}\frac{1}{n}\left[\left(\frac{r_<}{r_>}\right)^n - \left(\frac{R}{r}\right)^n\left(\frac{R}{r'}\right)^n\right]\cos n(\theta-\theta')\right\} \end{aligned} \tag{8.32}$$

where

$$r_> \equiv \max\,(r,r'), \quad r_< \equiv \min\,(r,r')$$

Equation 8.32 can also be rewritten in the form[43]

$$\psi_G(r,\theta;r',\theta') = \frac{-1}{2\pi k}\ln\left\{\frac{\left(\frac{r}{R}\right)^2 + \left(\frac{R}{r'}\right)^2 - 2\left(\frac{r}{r'}\right)\cos(\theta-\theta')}{1 + \left(\frac{r}{r'}\right)^2 - 2\left(\frac{r}{r'}\right)\cos(\theta-\theta')}\right\} \tag{8.33}$$

We now can use Green's theorem, Equation 8.28, to write

$$\varphi(\mathbf{r})=k\int_0^{2\pi}\left[\varphi_S\frac{\partial\psi_G}{\partial r'}\right]_{r'=R}R\,\mathrm{d}\theta'-\lim_{\rho\to\infty}k\int_0^{2\pi}\left[\varphi\frac{\partial\psi_G}{\partial r'}-\psi_G\frac{\partial\varphi}{\partial r'}\right]_{r'=\rho}\rho\,\mathrm{d}\theta' \qquad (8.34)$$

Generally, in two dimensions one cannot impose as boundary condition that $\varphi\sim 0(1/r)$ as r becomes very large (under which condition, we could neglect the second integral on the right hand side of Equation 8.34). In fact, as is well known, the potential in two dimensions has a logarithmic divergence as $r\to\infty$. It can be shown[44], however, that specifying a boundary condition of the form

$$\lim_{r\to\infty}\varphi(\mathbf{r})=\frac{\psi_\infty}{2\pi}\ln\frac{R}{r}+\varphi_\infty+0\left(\frac{1}{r}\right) \qquad (8.35)$$

leads to a well posed problem, i.e. one with a unique solution. In the above equation, ψ_∞ is a specified constant (viz. the negative of the total flux extending to infinity) and φ_∞ is an unknown constant to be determined. With a boundary condition of this type and the explicit form of ψ_G, it is easy to see that Equation 8.34 reduces to

$$\varphi(\mathbf{r})=k\int_0^{2\pi}\left[\varphi_S\frac{\partial\psi_G}{\partial r'}\right]_{r'=R}R\,\mathrm{d}\theta'-\frac{\psi_\infty}{2\pi}\ln\frac{r}{R} \qquad (8.36)$$

Equation 8.36, in this very simple case, corresponds to Equation 8.31 in our general discussion of the GBEM.

As mentioned earlier, the contribution from D_2 to the total 'energy' that must be minimized is given by

$$E_2[\varphi]=\frac{k}{2}\int_{D_2}(\vec{\nabla}\varphi)^2 r\,\mathrm{d}r\,\mathrm{d}\theta$$

We now apply Green's theorem and the fact that φ is harmonic in D_2 to obtain

$$E_2[\varphi]=\frac{k}{2}\int_0^{2\pi}\left[\varphi\frac{\partial\varphi}{\partial n}\right]_{r=R}R\,\mathrm{d}\theta+\frac{k}{2}\lim_{\rho\to\infty}\int_0^{2\pi}\left[\varphi\frac{\partial\varphi}{\partial n}\right]_{r=\rho}\rho\,\mathrm{d}\theta$$

Making use of the imposed boundary condition at large distance, Equation 8.35, the second integral yields a contribution to E_2 given by

$$\frac{k}{2}\int_0^{2\pi}\left[\varphi\frac{\partial\varphi}{\partial n}\right]_{r=\rho}\rho\,\mathrm{d}\theta=\frac{k}{2}\left[\frac{\psi_\infty}{2\pi}\ln\frac{R}{\rho}+\varphi_\infty\psi_\infty+0\left(\frac{1}{\rho}\right)\right]$$

Since ψ_∞ is given (and thus fixed), the first term can in fact be neglected (even though it is logarithmically divergent as $\rho \to \infty$) because additive constants are unimportant in the variational formulation. Therefore,

$$E_2[\varphi] = \frac{k}{2} \int_0^{2\pi} \left[\varphi \frac{\partial \varphi}{\partial n} \right]_{r=R} R \, d\theta + \frac{k}{2} \varphi_\infty \psi_\infty$$

Equation 8.36 allows us to explicitly evaluate $\partial\varphi/\partial n(\mathbf{r};\mathbf{r}\in S)$ as a functional of φ_S; it then becomes straightforward to compute E_2 itself as a functional of φ_S. To avoid possible difficulties with singularities, questionable interchange of order of integrations, etc. we note that E_2 being the total energy of the field φ in D_2 must naturally be finite, up to an unimportant additive constant. In particular, then, the total energy contained in a small region between the circle S and another one, S_ε, infinitesimally close to S, must also be finite and tend to zero as the two circles coalesce. Therefore, E_2 can be computed by first evaluating $\partial\varphi/\partial n$ on S_ε (rather than on S), then performing all the necessary integrations, summations, etc., and finally taking the limit $S_\varepsilon \to S$. This procedure gives mathematical rigour and meaning to all encountered expressions and operations. With this caveat in mind, we find

$$\frac{\partial \varphi}{\partial n_\varepsilon} = \int_0^{2\pi} \left[\varphi_S \frac{\partial^2 \psi_G}{\partial r \partial r'} \right]_{r'=R+\varepsilon} (R+\varepsilon) \, d\theta' - \frac{\psi_\infty}{2\pi(R+\varepsilon)}$$

Hence,

$$E_2 = \frac{k}{2} \lim_{\varepsilon \to 0} \int_0^{2\pi} \varphi_S \int_0^{2\pi} \left[\varphi_S \frac{\partial^2 \psi_G}{\partial r \partial r'} \right]_{r'=R+\varepsilon} R(R+\varepsilon) \, d\theta \, d\theta'$$

$$- \frac{k}{2} \cdot \frac{\psi_\infty}{2\pi R} \int_0^{2\pi} \varphi_S R \, d\theta + \frac{k}{2} \psi_\infty \varphi_\infty$$

Using the explicit form of ψ_G and Equation 8.36, it is easily proven that

$$\int_0^{2\pi} \varphi_S R \, d\theta = 2\pi R \varphi_\infty$$

Thus, we finally obtain

$$E_2 = \frac{k}{2} \lim_{\varepsilon \to 0} \int_0^{2\pi} \varphi_S \int_0^{2\pi} \left[\varphi_S \frac{\partial^2 \psi_G}{\partial r \partial r'} \right]_{r'=R+\varepsilon} R(R+\varepsilon) \, d\theta \, d\theta'$$

completely analogous to the standard result for three-dimensional problems.

The stiffness matrix corresponding to D_2 can now be readily calculated. If we let

$$\varphi_S(\mathbf{r}) \simeq \sum_{i=1}^{N} \varphi_i v_i(\mathbf{r}), \quad \mathbf{r} \in S$$

we find

$$E_2(\boldsymbol{\varphi}_S) = \frac{1}{2} \boldsymbol{\varphi}_S^T \mathbf{K}_2 \boldsymbol{\varphi}_S$$

where

$$(\mathbf{K}_2)_{ij} = k \lim_{\varepsilon \to 0} \int_0^{2\pi} v_i \int_0^{2\pi} \left[v_j \frac{\partial^2 \psi_G}{\partial r' \partial r} \right]_{r'=R+\varepsilon} R(R+\varepsilon)\, d\theta\, d\theta'$$

To illustrate, let us assume that a piecewise linear trial form was chosen, i.e. we approximate φ_S as follows:

$$\varphi_S = \varphi_i + \frac{\theta - \theta_i}{\Delta\theta}(\varphi_{i+1} - \varphi_i), \quad \theta_i \leqslant \theta \leqslant \theta_{i+1} \quad i = 1,2,\ldots N$$

$$\theta_i = (i-1)\Delta\theta, \quad i = 1,2,\ldots,N$$

$$\Delta\theta = \frac{2\pi}{N}$$

(recalling that $\theta_{N+1} = \theta_1$). It can then be shown[34] that the stiffness matrix elements are given by

$$(\mathbf{K}_2)_{ij} = \frac{k}{8\pi}\{(i-j+4)^2 \ln (i-j+4)^2 + (i-j-4)^2 \ln (i-j+4)^2$$
$$-4(i-j-2) \ln (i-j-2)^2 - 4(i-j+2)^2 \ln (i-j+2)^2 \qquad (8.37)$$
$$+6(i-j)^2 \ln (i-j)^2\}$$

From Equation 8.37 we notice that each $(K_2)_{ij}$ depends only on $(i-j)$. As a consequence, the stiffness matrix for D_2 is of Toeplitz form, i.e.

$$\mathbf{K}_2 = \begin{bmatrix} a_1 & a_2 & a_3 & \cdots & a_N \\ a_2 & a_1 & a_2 & \cdots & a_{N-1} \\ a_3 & a_2 & a_1 & \cdots & a_{N-2} \\ \vdots & \vdots & \vdots & & \vdots \\ a_N & a_{N-1} & & \cdots & a_1 \end{bmatrix}$$

It is also immediately obvious that $a_2 = a_N$, $a_3 = a_{N-1}$, etc. Hence, K_2 can be generated very easily and requires the computation of at most only $(N/2+1)$ numbers. Furthermore, its particular form and symmetry property result in significant savings, both in terms of storage and in terms of the resolution of the final system of linear equations. Once the solution has been found, the potential at an arbitrary point of D_2 can be evaluated easily. Using Equation 8.36 we find, in matrix notation:

$$\varphi(\mathbf{r}) = \boldsymbol{\varphi}_S^T \mathbf{M}(\mathbf{r}) - \frac{\psi_\infty}{2\pi} \ln \frac{R}{r} \qquad r \in D_2$$

where

$$M_i(\mathbf{r}) \equiv \int_0^{2\pi} \left[v_i \frac{\partial \psi_G}{\partial r'} \right]_{r'=R} R \, \mathrm{d}\theta'$$

The two methods of combining the BEM and the FEM described here are applicable only if the region in which the BEM is implemented has homogeneous material properties. This requirement can be relaxed in some special cases, the most prominent of which occurs when D_2 consists of two homogeneous regions (with different material properties) separated by a single plane interface. It is usually possible then to draw upon the theory of images (be they 'electrical' images or 'elastic' images[45]) to obtain a Green's function (free space or bounded) which satisfies the governing equations in D_2 and the required boundary conditions on the plane interface. Having obtained such a Green's function, either of the methods can be applied as described above; the only explicit change will then occur in that the particular form of ψ^* or ψ_G used in Equation 8.9 or Equation 8.28 must be modified to reflect the presence of 'image' sources.

8.3 APPLICATIONS

As mentioned at the beginning of this chapter, combination of the FEM and the BEM is particularly attractive when dealing with problems posed in infinite domains. Following are a few illustrations of applications of this kind.

Structural mechanics[36]

Consider the situation of a building standing of an infinite halfspace representing the soil (see Figure 8.1). The finite element domain represents a structure and the halfspace boundary element domain the infinite foundation. Note that since the plane representing the soil extends to 'infinity', it is in principle necessary to introduce an infinite number of boundary elements. This difficulty can be eliminated in practice in either of two ways:

(1) Truncate the plane at finite distance. (Approximate method, as shown in Figure 8.1 and used in the present example.)

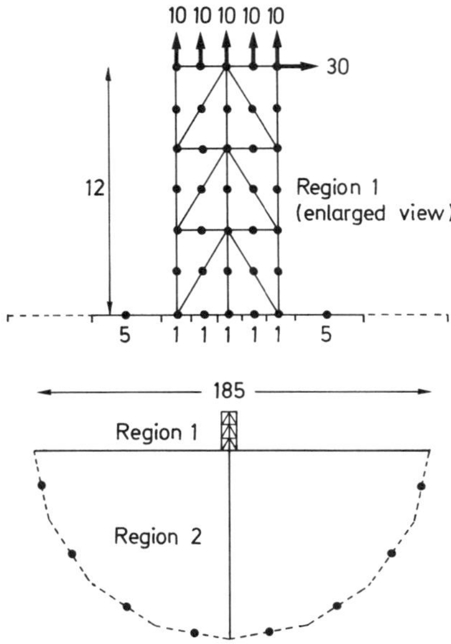

Figure 8.1 Finite element and boundary element combination mesh

(2) Use a fundamental solution appropriate to the halfspace problem (without the structure), rather than a free space-Green's function[46]. With this last approach, boundary elements are needed only at the interface between the structure and the soil, yet the remainder of the soil surface is rigorously taken into account.

Regions 1 and 2 shown in Figure 8.1 were then combined, where region 2 can be a boundary element on a finite element domain (Figure 8.2). The combination of the boundary element region 2 with the finite element region 1 was carried out by finding an equivalent symmetric stiffness matrix for the boundary element domain (similar to the first method shown in this chapter). Five concentrated vertical loads along the top and an additional horizontal load acting at a corner were considered. The results obtained are shown in Table 8.1.

The combination solution is in excellent agreement with the results obtained using finite elements for the whole domain (Figure 8.2), but it is interesting to note that the foundation was adequately represented using only 37 boundary element nodes as opposed to 163 for the finite element case.

Electrostatics

Coupling of the BEM and the FEM is used in this example to compute the electrical field surrounding a two dimensional charged parallel place capacitor[35]. The surface S separating the interior and exterior regions is shown in Figure 8.3. Combination of the two techniques was implemented by the FBEM. As can be seen, agreement with the solution derived by a different technique[11] is excellent.

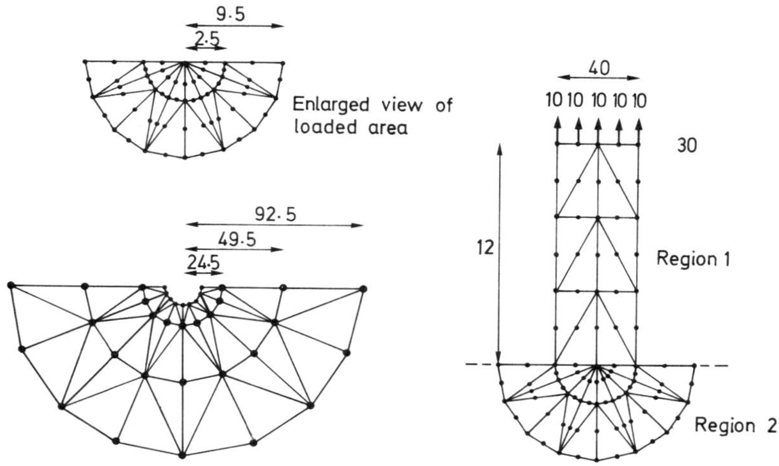

Figure 8.2 Finite element mesh

Table 8.1 VERTICAL DISPLACEMENTS ALONG LOADED TOP ($\times 10^{-6}$)

FE solution	*Combination solution*
−339	−355
−97	−105
135	135
361	370
600	617

Steady state heat conduction

This example[47] involves heat conduction (governed by Laplace's equation) in one quarter of a two-dimensional disc. The temperature is specified on the inner and outer circular boundaries and the flux on the radial edges is required to vanish (see Figure 8.4). Thermal properties are homogeneous and isotropic throughout the domain. A coupled discretization involving linear finite elements and boundary integral segments with linear temperature variation is illustrated (the surface S is chosen arbitrarily). Agreement with the known analytical solution is excellent. Note that this example is purely pedagogical as the BEM (or the FEM) could equally well have been used as a stand-alone solution method in this particular case.

Shallow water wave propagation

Wave propagation in shallow water is governed by the Helmholtz equation (for constant water depth). A typical situation is the interaction of an incident wave with an obstacle. Figure 8.5 shows an exterior problem of wave

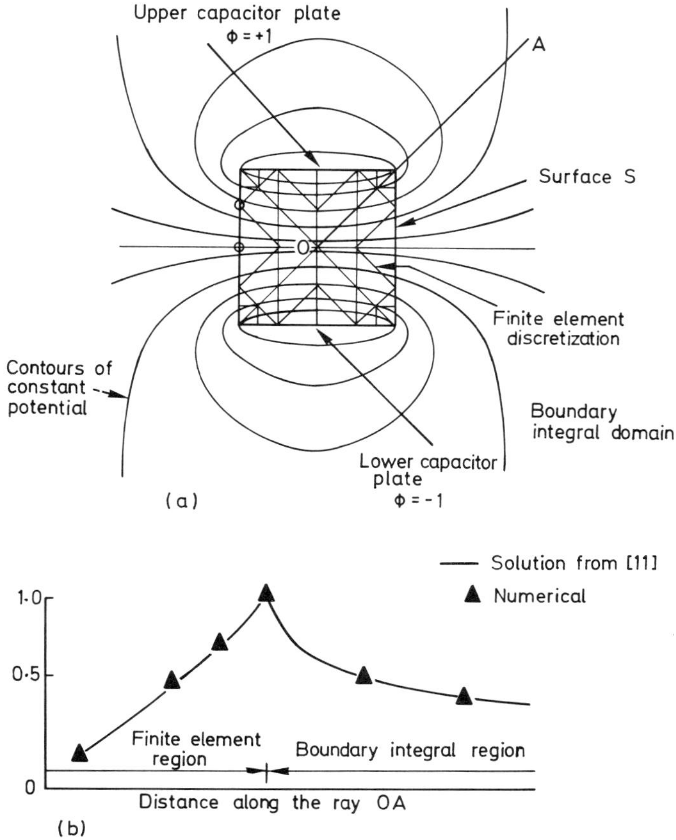

Figure 8.3

refraction caused by a cylinder placed in a field of uniform waves[47]. An exact solution is available, and comparison between it and the numerical results shows excellent accuracy. Although not particularly advantageous here, the field was divided into a set of conventional (isoparametric parabolic) elements and a single exterior boundary element.

A similar technique was used to investigate the more realistic problem of the resonance of a harbour of variable depth on a straight coastline of infinite length[24]. Resonant wavelengths and amplitudes are found to correlate well with those obtained by different approaches.

Another class of applications consists in the analysis of fracture mechanics. Two approaches have been used to deal with this problem. In the first[48], the crack itself is used as part of the boundary and the free space fundamental solution is used to derive an integral equation for displacements and tractions (both finite) on the crack surface and the surface adjacent to the FEM region. Solution of this integral equation then allows one to compute stress intensity

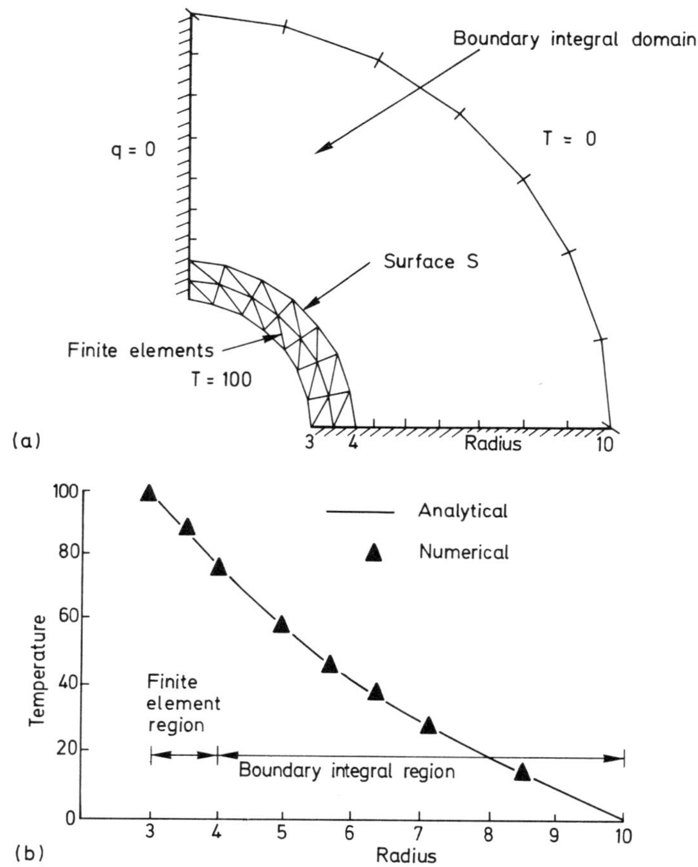

Figure 8.4

factors K_I, K_{II} and K_{III} which are the quantities of interest when analysing the possibility of crack propagation. In the second method, a fundamental solution corresponding to a crack in an infinite homogeneous medium is used[49,50]. A domain D_2 surrounding the crack is introduced as a single element in the FEM and the representation for the relevant fields in that domain is derived using the fundamental solution and a boundary integral equation on the surface separating the domain containing the crack from the rest of the solid. Once the solution has been found on that surface, stresses and displacements can be computed at every point of D_2 and thence stress intensity factors derived.

Geophysical modelling

A typical example in geophysical modelling[15,16] is presented as the final illustration. The problem is to compute the potential at any point on the

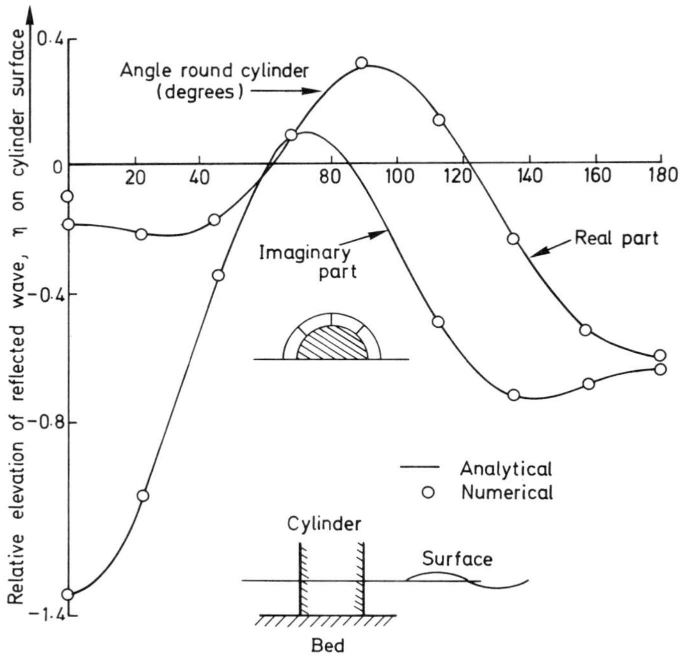

Figure 8.5

symmetry axis of a vertical borehole due to a moving current source, located on that axis as well. As shown in Figure 8.6, the geological formation consists of many horizontal layers each having different electrical conductivities. Because the electrical potential has a singularity at the location of the point source, it is very difficult if not impossible to compute the potential with accuracy near the current source, no matter how fine the mesh chosen. Furthermore, since the solution is sought for many source locations, even if that were an acceptable method it would rapidly become very combersome to redefine the fine grid for each new case. Both of these difficulties can be eliminated at one fell swoop by using the BEM inside the borehole and connecting it to the FEM in the formation (thus preserving the possibility of modelling general, complicated geological strcutres). In order to insure perfect matching between the BEM and FEM solutions, this problem was solved using a bounded Green's function appropriate to the inside of an infinitely long cylindrical region. Details of the procedure are given in Margulies[16]; the results yielded by this approach were verified in simple cases for which the exact solution was known. Agreement between the numerical method and the analytical or analog one is excellent (see Figures 8.7 and 8.8).

8.4 CONCLUSIONS

In this chapter we have presented general classes of problems, often encountered in practice, for which coupling of the BEM and the FEM results

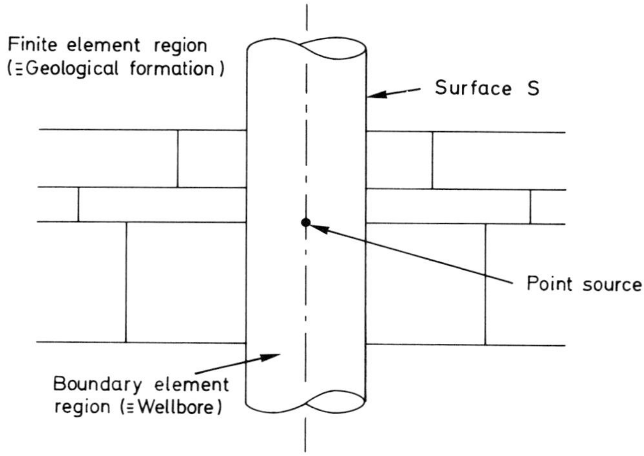

Figure 8.6

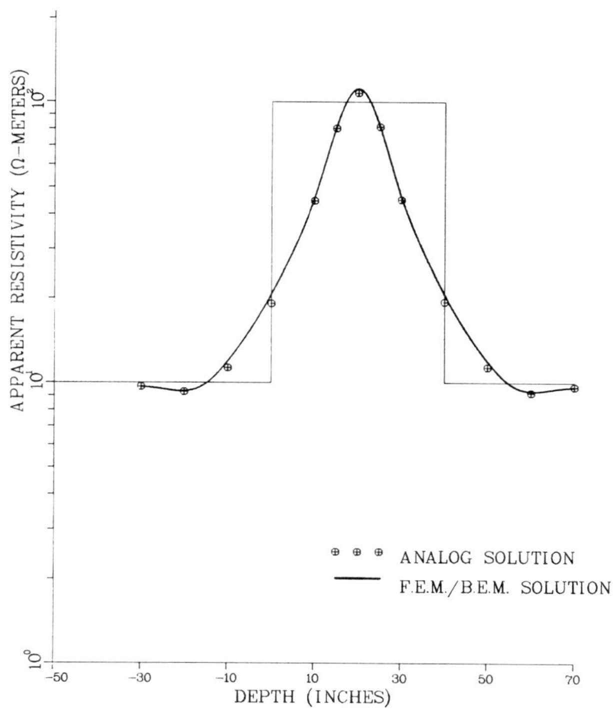

Figure 8.7 Comparison of analog solution with FEM/BEM solution

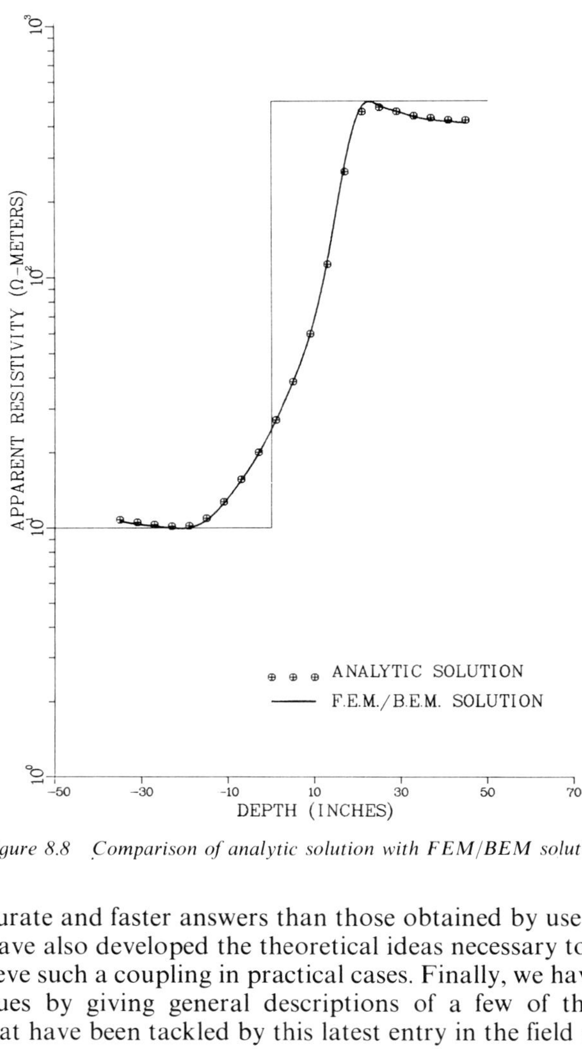

Figure 8.8 Comparison of analytic solution with FEM/BEM solution

in more accurate and faster answers than those obtained by use of the FEM alone. We have also developed the theoretical ideas necessary to understand how to achieve such a coupling in practical cases. Finally, we have illustrated the techniques by giving general descriptions of a few of the numerous problems that have been tackled by this latest entry in the field of numerical methods.

REFERENCES

1 Martin, H. C., 'Finite element analysis of fluid flows', *Proc. 2nd Conf. on Matrix Methods in Structural Mechanics*, Air Force Inst. Tech., Wright-Patterson A. F. Base, Ohio (1968)

2 de Vries, G. and Norrie, D. H., 'The application of the finite element technique to potential flow problems', *J. Appl. Mech., Am. Soc. Mech. Eng.*, **38**, pp. 798–802 (1971)

3 Huebner, K. H., 'Application of finite element methods to thermo-hydrodynamic lubrication', *Int. J. Num. Meth. Engng*, **8**, pp. 139–168 (1974)

4 Argyris, J. H., 'Two and three dimensional flow using finite elements', *J. Roy. Aero. Soc.*, **73**, pp. 961–964 (1969)

5 Oden, J. T. *et al.* (Eds.), *Proc. 1st Symp. Finite Element Methods in Flow Problems*, Swansea, Univ. of Alabama Press (1974)
6 Oden, J. T. (Ed.), *Finite Elements in Fluids*, Vols. I and II, J. Wiley and Sons, New York (1975)
7 Connor, J. J. and Brebbia, C. A., *Finite Element Techniques for Fluid Flow*, Newnes-Butterworths (1976)
8 Desai, C. S., 'Finite element methods for flow in porous media', in *Finite Elements in Fluids, Vol. 1*, R. H. Gallagher (Ed.), pp. 157–182 (1975)
9 Javandel, I. and Witherspoon, P. A., 'Applications of the finite element method to transient flow in porous media', *Trans. Soc. Petr. Eng.*, **243**, pp. 241–251 (1968)
10 Taylor, R. L. and Brown, C. A., 'Darcy flow solutions with a free surface', *Proc. Am. Soc. Civ. Eng.*, **93**, HY2, pp. 25–33 (1967)
11 Silvester, P. and Hsieh, M. S., 'Finite element solution of two dimensional exterior field problems', *Proc. IEE*, **118**, pp. 1743–1747 (1971)
12 McDonald, B. H. and Wexler, A., 'Finite element solution of unbounded field problems', *Proc. IEEE*, MTT–20, pp. 841–847 (1972)
13 Chari, M. V. K. and Silvester, P. (Eds.), *Finite Elements in Electrical and Magnetic Field Problems*, J. Wiley and Sons (1978)
14 Zienkiewicz, O. C., 'The finite element method and the solution of some geophysical problems', *Phil. Trans. R. Soc.* (London), **A283**, pp. 139–151 (1976)
15 Margulies, M., 'Modelling of resistivity sondes: a numerical approach', Schlumberger-Doll Research Report, (unpublished) (1980)
16 Margulies, M., 'Electrical well logging modelling by a combination of the boundary element and finite element methods', *1st Chautauqua on Finite Element Modelling*, J. H. Conaway (Ed.), pp. 353–366 (1980)
17 Douglas, A., 'Finite elements in geological modelling', *Nature*, **222**, pp. 630–631 (1970)
18 Coggon, J. H., 'Electromagnetic and electrical modelling by the finite element method', *Geophysics*, **36**, pp. 132–155 (1971)
19 Ruiz, L., 'Modelling of electric and electromagnetic data', Ph.D. thesis, Univ. of Utah (1977)
20 Chen, H. S. and Mei, C. C., 'Oscillations and wave forces in a man-made harbour in open sea', *10th Naval Hydrodynamics Symposium*, Dept. of Civil Eng., M.I.T., Cambridge, Mass. (1974)
21 Zienkiewicz, O. C. and Bettess, P., 'Infinite elements in the study of fluid-structure interaction problems', *2nd Int. Symp. on Computing Methods in Applied Science and Engineering*, IRIA, Versailles, France (1975)
22 Bettess, P., 'Infinite elements', *Int. J. Num. Meth. Engng*, **11**, pp. 53–64 (1977)
23 Bettess, P. and Zienkiewicz, O. C., 'Diffraction and refraction of surface waves using finite and infinite elements', *Int. J. Num. Meth. Eng.*, **11**, pp. 1271–1290 (1977)
24 Shaw, R. P. and Falby, W., 'FEBIE — A combined finite element-boundary integral equation method', *Computers and Fluids*, **6**, pp. 153–160 (1978)
25 Beer, G. and Meek, J. L., "A 'Boundary Finite Element' for underground mining applications", in *New Developments in Boundary Element Methods*, C. A. Brebbia (Ed.), CML Publications, Southampton (1980)
26 Irwin, G. R., 'Fracture', *Handbuch der Physik*, Vol. 6, pp. 551–590, Springer, Berlin (1958)
27 Griffiths, A. A., 'The Phenomena of flow and rupture in solids', *Phil. Trans. Roy. Soc.*, **A221**, pp. 163–198 (1920)
28 Anderson, D. L. and Ungless, R. L., 'Infinite finite elements', *Int. Symp. Innovation in Numerical Analysis in Engng Sciences*, CETIM, Senlis, France, pp. 2.13–2.16 (1977)
29 Gartling, D. K. and Becker, E. B., 'Finite element analysis of viscous incompressible flow', *Comp. Meth. Appl. Mech. Engng*, **8**, pp. 51–60 (1976)
30 Tong, P. and Pian, T. H. H., 'A hybrid-element approach to crack problems in plane elasticity', *Int. J. Num. Meth. Eng.*, **7**, pp. 297–308 (1973)
31 Tong, P. and Pian, T. H. H., 'On the convergence of the finite element method for problems with singularity', *Int. J. Solids Struct.*, **9**, pp. 313–321 (1972)
32 Hormander, H., 'On the theory of general partial differential operators', *Acta Math.*, **94**, pp. 162–248 (1955)
33 Kellogg, O. D., *Foundations of Potential Theory*, F. Ungar Publishing Co., New York

34 Margulies, M., 'Exact treatment of the exterior problem in the combined FEM–BEM', in *New Developments in Boundary Element Methods*, C. A. Brebbia (Ed.), CML Publications, Southampton (1980)

35 Zienkiewizc, O. C., Kelly, D. W. and Bettess, P., 'The coupling of the finite element method and boundary solution procedures', *Int. J. Num. Meth. Engng*, **11**, 355–375

36 Brebbia, C. A. and Georgiou, P., 'Combination of boundary and finite elements in elastostatics', *Appl. Math. Modelling*, **3**, No. 2, (1979)

37 Brebbia, C. A., *The Boundary Element Method for Engineers* 2nd rev. edn., Pentech Press, London (1980)

38 Jeng, G. and Wexler, A., 'Isoparametric, finite element, variational solution of integral equations for three-dimensional fields', *Int. J. Num. Meth. Eng.*, **11**, pp. 1455–1471 (1977)

39 Lachat, J. C., 'A further development of the boundary integral technique for elastostatics', Ph.D. thesis, University of Southampton, (unpublished) (1975)

40 Lachat, J. C. and Watson, J. O., 'Effective numerical treatment of boundary integral equations: a formulation for three-dimensional elastostatics', *Int. J. Num. Meth. Engng*, **10**, pp. 991–1005 (1976)

41 Cristescu, M. and Loubignac, G., 'Gaussian quadrature formulas for functions with singularities in $1/R$ over triangles and quadrangles', in *Recent Advances in Boundary Element Methods*, C. A. Brebbia (Ed.), Pentech Press, London (1978)

42 Rao, S. M. *et al.*, 'A simple numerical solution procedure for static problems involving arbitrary-shaped surfaces', *IEEE Trans.*, **AP–27**, pp. 604–607 (1979)

43 Morse, P. and Feshbach, H., *Methods of Theoretical Physics*, McGraw-Hill, New York (1953)

44 Hsiao, G. and MacCramy, R. C., 'Solution of boundary value problems by integral equations of the first kind', *SIAM Review*, **15**, pp. 687–705 (1973)

45 Mindlin, R. D., 'Force at a point in the interior of a semi-infinite solid', *Physics*, **7**, pp. 195–202 (1936)

46 Pircher, H. and Beer, G., "On the treatment of 'infinite' boundaries in the finite element method", *Int. J. Num. Meth. Engng*, **11**, pp. 1194–1197 (1977)

47 Zienkiewicz, O. C., *The Finite Element Method*, McGraw-Hill, London (1977)

48 Cruse, T. A. and Vanburen, W., 'Three dimensional elastic stress analysis of a fracture specimen with edge crack', *Int. J. Fract. Mech.*, **7**, pp. 1–15 (1971)

49 Snyder, M. D. and Cruse, T. A., 'Boundary-integral equation analysis of crack anisotropic plates', *Int. J. Fract.*, **11**, pp. 315–328 (1975)

50 Rao, A. K., 'A powerful hybrid method in finite element analysis', *Int. J. Num. Meth. Engng*, **3**, pp. 389–403 (1971)

APPENDIX

Helmholtz equation

(Electromagnetic wave propagation, shallow water wave propagation, sound propagation in acoustical medium, etc.)

$$k\nabla^2\varphi + w^2\varphi = 0$$

Green's theorem (Equation 8.7) and the resulting boundary integral equation (Equation 8.9) are unchanged in form. The free space Green' functions, however, are now given by

$$\psi^*(\mathbf{r};\mathbf{r}') = -\frac{1}{4ik}H_0^{(2)}\left(\frac{w}{k^{1/2}}|\mathbf{r}-\mathbf{r}'|\right) \qquad \text{in 2-D}$$

$$\psi^*(\mathbf{r};\mathbf{r}')=\frac{-1}{4\pi k|\mathbf{r}-\mathbf{r}'|}\exp\left[-i\frac{w}{k^{1/2}}|\mathbf{r}-\mathbf{r}'|\right] \qquad \text{in 3-D}$$

where $H_0^{(2)}$ is the zeroth order Hankel function of the second kind.

Isotropic elastostatics

The governing field equations are written in the form

$$\frac{\partial}{\partial x_j}\sigma_{ij}[\mathbf{v}]-F_i(\mathbf{r})=0$$

where $\mathbf{v}(\mathbf{r})$ is the displacement vector at a point $\mathbf{r}$, σ_{ij} denotes the stress field corresponding to $\mathbf{v}$ and $\mathbf{F}$ represents the body forces. Green's theorem (Equation 8.7) is now replaced by Betti's theorem

$$\int_v\left[b_j\frac{\partial\sigma_{jl}[\mathbf{a}]}{\partial x_l}-a_j\frac{\partial\sigma_{jl}[\mathbf{b}]}{\partial x_l}\right]\mathrm{d}v$$

$$=\int_S\{b_j\sigma_{jl}[\mathbf{a}]-a_j\sigma_{jl}[\mathbf{b}]\}n_l\,\mathrm{d}a$$

Upon replacing in the above equation a_j by u_j and b_j by $u^*_{ij}(\mathbf{r};\mathbf{r}')$, the fundamental solution, we obtain Betti's reciprocal work theorem. Since u^*_{ij} is not differentiable at $\mathbf{r}$, the theorem must be written for $D-D_\varepsilon$, where D_ε isolates the point $\mathbf{r}$. Thus,

$$\int_{S+S_\varepsilon}\{u_i t^*_{ij}-t_i u^*_{ij}\}\,\mathrm{d}S=0$$

the tractions t_i and t^*_{ij} being defined by

$$t_i=\sigma_{ij}n_j$$

$$t^*_{ij}=\sigma_{il}[\mathbf{u}^*_j]n_l$$

Now, let the observation point tend to the surface S and take the limit $\varepsilon\to 0$. This yields an integral equation

$$c_{ij}(\mathbf{r})u_j(\mathbf{r})=\fint_S\{t^*_{ij}(\mathbf{r};\mathbf{r}')u_j(\mathbf{r}')-u^*_{ij}(\mathbf{r};\mathbf{r}')\,t_j(\mathbf{r}')\}\,\mathrm{d}a \qquad \text{(A8.1)}$$

with

$$c_{ij}(\mathbf{r}) \equiv \lim_{\varepsilon \to 0} \int_{S_\varepsilon} t_{ij}^*(\mathbf{r};\mathbf{r}') \, \mathrm{d}a$$

Equation (A8.1) is the analog of Equation 8.9 in the main text. Upon solving it, displacements and stresses can be computed at any point of D by using Somigliana's identity, derived from Betti's theorem on taking the limit $\varepsilon \to 0$ for $\mathbf{r} \in D$:

$$u_i(\mathbf{r}) = \int_S \{u_{ij}^*(\mathbf{r};\mathbf{r}')t_j(\mathbf{r}') - t_{ij}^*(\mathbf{r};\mathbf{r}')u_j(\mathbf{r}')\} \, \mathrm{d}a$$

The fundamental solutions u_{ij}^* are explicitly given by:

$$u_{ij}^* = \frac{-1}{8\pi\mu(1-v)} \left[(3-4v)\ln\frac{1}{|\mathbf{r}-\mathbf{r}'|}\delta_{ij} + |\mathbf{r}-\mathbf{r}'|_{,i}|\mathbf{r}-\mathbf{r}'|_{,j} \right], \quad \text{in 2-D}$$

$$u_{ij}^* = \frac{-1}{16\pi\mu(1-v)|\mathbf{r}-\mathbf{r}'|}[(3-4v)\delta_{ij} + |\mathbf{r}-\mathbf{r}'|_{,i}|\mathbf{r}-\mathbf{r}'|_{,j}], \quad \text{in 3-D}$$

where v is Poisson's ratio and μ Lame's constant.

9

Asymptotic accuracy and convergence

by W. L. Wendland

9.1 INTRODUCTION

Here we are concerned with boundary element methods governed by a class of boundary integral equations satisfying a coerciveness inequality in the form of Garding's inequality and having convolutional operators as principal parts. This class is characterized in Section 9.2. It turns out that most problems of the applications belong to this class[16,40]. Since it seems to be a too pretentious task to find computable error bounds, we are concerned with asymptotic results in terms of the order of convergence with respect to the meshwidth h of the boundary elements. Such results are well known if one uses Galerkin's method for approximating the original equations. In Section 9.3 we formulate Galerkin's method and in Section 9.4 we use finite element functions on the bounary manifold, i.e. boundary elements for the trial and test functions. Here we find optimal order of convergence following Stephan and Wendland[37], including super approximation due to Hsiao and Wendland[20].

Since Galerkin's method requires a double integration for the calculation of the stiffness matrix, we replace the integrations by suitable numerical quadratures. To this end we assume that the principal part of the equation is given by a convolutional integral. After converting the principal part into a standard form, the corresponding weights in the stiffness matrix can be evaluated from two vectors of weights forming a Töplitz matrix provided the finite elements are defined by shifting and stretching subject to a regular uniform grid. The vector of weights can be evaluated independently of the boundary and independently of the meshwidth. For all remaining smooth terms we use numerical integrations with node points connected with the finite element grid. For $n=2$, i.e. boundary elements on a curve we perform this method in Section 9.5. It results in a very simple and effective modified collocation scheme which we call *Galerkin-collocation*. In Section 9.6 we execute the asymptotic error analysis for the Galerkin-collocation provided one uses enough node points.

For some examples the Galerkin-collocation has been used also for numerical computations[16,17,40] but usually the boundary element method is discretized numerically with the standard collocation method. In Section 9.7

we announce some yet unpublished recent asymptotic convergence results for the standard collocation of the same integral equations as in Section 9.5.

For convenience, in the asymptotic error analysis we are using $c,c',\ldots$ as generic constants which might change their size and meaning at different places.

9.2 STRONGLY ELLIPTIC INTEGRAL EQUATIONS AND COERCIVITY

All boundary element methods can be considered as to be approximations of corresponding boundary integral equations. Conversely, all boundary value problems of mathematical physics with elliptic differential equations can also be transformed into equivalent boundary integral equations provided the fundamental solution of the differential operator is available. For practical purposes this restricts us essentially to problems with constant (or piecewise constant) coefficients. The boundary integral equations are by no means uniquely determined by the boundary value problem. In general there are several different integral equations being equivalent to the original problem. For example, to the Dirichlet problem of the Laplacian in a bounded domain $\Omega \in R^n$ with boundary Γ one may associate Carl Neumann's classical integral equation

$$Au(x)=u(x)+\frac{2}{\omega_n}\int\limits_{y\in\Gamma} u(y)\frac{\nu(y)(y-x)}{|y-x|^n}\,\mathrm{d}o(y)=f(x),\ x\in\Gamma \tag{9.1}$$

for the desired double layer $u(x)$, if the original solution is required as a double layer potential. Here ω_n denotes the surface of the unit sphere in R^n, $\nu(y)$ is the outer normal to Γ at the integration point y and $\mathrm{d}o(y)$ denotes the corresponding surface element; $|\cdot|$ denotes the Euclidean length. The integral Equation 9.1 is *the* Fredholm integral equation of the *second* kind (see, e.g. Kral and Netuka[23]).

If the solution of the Dirichlet problem is required to be a simple layer potential with charge $u(y)$ then one obtains a different integral equation, namely

$$Au(x)=\int\limits_{y\in\Gamma} u(y)\gamma(x,y)\,\mathrm{d}o(y)=f(x),\ x\in\Gamma \tag{9.2}$$

where γ denotes the fundamental solution of the Laplacian,

$$\gamma(x,y)=\begin{cases}-\dfrac{1}{2\pi}\log|x-y|, & \text{for } n=2,\\ \dfrac{1}{\omega_n}|x-y|^{-n+2}, & \text{for } n\geqslant 3.\end{cases}$$

This is a Fredholm integral equation of the *first* kind. A third type of integral equation often comes up in connection with oblique problems and is of the form

$$Au(x)=a(x)u(x)+\frac{2}{\omega_n}p.v.\int\limits_{y\in\Gamma}\frac{b(x,y)}{|x-y|^{n-1}}u(y)\,\mathrm{d}o(y)=f(x) \tag{9.3}$$

where the integral operator is defined by the *Cauchy principal value* integral. (For the appropriate definition of a singular integral operator see, e.g. Michlin and Prössdorf[25], chapter XIII). For $n=2$, Equation 9.3 is a singular integral operator with Cauchy kernel on a curve Γ and with additional smoother kernels and $b(x,x+0)=-b(x,x-0)$.

As the fourth example let us consider the classical Neumann problem for the Laplacian. If one uses Green's identity or if the solution is required as to be a simple layer potential, then one again finds a Fredholm integral equation of the second kind which is very similar to Equation 9.1, (see e.g. Haack and Wendland[13] formulas 4.11.13, 4.11.21, 22.2.3) or its adjoint (see e.g. Riesz and Nagy[33] Chapter 81). But if the solution is required to be a double layer potential then the Neumann boundary condition yields the integro differential equation

$$Au(x)=-\frac{\partial}{\partial \nu_x}\frac{2}{\omega_n}\int\limits_{\Gamma}u(y)\frac{\nu(y)\cdot(y-x)}{|y-x|^n}\,\mathrm{d}o(y)=f(x),\ x\in\Gamma \tag{9.4}$$

where the normal derivative $\partial/\partial\nu_x\ldots$ of the double layer potential on Γ can also be expressed by the bounded part of a principal value integral.

All these types of equations or corresponding systems appear in applications as Equation 9.1 in classical potential theory, scattering theory and inviscid incompressible ideal flows (see e.g. references in Kleinman and Wendland[21]), Equation 9.2 in conformal mappings, viscous flows, electrostatics and elastomechanics, Equation 9.3 in elasticity, thermoelasticity and electromagnetic field theory, and Equation 9.4 in all the above mentioned fields. (For references see, e.g. Wendland[40], Section 5.)

Although these four types of equations have very different properties in the classical theory of integral equations it turns out that if they are considered as so called *pseudodifferential operators* they have a very strong, common property. Namely the equations of practical interest are 'strongly elliptic'. In order to formulate this property we need to introduce the Sobolev spaces $H^s(\Gamma)$ of generalized functions on Γ, their interpolation spaces and their dual spaces. For simplicity we assume that the $(n-1)$-dimensional boundary manifold Γ is sufficiently smooth.

Following Adams[3] and Triebel[38], we define the real function spaces

$$H^s(\Gamma):=\left\{f\,|\ \|f\|^2_{H^s}:=\sum_{|\alpha|\leqslant s}\int\limits_{\Gamma}|D^\alpha f|^2\,\mathrm{d}o<\infty\right\} \tag{9.5}$$

for $s \geqslant 0$ an integer and

$$H^s(\Gamma) := \{f \mid \|f\|^2_{H^s} := \|f\|^2_{H^{[s]}} + \sum_{|\alpha|=[s]} \int_\Gamma \int_\Gamma \frac{|D^\alpha f(x) - D^\alpha f(y)|^2}{|x-y|^{n-1+2(s-[s])}} \, do\, do < \infty\} \tag{9.6}$$

for non integer s with $[s] < s < [s]+1$, where $[s]$ denotes the integer part of s. Here

$$D^\alpha = \frac{\partial^{\alpha_1}}{\partial u_1^{\alpha_1}} \cdots \frac{\partial^{\alpha_{n-1}}}{\partial u_{n-1}^{\alpha_{n-1}}}$$

denote covariant differentiations of order α within Γ. For $s<0$ we define $H^s(\Gamma)$ by duality with respect to the L_2 scalar product

$$(f,g) := \int_\Gamma fg \, do$$

$$H^s(\Gamma) := \left\{ F \text{ distribution on } \Gamma \text{ with } \|F\|_{H^s} := \sup_{\|f\|_{H^{-s}} \leqslant 1} \left| \int_\Gamma Ffdo \right| < \infty \right\}$$

Then each of the above operators A (9.1–9.4) defines a continuous linear mapping $A: H^s \rightarrow H^{s-2\alpha}$ for a whole scale of real s (depending on the smoothness of Γ). 2α is called the *order* of the pseudodifferential operator A, Seeky[36]. (Richter[32] calls -2α the 'smoothing index'.) For our examples we have

(1) $2\alpha = 0$, (2) $2\alpha = -1$, (3) $2\alpha = 0$, (4) $2\alpha = 1$.

The announced common property of (1), (2) and (4) is the Garding inequality or

Coerciveness:
$$Re(Av,v) = Re \int_\Gamma vAv \, do \geqslant \gamma \|v\|^2_{H^\alpha} - |k[v,v]|, \quad \text{for all } v \in H^\alpha(\Gamma) \tag{9.8}$$

where $\gamma > 0$ is a constant independent of v and where $k[u,v]$ denotes a compact bilinear form on $H^\alpha \times H^\alpha$.

In some cases Equation 9.8 can be proved directly as in Le Roux[34] and Hsiao and Wendland[19] or it is equivalent to the energy estimates for the corresponding solution of the original boundary value problem and Equation 9.8 holds even with $k=0$, as in Giroire and Nedelec[12] and Nedelec[28].

In general, Equation 9.8 is a consequence of a simple property of the principal part of A which has a so called 'principal symbol' $a_0(x,\xi)$ for $x\in\Gamma$ and $\xi\in R^{n-1}$. a_0 is not uniquely determined, it depends on the local representation of Γ (Seeley[36]). If the principal part of A is given by a convolution as in Equations 9.2 and 9.4, one may replace Γ by its tangential plane in x and find a_0 by the $(n-1)$ dimensional Fourier transformed kernel function. For our examples the principal symbols are given by

(1) $a_0=1$,

(2) $a_0(x,\xi)=\dfrac{1}{|\xi|}$,

(3) and $n=2{:}a_0(x,\xi)=a(x)+ib(x,x+0)\dfrac{\xi}{|\xi|}$, for $n>2$ see Michlin and Prössdorf[25]

(4) $a_0(x,\xi)=|\xi|$.

An integral operator A, or more general, pseudodifferential operator A of order 2α is called *strongly elliptic*, if

$$Re\ a_0(x,\xi)\geqslant\gamma'>0 \text{ for all } |\xi|=1 \text{ and } x\in\Gamma \tag{9.9}$$

where γ' is independent of x and ξ.

Obviously, the operators in (1), (2) and (3) are strongly elliptic. Only for (3) strong ellipticity, Equation 9.9, i.e.

$$Re\{a(x)+ib(x,x)\ \text{sign}\ \xi\}\geqslant\gamma'>0 \text{ for } \xi=\pm1 \tag{9.10}$$

for $n=2$ (and for $n>2$ correspondingly) is an additional requirement on A.

Due to Kohn and Nirenberg[22] *strong ellipticity*, Inequality 9.9, *implies coerciveness*, Inequality 9.8.

For most boundary integral equations of practical interest it turns out that they are strongly elliptic (see Wendland[40]). However, most problems of practical applications yield more than just one boundary integral equation but systems, i.e. u have to be replaced by a vector of unknowns $\vec{u}(x)=\{u_1(x),\ldots,u_p(x)\}$. Then the boundary integral equations form a *system*

$$A\vec{u}=\left(\sum_{k=1}^{p}A_{jk}u_k\right)_{j=1,\ldots,p}=\vec{f} \text{ on } \Gamma \tag{9.11}$$

and our concept of strong ellipticity has to be modified. To this end we assume that to A there exists an index vector $\alpha=(\alpha_1,\ldots\alpha_p)\in IR^p$ such that the numbers

$$\alpha_{qr}=\alpha_q+\alpha_r,\quad q,r=1,\ldots,p \tag{9.12}$$

are the orders of A_{jk} with corresponding principal symbols $a_{ojk}(x,\xi)$. Then the *system*, Equation 9.11, is called *strongly elliptic* if

$$Re \sum_{j,k=1}^{p} \zeta_j a_{ojk}(x,\xi)\bar{\zeta}_k \geqslant \gamma'|\zeta|^2 \tag{9.13}$$

for all $\xi \in C^p$, all $x \in \Gamma$ and all $\xi \in R^{n-1}$ with $|\xi| = 1$, where $\gamma' > 0$ is independent of ζ, x and ξ. Now

$$H^\alpha(\Gamma) - H^{\alpha_1}(\Gamma) \times H^{\alpha_2}(\Gamma) \times \ldots \times H^{\alpha_p}(\Gamma)$$

and $\vec{v} \in H^\alpha$ means

$$v_j \in H^{\alpha_j} \text{ for } j = 1,\ldots,p, \quad ||\vec{v}||^2_{H^\alpha} := \sum_{j=1}^{p} ||v_j||^2_{H^{\alpha_j}}.$$

Again *strong ellipticity*, Inequality 9.13, *implies coerciveness*[22]:

$$Re(Av,v)_0 \geqslant \gamma ||\vec{v}||^2_{H^\alpha} - |k[\vec{v},\vec{v}]| \quad \text{for all } \vec{v} \in H^\alpha \tag{9.14}$$

To our knowledge, all examples of boundary integral equations in applications belong to the class of strongly elliptic systems[16,40]. In the sequel we shall call these systems *strongly elliptic integral equations*.

9.3 CONVERGENCE OF GALERKIN'S METHOD

Now let us consider a strongly elliptic integral equation

$$Au = f, \quad \text{on } \Gamma \tag{9.15}$$

If Equation 9.15 has eigensolutions, i.e. there exist solutions $0 \not\equiv u_0 \in H^\alpha$ satisfying

$$Au_0 = 0$$

then let us introduce appropriate new unknowns as well as a finite number of side conditions such that the extended problem is uniquely solvable. (For specific problems of this type see, e.g. Hsiao and MacCamy[18]). Then the following arguments need a very simple modification in order to remain valid for the modified problem. Thus, for simplicity we assume, that Equation 9.15 is *uniquely* solvable.

If Equation 9.15 is not one equation but a system, again there is no difficulty to extend our approach and our results to the system. Hence let us consider just one equation in 9.15. We follow closely Stephan and Wendland[37] and Wendland[40].

For the approximation of Equation 9.15 let us introduce a family of finite dimensional nested subspaces $\tilde{H}_h \subset \tilde{H}_{h/2} \subset \ldots \subset (H^\alpha(\Gamma) \cap L_2)$ where h denotes a parameter with h^{-1} proportional to the dimension of $\tilde{H}_h$. Later on h will denote the meshwidth of the boundary elements. Since $\tilde{H}_h$ ought to approximate H^α we required that $\bigcup_{0<h} \tilde{H}_h$ is dense in H^α. Let μ_j, $j=0,1,\ldots,N$ denote a basis of $\tilde{H}_h$. Then the well known Galerkin procedure for Equation 9.15 is to find the coefficients γ_j of the approximate solution

$$\hat{v}(x) = \sum_{j=0}^{N} \gamma_j \mu_j(x), \quad x \in \Gamma \tag{9.16}$$

by solving the finite system of linear equations,

$$\sum_{j=0}^{N} (A\mu_j, \mu_k)\gamma_j = (f, \mu_k), \quad k=0,\ldots,N \tag{9.17}$$

For the convergence of this procedure we have well known results going back to Michlin[24], Hildebrandt and Wienholtz[15]. Here we use the version known as Cea's lemma (Ciarlet[11], p. 104).

For the application to our problem we need an additional property of our approximation. Let us denote by P_h the L_2 orthogonal projection onto $\tilde{H}_h$. Then we require

$$\|P_h g - g\|_\alpha \to 0 \text{ for } h \to 0 \text{ and any } g \in H^\alpha \tag{9.18}$$

This assumption implies with the Banach-Steinhaus theorem the stability

$$\|P_h\|_{H^\alpha, H^\alpha} \leqslant c \tag{9.19}$$

uniformly for all h and by duality also

$$\|P_h\|_{H^{-\alpha}, H^{-\alpha}} \leqslant c \tag{9.20}$$

This requirement is satisfied for the Fourier expansion with trigonometric polynomials, spherical harmonics and also with regular finite elements. Now we are in the position to state the convergence result.

Theorem 3.1: *Let* Equation 9.15 *with A be a strongly elliptic equation with unique solution $u \in H^\alpha$ to any $f \in H^{-\alpha}$. Then there exists $h_0 > 0$ such that* Equation 9.17 *is uniquely solvable for every $0 < h \leqslant h_0$. Moreover there exists a constant c independent of h and f such that*

$$\|\hat{v} - u\|_\alpha \leqslant c \inf_{\chi \in \tilde{H}_h} \|u - \chi\|_\alpha \tag{9.21}$$

Proof: Since A is strongly elliptic we have coerciveness, Inequality 9.8, which also holds for $v \in \tilde{H}_h$. That means that A can be decomposed as

$$A = D + C \tag{9.22}$$

where D is a positive definite operator and $C:H^{\alpha}\to H^{-\alpha}$ is compact. (The construction of D and C needs the spectrum of $(Au,v)=\lambda k[u,v]$ and is in general rather involved.) Equation 9.17 is equivalent to finding $\hat{v}\in\tilde{H}_h$ with

$$P_hAP_h\hat{v}=P_hDP_h\hat{v}+P_hCP_h\hat{v}=P_hAu \tag{9.23}$$

From the definiteness of D we have

$$Re(Dv,v)\geqslant\gamma\|v\|_{\alpha}^2$$

and also

$$Re(DP_hv,P_hv)\geqslant\gamma\|P_hv\|_{\alpha}^2$$

which yields

$$\|\chi\|_{\alpha}\leqslant\frac{1}{\gamma}Re(D\chi,\chi)\frac{1}{\|\chi\|_{\alpha}}\leqslant c\|P_hDP_h\chi\|_{-\alpha} \tag{9.24}$$

for all $\chi\in\tilde{H}_h$. In other words, Inequality 9.24 means

$$\|(P_hDP_h)^{-1}\|_{H^{-\alpha},H^{-\alpha}}\leqslant c \tag{9.25}$$

where c is independent of h. Now we have

$$P_hAP_h=P_hDP_h(I+(P_hDP_h)^{-1}P_hCP_h)$$

The sequence of operators

$$(P_hDP_h)^{-1}P_hCP_h$$

as compositions of a stable sequence 9.25, of P_h with Inequalities 9.19 and 9.20 and the compact operator C, defines a family of collectively compact operators in the sense of Aneslone[5], Chapter 1. Moreover, for any $g\in H^{\alpha}$ we have

$$(P_hDP_h)^{-1}P_hCP_hg\to D^{-1}Cg \text{ as } h\to 0$$

where $D^{-1}C:H^{\alpha}\to H^{\alpha}$ is compact. Hence

$$(I+(P_hDP_h)^{-1}P_hCP_h)^{-1}g\to(I+D^{-1}C)^{-1}g \text{ as } h\to 0$$

for any $g\in H^{\alpha}$, and the approximating inverses exist for $h\leqslant h_0$ (Anselone[5], p. 10) with some $h_0>0$, since

$$(I+D^{-1}C)^{-1}=A^{-1}D$$

exists due to our assumptions. This implies the uniform boundedness

$$\|(I+(P_hDP_h)^{-1}P_hCP_h)^{-1}\|_{H^\alpha,H^\alpha}\leqslant c$$

for all $h\leqslant h_0$. With Equation 9.25 we find the uniform boundedness of

$$G_h:=(P_hAP_h)^{-1}P_hA, \|G_h\|_{H^\alpha,H^\alpha}\leqslant c \text{ for } h\leqslant h_0 \tag{9.26}$$

where G_h is the *Galerkin projection* (often called 'elliptic projection'). Now Equation 9.26 implies 9.21, since

$$G_h\chi=(P_hAP_h)^{-1}P_hS\chi=(P_hAP_h)^{-1}P_hAP_h\chi=\chi, \text{ for any } \chi\in\tilde{H}_h$$

and, hence

$$\|\hat{v}-u\|_\alpha=\|G_hu-u\|_\alpha=\|G_h(u-\chi)+\chi-u\|_\alpha\leqslant(c+1)\|x-u\|_\alpha$$

for every $\chi\in\tilde{H}_h$. This completes the proof.

9.4 BOUNDARY ELEMENTS AND GALERKIN'S METHOD

In the following we shall approximate functions on the boundary Γ by finite elements called *boundary elements*. To this end we assume that Γ is given by local representations such that regular triangular partitions in the parameter domains are mapped onto a corresponding partition of Γ. On the partitions in the parameter domains we use a regular $(m+1,m)$ system of finite elements (e.g. piecewise polynomials)[8]. Then the local representation of Γ transplants these finite element functions on to Γ. For calculations, the integrals can be evaluated by using the local coordinates. In these the finite elements appear as simple functions over the parameter domains. This construction of finite elements on Γ requires that the parameter representations are fully available. For two dimensions, i.e. if Γ is a curve, this is a sensible requirement whereas in higher dimensions, as $n=3$, for the boundary surface Γ the parameter representations can become rather involved. Hence in this case an additional approximation of the surface representation becomes necessary leading to isoparametric elements as well as to a boundary approximation. These approximations have been investigated by Nedelec[27,28] for special cases of Equation 9.15. But this approach is yet to be generalized to our more general class of problems.

If we use the transplanted finite elements subject to regular $(m+1,m)$ systems in the parameter domains in connection with regular representations of the surface then they provide the *approximation property*:

Let $-m-1\leqslant t\leqslant s\leqslant m+1$, $-m\leqslant s,t\leqslant m$. Then to any given $u\in H^s(\Gamma)$ and any $\tilde{H}_h$ of our family there exists a finite element function $u_h\in\tilde{H}_h$ such that[10]

$$\|u-u_h\|_{H^t}=ch^{s-t}\|u\|_{H^s} \tag{9.27}$$

The constant c is independent of u, h and u_h. Moreover, since we use a regular finite element approximation, it provides the *inverse assumption*:

For $-m \leqslant t \leqslant s \leqslant m$ *and any* $\mu \in \tilde{H}_h$ *there holds an estimate*

$$\|\mu\|_{H^s} \leqslant ch^{t-s}\|\mu\|_{H^t} \tag{9.28}$$

where the constant c is independent of μ and h, Nitsche[29].

In addition, our stability properties Equations 9.19 and 9.20 hold for $-m \leqslant \alpha \leqslant m$ since Equation 9.27 implies[8]

$$\|P_h\|_{H^s,H^s} \leqslant c, \quad -m \leqslant s \leqslant m \tag{9.29}$$

Using the regular finite elements for Galerkin's method, Equations 9.16, 9.17 and inserting Equation 9.27 into 9.21 we immediately find the optimal order of convergence. The result can easily be improved with the inverse assumption, Equation 9.28. But with all our assumptions, the Aubin-Nitsche lemma can be applied also and yields super approximation[20]. Collecting all these results we find the following convergence theorem[20,26,32,37].

Theorem 4.1: *Let A in* Equation 9.15 *be strongly elliptic and* Equation 9.15 *be uniquely solvable. Let* $\hat{v}$ *denote the Galerkin solution of* Equations 9.16, 9.17 *subject to regular finite element spaces providing* Equations 9.27 *and* 9.28. *Define* $\alpha' := \min\{\alpha, 0\}$ *and suppose*

$$2\alpha - m - 1 \leqslant t \leqslant s \leqslant m+1, \quad -m \leqslant \alpha \leqslant m, \ \alpha \leqslant s, \ t \leqslant m$$

Then we have the asymptotic error estimate

$$\|u - \hat{v}\|_{H^t} \leqslant ch^{s-t}\|u\|_{H^s} \tag{9.30}$$

In addition, if we consider the discrete Equations 9.17 *in* $L_2(\Gamma)$ *then we find for the conditioning the asymptotic estimate*

$$\|\hat{v}\|_{L_2(\Gamma)} \tag{9.31}$$

Remark: The super approximation in Equation 9.30 with $t = -m - 1 + 2\alpha$ and $s = m + 1$ implies for the desired potential U, the solution of the underlying boundary value problem and its approximation $\hat{V}$ generated by the approximate layer $\hat{v}$ *inner super approximation*,

$$\|U - \hat{V}\|_{X(\tilde{\Omega})} \leqslant ch^{2m+2-2\alpha}\|v\|_{H^{m+1}(\Gamma)} \tag{9.32}$$

where $\tilde{\Omega}$ is any compact subdomain in the interior (or exterior) of Γ and $X(\tilde{\Omega})$ denotes *any* norm. Clearly, c depends here on the distance between $\tilde{\Omega}$ and Γ and on $X(\tilde{\Omega})$.

Proof: Equation 9.27 with $t = \alpha$ and Equation 9.21 imply Equation 9.32 with $t = \alpha$. Next use Equation 9.28 for $\alpha < t \leqslant s$ and Equation 9.30 with $t = \alpha$:

$$\begin{aligned} \|u - \hat{v}\|_{H^t} &\leqslant \|u - u_h\|_{H^t} + \|u_h - \hat{v}\|_{H^t} \\ &\leqslant \|u - u_h\|_{H^t} + c\|u_h - \hat{v}\|_{H^\alpha} h^{\alpha - t} \end{aligned}$$

$$\leqslant \|u-u_h\|_{H^t}+ch^{\alpha-t}(u-\hat{v}\|_{H^\alpha}+\|u-u_h\|_{H^\alpha})$$

$$\leqslant ch^{\alpha-t+s-\alpha}\|u\|_{H^s}+\|u-u_h\|_{H^t}+ch^{\alpha-t}\|u-u_h\|_{H^\alpha}$$

Using Equation 9.27 for each of the last terms implies Equation 9.30 for $\alpha \leqslant t \leqslant s$.

The remaining case can be proved by using an extended version of the Aubin-Nitsche lemma, see Hsiao and Wendland[20]. We omit the proof.

9.5 NUMERICAL INTEGRATION AND GALERKIN-COLLOCATION

For the numerical implementation of Galerkin's procedure, the weights

$$a_{jk}:=(A\mu_j,\mu_k), \quad j,k=0,\ldots,N \tag{9.33}$$

have to be evaluated. Since A is given by an integral operator (in the usual or the generalized) sense, the computation of a_{jk} requires a double integration over $\Gamma \times \Gamma$. If this is done numerically, the kernels of the integral operators must be computed at all combinations of grid points on Γ. In addition, special care must be taken of the singular integrals. All these facts result in a rather high computing time for computing the stiffness matrix a_{jk}. Therefore we shall adapt numerical integration to the special integrals in Equations 9.33. Provided the principal part is given by a convolutional operator as in Equations 9.1–9.4 we split A into two parts, a principal part in a standard form and a remaining part with smooth kernel. The principal part in standard form will be handled independently of the special boundary Γ yielding a Toeplitz matrix whose elements are given by a vector. This vector can be computed exactly up to the desired accuracy once for all independent of Γ as well as of h for any fixed type of elements. It should be pointed out that the accuracy of the numerical results depends significantly on how to compute the approximate principal part.

The Galerkin weights due to the smooth remaining parts will be treated numerically by appropriate quadrature formulas depending on the particular finite elements to be used. In them we use only grid points in a regular grid connected with the finite elements such that the kernel functions are to be evaluated as seldom as necessary. This leads to simple modified collocation formulas and the computation of the corresponding stiffness matrix is extremely fast.

The advantages of this Galerkin-collocation are most apparent for two-dimensional problems, i.e. $n=2$ and Γ a curve. Thus let us restrict ourselves here to this case. Moreover for brevity, let us consider for Γ only one simple closed boundary curve.

Let Γ be given by a regular parameter representation

$$\Gamma: x=x(t), \quad t\in[0,1] \tag{9.34}$$

with $x(t)$ a 1-periodic sufficiently smooth vector valued function satisfying

$$\left|\frac{\mathrm{d}x}{\mathrm{d}t}\right| = \rho(t) \geqslant \rho_0 \geqslant 0 \text{ for all } t \tag{9.35}$$

where ρ denotes the Jacobian. The principal part of A with a convolution operator has the form[36]

$$p.v. \int_{|t-\tau|<\frac{1}{2}} [p_1(t-\tau) + \log|t-\tau| p_2(t-\tau)] u(t)\rho(t)\, \mathrm{d}t \tag{9.36}$$

where $p_1(\zeta)$ and $p_2(\zeta)$ for $\zeta \neq 0$ are homogeneous functions of degree $\beta = 1 - 2\alpha - n$. The kernel of Equation 9.36 and the principal symbol of A subject to Equation 9.34 are connected by the Fourier transformation F,

$$a_0(\xi) = F(p_1(\cdot) + \log|\cdot| p_2(\cdot))|_\xi \tag{9.37}$$

With Equation 9.36, Equation 9.15 takes the form

$$\begin{aligned} p.v. &\int_{|t-\tau|<\frac{1}{2}} [p_1(t-\tau) + \log|t-\tau| p_2(t-\tau)](u(t)\rho(t))\, \mathrm{d}t \\ &+ \int_{|t-\tau|<\frac{1}{2}} L(\tau,t)(u(t)\rho(t))\, \mathrm{d}t = f(t) \end{aligned} \tag{9.38}$$

Hence we consider Equation 9.38 as an integral equation over [0,1] for the 1-periodic new unknown function

$$v(t) := \rho(t)u(t) \tag{9.39}$$

Note that the principal part in Equation 9.38 then becomes independent of the special choice of the curve Γ. In $L(\tau,t)$ we have collected all the remaining terms. For convenience we assume that $L$is a sufficiently smooth function of τ and t. Otherwise we again split into two terms, where the first contains the singularities and has to be treated similarly to the principal part of Equation 9.38.

In order to utilize the convolution in the principal part we use regular finite elements on a uniform grid of [0,1] defined with shifts and stretched variables from one shape function $\mu(\eta)$. The latter we define as in Aubin[7], Chapter 4 by suitable piecewise polynomials of order m with $\mu \in C^{m-1}$. For $m = 0,1,2$ e.g. we have

	$m=0$	$m=1$	$m=2$	*for*	
	1	η	$\frac{1}{2}\eta^2$	$0\leqslant\eta<1$	
$\mu(\eta)=$	0	$2-\eta$	$-\eta^2+3\eta-3/2$	$1\leqslant\eta<2$	(9.40)
	0	0	$\frac{1}{2}\eta^2-3\eta+9/2$	$2\leqslant\eta<3$	
	0	0	0	elsewhere	

With μ we define a basis of $\tilde{H}_h$ by

$$\mu_j(t):=\mu\left(\frac{t}{h}-j\right) \text{ for } hj\leqslant t\leqslant 1+hj,\ j=0,\ldots,N,\ h=1/(N+1) \tag{9.41}$$

and their 1-periodic extensions

$$\mu_j(t+l):=\mu_j(t) \quad \text{for integer } l$$

For v in Equation 9.39 we use the approximation

$$v_h(t):=\sum_{j=0}^{N}\gamma_j\mu_j(t) \tag{9.42}$$

Inserting Equation 9.41 into 9.33 we find for the terms due to the first expression in 9.38

$$d_{jk}=\int_0^1 pv \int_{|t-\tau|<\frac{1}{2}} [p_1(t-\tau)+\log|t-\tau|p_2(t-\tau)]\mu_j(t)\,\mathrm{d}t\mu_k(\tau)\,\mathrm{d}\tau$$

$$=h^{2+\beta}\Bigg\{\int_{\tau'=0}^{m+1} p.v.\int_{t'=0}^{m+1}[p_1(t'-\tau'+(j-k))+p_2\log|t'-\tau'+(j-k)|]\mu(t')\mu(\tau')\,\mathrm{d}t'\,\mathrm{d}\tau'$$

$$+\log h\int_{\tau'=0}^{m+1} p.v.\int_{t'=0}^{m+1} p_2(t'-\tau'+(j-k))\mu(t')\mu(\tau')\,\mathrm{d}t'\,\mathrm{d}\tau'\Bigg\},$$

$$\mathrm{d}_{jk}=h^{2+\beta}\{W_1{}^{\rho}+W_{2\rho}\log h\} \text{ with } \rho=j-k\in Z \tag{9.43}$$

Here the two vectors of weights

$$W_{1\rho} = \sum_{\tau'=0}^{m+1} pv \int_{t'=0}^{m+1} [p_1(t'-\tau'+\rho) + p_2 \log|t'-\tau'+\rho|]\mu(t')\mu(\tau')\, dt'\, d\tau' \quad (9.44)$$

$$W_{2\rho} = \int_{\tau'=0}^{m+1} pv \int_{t'=0}^{m+1} p_2(t'-\tau'+\rho)\mu(t')\mu(\tau')\, dt'\, d\tau',\ \rho \in Z \quad (9.45)$$

can be computed once for all independent of Γ and h. For their numerical evaluation one should rotate the t',τ' coordinate system by introducing the new variables

$$\sigma := t' - \tau' + \rho,\ \theta := t' + \tau' \quad (9.46)$$

Since then the integrands are piecewise polynomials in θ for any fixed σ, the integrations over θ can be done *exactly* by the use of appropriate integration formulae as Gaussian formulae or Newton-Cotes formulae. The remaining integrals take the form

$$W_{1\rho} = \int_{\rho-m-1}^{\rho+m+1} [p_1(\sigma) + \log|\sigma| p_2(\sigma)] F(\sigma)\, d\sigma \quad (9.47)$$

and the corresponding formula for $W_{2\rho}$. With the exactly computable integrand $F(\sigma)$ one can use weighted Gaussian quadrature in Equation 9.47. To this end one observes that $F(\sigma)$ itself is a spline function and one has to care for its knots. All this has been done explicitly for $p_1 \equiv 0$ and $p_2 \equiv 1$ in Hisiao *et al.*[16] and there one also can find more details. For the more general case, Equations 9.44 and 9.45, one has to use appropriate Gaussian quadrature rules for the more general p_1, p_2.

It seems to be possible to extend the above approach also to equations with $p_1 = p_1(\tau,\zeta)$ and $p_2 = p_2(\tau,\zeta)$, i.e. general strongly elliptic pseudodifferential operators, by using Equation 9.43 only *locally*. But this is yet to be done.

For all the remaining smooth terms in the Galerkin equations to 9.38 we use numerical integration.

Since in the corresponding integrals

$$\int_{\text{supp}\,\mu_j} f(t)\mu_j(t)\, dt = h \int_{\sigma=0}^{m+1} f(h(j+\sigma))\mu(\sigma)\, d\sigma \quad (9.48)$$

the finite element functions appear as factors, the numerical integrations are chosen accordingly to the respective reference function μ such that poly-

Table 9.1

$m=0$			$m=1$				$m=2$			
M	b_0	b_1	M	b_0	b_1	b_2	M	b_0	b_1	b_2
0	1	0	0	1	0	0	0	$\frac{1}{3}$	0	0
1	$\frac{11}{12}$	$\frac{1}{24}$	1	$\frac{5}{6}$	$\frac{1}{12}$	0	1	$\frac{3}{4}$	$\frac{1}{8}$	0
	$b_l=b_{-l}$		2	$\frac{97}{120}$	$\frac{1}{10}$	$-\frac{1}{240}$	2	$\frac{59}{960}$	$\frac{67}{480}$	$-\frac{7}{1920}$

nomials f up to the order $2M+1$ are integrated exactly. This leads to formulas like

$$\int\limits_{\text{supp }\mu_j} f(t)\mu_j(t)\,\mathrm{d}t = h\sum_{l=-M}^{M} b_l f(z_{jl}) + R \tag{9.49}$$

where

$$z_k := z\left(h\left(k+\frac{m+1}{2}\right)\right) \text{ and } z_{jl} := z\left(h\left(j+\frac{m+1}{2}+l\cdot\gamma_l\right)\right) \tag{9.50}$$

are the gridpoints subject to the boundary elements and, correspondingly subject to the integration formula. The error term R satisfies an estimate of the form

$$|R| \leqslant ch^{s+1/2}\left\{\int\limits_{h\cdot\min\{j,j+(m+1)/2-M\gamma_M\}}^{h\cdot\max\{j+m+1,\,j+(m+1)/2+M\gamma_M\}} |f^{(s)}(t)|^2\,\mathrm{d}t\right\}^{1/2} \quad \text{for } 1\leqslant s\leqslant 2M+2 \tag{9.51}$$

The simplest choice $\gamma_l=1$ yields $z_{jl}=z_{j+l}$ and weights b_l as shown in Table 9.1. As one can see, for higher order formulas one obtains in this also negative b_l.

This can be avoided by using only gridpoints $z_{jl}\in\text{supp }\mu_j$. E.g. for $\gamma_l=\frac{1}{2}$ and $M=2$ we find the following weights ($b_l=b_{-l}$ for $M=2$ and $\gamma_1=\frac{1}{2}$):

m	b_0	b_1	b_2
1	$\frac{13}{30}$	$\frac{4}{15}$	$\frac{1}{60}$
2	$\frac{2}{5}$	$\frac{7}{30}$	$\frac{1}{15}$

Using Equation 9.49 for the smooth terms of the weights in Equation 9.33 we obtain

$$\int_{\tau=0}^{1} \int_{|\tau-t|\leqslant\frac{1}{2}} L(\tau,t)\mu_j(t)\,\mathrm{d}t\ \mu_k(\tau)\,\mathrm{d}\tau = h^2 \sum_{l,i=-M}^{M} b_i b_l L(z_{ki},z_{jl}) + R \tag{9.52}$$

with the error term

$$|R| \leqslant h^{s+2} c \left\{ \max\left|\frac{\partial^s L}{\partial\tau^s}\right| + \max\left|\frac{\partial^s L}{\partial t^s}\right| \right\},\ 0\leqslant s\leqslant 2M+2 \tag{9.53}$$

Now we are ready to formulate the *Galerkin-collocation equations* by using Equations 9.39, 9.42, 9.43 and 9.52. They read as

$$\begin{aligned}\sum_{j=0}^{N} \Big\{ & h^{2+\beta}(W_{1,\rho(j,k)} + \log h\ W_{2,\rho(j,k)}) \\ & + h^2 \sum_{l,i=-M}^{M} b_i b_l L(z_{ki},z_{jl}) \Big\} \gamma_j = h \sum_{i=-M}^{M} b_i f(z_{ki}),\ k=0,\ldots,N\end{aligned} \tag{9.54}$$

Remarks

(1) As we shall see in the next section we must choose $M \geqslant (m-1/2)-\alpha'$ in order to maintain the order h^{m+1} of the asymptotic convergence in Equation 9.30 with $t=0$, also for the fully discretized scheme, Equation 9.54. In this case the weights $W_{1,\rho}$ and $W_{2,\rho}$ have to be accurate up to an order h^{a} with $a > m+2 + 2(\alpha-\alpha')$.

In case one wants to maintain the super approximation, Equation 9.32, one has to choose $M \geqslant m-\alpha'-\alpha$ and the weights $W_{1,\rho}$ and $W_{2,\rho}$ have to be accurate with $a > 2m+3-2\alpha'$.

(2) Since p_1, p_2 and $\log|t-\tau|$ are no longer periodic, *L will no longer be periodic*. But since μ_j and the problem are 1-periodic, one should use for ρ in Equation 9.54 the 'modified discrete difference':

$$\rho(j,k) := \begin{cases} j-k, & \text{if } |j-k| \leqslant |j-k\pm N| \\ j-k+N, & \text{if } |j-k+N| < |j-k| \\ j-k-N, & \text{if } |j-k-N| < |j-k| \end{cases} \tag{9.55}$$

and choose for L in Equation 9.54 that branch of the function which is continuous in $[z_{k,-M},z_{k,+M}][z_{j,-M},z_{j,+M}]$.

(3) In order to save computing time, the values of L and f at the grid points should be evaluated *once only* at the beginning and then be stored for further use so as to build up the stiffness matrix in Equation 9.54. This suggests a choice $\gamma_l = 1$ or $\frac{1}{2}$ or $\frac{1}{3}$ etc. in the numerical integration formulae.

For the examples treated in Hsiao *et al.*[16] ánd Wendland[40] with $\alpha = -\frac{1}{2} = \alpha'$ the choise $m=2$, $M=1$ and $\gamma_l = 1$ provided excellent numerical results in combination with short computing times.

It seems that the Galerkin-collocation, Equation 9.54, combines the theoretical advantages of Galerkin's method with the practical advantages of the collocation methods.

9.6 ASYMPTOTIC ERROR ANALYSIS FOR THE GALERKIN-COLLOCATION METHOD

For the asymptotic error due to the Galerkin-collocation we shall use the already established error estimates, Equation 9.35, for Galerkin's method. To this end we first consider the Estimate 9.51 abbreviated by

$$\sum_{j=0}^{N} a_{hjk}\gamma_j = F_k, \quad k=0,\ldots,N \tag{9.56}$$

as mappings in $\tilde{H}_h$. If

$$W_h = \sum_{j=0}^{N} \alpha_j \mu_j \tag{9.57}$$

then the mapping $\tilde{A}$ associated with Equation 9.56,

$$\sum_{l=0}^{N} \beta_l \mu_l = \tilde{A} w_h \tag{9.58}$$

will be defined by the linear equations for the coefficients β_l,

$$\sum_{l=0}^{N} \beta_l(\mu_l,\mu_k) = \sum_{j=0}^{N} a_{hjk}\alpha_j, \quad k=0,\ldots,N \tag{9.59}$$

An elementary explicit computation of the Gram matrix

$$M_{lk} = M_{kl} := \frac{1}{h}(\mu_l,\mu_k) \geqslant 0 \tag{9.60}$$

shows that it is symmetric, positive, positive definite, sparse and diagonally dominant. Hence, its inverse $((P_{qr}))$ can be computed explicitly by the Neumann series,

$$((P_{qr})) = ((M_{lk}))^{-1} = \frac{1}{M_{00}}\left\{I + \sum_{j=1}^{\infty}\left(\left(-\frac{M_{kl}}{M_{00}}(1-\delta_{kl})\right)\right)^j\right\} \tag{9.61}$$

The convergence is uniform for all N.

Hence, $\tilde{A}$ in Equation 9.58 is well defined. Correspondingly we define $\tilde{F} \in \tilde{H}_h$ by

$$(\tilde{F},\mu_k) = F_k, \text{ for } k=0,\ldots,N \tag{9.62}$$

Then the Galerkin Equations 9.17 and the Galerkin-collocation Equations 9.54 take the form

$$P_h A P_h \hat{v} = P_h f \text{ and } \tilde{A} v_h = \tilde{F},\ \hat{v}, v_h \in \tilde{H}_h \subset L_2 \cap H^\alpha \tag{9.63}$$

respectively. Thus, the identity

$$\tilde{A}(\hat{v} - v_h) = (\tilde{A} - P_h A P_h)\hat{v} + (P_h f - \tilde{F})$$

yields the estimate

$$\|\hat{v} - v_h\|_{L_2} \leqslant \|\tilde{A}^{-1}\|_{L_2 L_2} \{\|(\tilde{A} - P_h A P_h)\hat{v}\|_{L_2} + \|P_h f - \tilde{F}\|_{L_2}\} \tag{9.64}$$

This estimate shows clearly that we need estimates for stability, i.e. $\|\tilde{A}^{-1}\|_{L_2 L_2}$, consistency, i.e. $\|(\tilde{A} - P_h A P_h)\hat{v}\|_{L_2}$ and the truncation error $\|P_h f - \tilde{F}\|_{L_2}$. Let us begin with the consistency.

Theorem 6.1: *Let the weights $W_{1\rho}, W_{2\rho}$ be accurate to an order h^a and let $(\partial/\partial\tau)^{2M+2} L$ be continuous. Then we have the consistency*

$$|(\tilde{A}\mu, v) - (A\mu, v)| \leqslant \lambda(h) \|\mu\|_{L_2} \|v\|_{L_2}, \text{ for all } \mu, v \in \tilde{H}_h \tag{9.65}$$

where

$$\lambda(h) \leqslant c_1 |\log h| h^{a+\beta} + c_2 h^{2M+2} \tag{9.66}$$

Here $\beta = -2\alpha - 1$.

Proof: With any

$$\mu = \sum_{j=0}^{N} \alpha_j \mu_j,\ v = \sum_{j=0}^{N} \beta_j \mu_j \tag{9.67}$$

we have

$$\|\mu\|_{L_2} \leqslant c\sqrt{h} \left\{ \sum_{j=0}^{N} \alpha_j^2 \right\}^{1/2} \leqslant c' \|\mu\|_{L_2} \tag{9.68}$$

and consequently

$$\begin{aligned} |(\tilde{A}\mu, v) - (A\mu, v)|^2 &= \left| \sum_{j,k=0}^{N} (a_{hjk} - a_{jk}) \alpha_j \beta_k \right|^2 \\ &\leqslant \sum_{j,k=0}^{N} |a_{hjk} - a_{jk}|^2 \sum_{l=0}^{N} \alpha_l^2 \sum_{p=0}^{N} \beta_p^2 \\ &\leqslant c h^{-2} \sum_{j,k=0}^{N} |a_{hjk} - a_{jk}|^2 \|\mu\|_{L_2}^2 \|v\|_{L_2}^2 \end{aligned} \tag{9.69}$$

Our assumptions concerning the numerical errors for the $W_{1\rho}, W_{2\rho}$ in Equation 9.43 and the estimate, Equation 9.53, yield

$$|a_{hjk} - a_{jk}|^2 \leqslant c_1^2 h^{4+2\beta} |\log h|^2 h^{2a} + c_2^2 h^{2M+8} \tag{9.70}$$

Inserting Equation 9.70 into Equation 9.69 $(N+1)^2 = h^{-2}$ times, we obtain

$$|(\tilde{A}\mu, v) - (A\mu, v)|^2 \leqslant \|\mu\|_{L_2}^2 \|v\|_{L_2}^2 \cdot \frac{1}{h^4}(c_1^2 h^{4+2\beta+2a} |\log h|^2 + c_2^2 h^{8+4M})$$

$$\leqslant \|\mu\|_{L_2}^2 \|v\|_{L_2}^2 (c_1 |\log h| h^{a+\beta} + c_2 h^{2M+2})^2$$

that is Estimate 9.65 with Estimate 9.66.

Now stability in Inequality 9.64 can easily be obtained from Estimate 9.65.

Theorem 6.2: Let the assumptions of Theorem 6.1 be fulfilled and in addition let $a > 1 + 2(\alpha - \alpha')$, $M > -\alpha' - 1$. Then we have *stability, i.e. there exists* $h_0 > 0$ *such that*

$$\|\tilde{A}^{-1}\|_{L_2 L_2} \leqslant c h^{2\alpha'} \tag{9.71}$$

where c is independent of h for all $0 < h \leqslant h_0$.

Proof: In the identity

$$\tilde{A} = P_h A P_h (I + (P_h A P_h)^{-1} (\tilde{A} - P_h A P_h))$$

we use Inequality 9.30, i.e.

$$\|(P_h A P_h)^{-1}\|_{L_2 L_2} \leqslant c h^{2\alpha'} \tag{9.72}$$

together with Inequalities 9.65 and 9.66, obtaining

$$\|(P_h A P_h)^{-1} (\tilde{A} - P_h A P_h)\|_{L_2 L_2} \leqslant c h^{2\alpha'} \{c_1 |\log h| h^{a+\beta} + c_2 h^{2M+2}\}$$

$$\leqslant c_1' |\log h| h^{a+2\alpha'-1-2\alpha} + c_2' h^{2+2M+2\alpha'} \to 0$$

for $h \to 0$. Hence, there is a $h_0 > 0$ such that the above expression is less than one for all $h \leqslant h_0$ and correspondingly Neumann's series yields

$$\|(I + (P_h A P_h)^{-1} (\tilde{A} - P_h A P_h))^{-1}\|_{L_2 L_2} \leqslant c$$

This together with Inequality 9.72 and

$$\tilde{A}^{-1} = (I + (P_h A P_h)^{-1} (\tilde{A} - P_h A P_h))^{-1} (P_h A P_h)^{-1}$$

yields Inequality 9.71.

For the truncation error we have

Theorem 6.3: *For*

$$F_k = h \sum_{l=-M}^{M} b_l f(z_{kl})$$

in Equation 9.62 *there holds*

$$\|P_h f - \tilde{F}\|_{L_2} \leqslant c h^{\sigma} \|f\|_{H^{\sigma}} \text{ with } 1 \leqslant \sigma \leqslant 2M+2 \tag{9.73}$$

Proof: With

$$P_h f - \tilde{F} = \sum_{k=0}^{N} \varepsilon_k \mu_k$$

we find

$$R_j = (f, \mu_j) - F_j = \sum_{k=0}^{N} (\mu_k, \mu_j) \varepsilon_k, \quad j = 0, \ldots, N \tag{9.74}$$

where R_j is given by Equation 9.49 satisfying Estimate 9.51. Using Equations 9.60 and 9.61 for Equation 9.74 we find

$$\varepsilon_k = \frac{1}{h} \sum_{j=0}^{N} P_{kj} R_j$$

and

$$\sqrt{h} \left\{ \sum_{k=0}^{N} \varepsilon_k^2 \right\}^{1/2} \leqslant \frac{1}{\sqrt{h}} c \left\{ \sum_{j=0}^{N} R_j^2 \right\}^{1/2}$$

Inserting Estimate 9.51 and using Equation 9.74 we obtain the desired estimate

$$\|P_h f - \tilde{F}\|_{L_2} \leqslant c \sqrt{h} \left\{ \sum_{k=0}^{N} \varepsilon_k^2 \right\}^{1/2} \leqslant c' \frac{1}{\sqrt{h}} h^{\sigma + 1/2} \|f\|_{H^{\sigma}}$$

Collecting these estimates and using Estimates 9.64 and 9.30 we find the following estimates for our Galerkin-collocation.

Theorem 6.4: *For* $a > m + 2 + 2(\alpha' - \alpha)$ *and* $M \geqslant (m - 1/2) - \alpha'$ *we find an error estimate*

$$\|v - v_h\|_{L_2} \leqslant c h^{s} \{ \|u\|_{H^s} + \|f\|_{H^{s-2\alpha'}} \} \tag{9.75}$$

with $0 \leqslant s \leqslant m + 1$ *and* $s \geqslant 1 + 2\alpha'$.

For $a > 2m+3-2\alpha'$ and $M \geqslant m-\alpha-\alpha'$ we have even the super approximation

$$\|v-v_h\|_{H^t} \leqslant ch^{s-t}\{\|u\|_{H^s}+\|u\|_{L_2}+\|f\|_{H^{s-t-2\alpha'}}\} \tag{9.76}$$

provided $2\alpha-m-1 \leqslant t \leqslant s \leqslant m+1$, $s-t \geqslant 1-2\alpha'$.

Proof: First we observe that in both cases our assumptions imply the inequalities $a > 1+2(\alpha-\alpha')$ and $M > -1-\alpha'$. Hence, we have Inequality 9.71 in 9.64 yielding with Inequality 9.30 the estimate

$$\begin{aligned}\|v-v_h\|_{L_2} &\leqslant \|v-\hat{v}\|_{L_2}+\|\hat{v}-v_h\|_{L_2}\\ &\leqslant ch^s\|v\|_{H^s}+c'h^{2\alpha'}\{\|\tilde{A}-P_hAP_h\|_{L_2}\|\hat{v}\|_{L_2}+\|P_hf-\tilde{F}\|_{L_2}\}\end{aligned}$$

For further estimation we use Inequalities 9.65, 9.66 and 9.73 finding with Estimate 9.71 the estimate

$$\begin{aligned}\|v-v_h\|_{L_2} \leqslant ch^s\|u\|_{H^s}&+c'h^{2\alpha'}|\log h|h^{a-2\alpha-1}\|u\|_{L_2}\\ &+c''h^{2\alpha'+\sigma}\|f\|_{H^\sigma}, \quad 1 \leqslant \sigma \leqslant 2M+2\end{aligned}$$

Since $a+2(\alpha'-\alpha)-1 > m+1$ we have with $1 \leqslant \sigma = s-2\alpha' \leqslant 2M+2$

$$\|v-v_h\|_{L_2} \leqslant ch^s(\|u\|_{H^s}+\|u\|_{L_2})+c'h^s\|f\|_{H^{s-2\alpha'}}$$

This is Inequality 9.75 if we guarantee $s-2\alpha' \leqslant 2M+2$ up to $s=m+1$, yielding $M \geqslant (m-1/2)-\alpha'$.

For the super approximation, Estimate 9.30 yields

$$\|v-v_h\|_{H^t} \leqslant ch^{s-t}\|u\|_{H^s}+\|\hat{v}-v_h\|_{H^t}$$

In the case $t > 0$ we use the inverse assumption 9.28 to find

$$\|v-v_h\|_{H^t} \leqslant ch^{s-t}\|u\|_{H^s}+c'h^{-t}\|\hat{v}-v_h\|_{L_2}$$

The same estimates as above yield Inequality 9.76.

For $t < 0$ we use the cruder inequality

$$\begin{aligned}\|v-v_h\|_{H^t} &\leqslant ch^{s-t}\|u\|_{H^s}+\|\hat{v}-v_h\|_{L_2}\\ &\leqslant ch^{s-t}\|u\|_{H^s}+c'|\log h|h^{a-1+2(\alpha'-\alpha)}\|u\|_{L_2}\\ &\qquad +c''h^{2\alpha'+\sigma}\|f\|_{H^\sigma}, \quad 1 \leqslant \sigma \leqslant 2M+2\end{aligned}$$

Since

$$s-t \leqslant 2m+2-2\alpha < a-1+2\alpha'-2\alpha$$

the choice

$$\sigma = s - t - 2\alpha' \leqslant 2m + 2 - 2(\alpha' + \alpha) \leqslant 2M + 2$$

yields Inequality 9.76.

9.7 REMARKS ON THE STANDARD COLLOCATION METHOD

Although most numerical implementations of boundary integral methods are done with the standard collocation yet there are known only few results on the asymptotic errors except in the case of Fredholm integral equations of the second kind as Equation 9.1. Here we refer to the extensive bibliography by Noble[30], the surveys by Atkinson[6], Baker[9] and results on superconvergence[14,35].

For our more general equations there are only preliminary results available for the special Equation 9.2 for $n=2$ with the logarithmic kernel[1,2,39] and Aleksidze[4], p. 119 ff.

For the singular integral Equation 9.3 on a curve ($n=2$) Prössdorf and Schmidt have proved that the collocation method with piecewise linear functions ($m=1$) only converges if Equation 9.3 is strongly elliptic[31].

According to very recent results[41] we are able to announce a more general result for our wider class of problems in the case $n=2$. Let us consider again strongly elliptic problems where the principal part is a convolutional operator as in Section 9.5.

The standard collocation method can be formulated as to find

$$\tilde{v} = \sum_{j=0}^{N} \tilde{\gamma}_j \mu_j \in \tilde{H}_h \tag{9.77}$$

such that the collocation equations at the grid points,

$$\sum_{j=0}^{N} \tilde{\gamma}_j A\mu_j(z_k) = f(z_k), \quad k = 0,\ldots,N \tag{9.78}$$

are satisfied. In order to formulate the convergence result let us introduce the interpolation operator I_h mapping continuous functions on to $\tilde{H}_h$:

$$I_h f = \sum_{j=0}^{N} \alpha_j \mu_j \text{ with } \sum_{j=0}^{N} \alpha_j \mu_j(z_k) = f(z_k), \quad k = 0,\ldots,N \tag{9.80}$$

For I_h we have the approximation property[8]

$$\|I_h f - f\|_{H^t} \leqslant c h^{s-t} \|f\|_{H^s} \text{ for } 0 \leqslant t \leqslant s \leqslant m+1,\ \tfrac{1}{2} < s,\ t \leqslant m \tag{9.81}$$

Then Equation 9.78 can also be written as

$$I_h A P_h \tilde{v} = I_h f = \tilde{f} \in \tilde{H}_h \tag{9.82}$$

Based on the special form of Equation 9.38 and on the general results by Stephan and Wendland[37] and by making use of the collectively compact operator theory by Anselone[5] we can prove asymptotic error estimates. The key result is the stability.

Theorem 7.1[41]: *The Equations* 9.82 *in* $\tilde{H}_h$ *are stable with*

$$\|(I_h A P_h)^{-1}\|_{L_2,L_2} \leqslant ch^{2\alpha'} \tag{9.83}$$

where c is independent of $h \leqslant h_0$ *with a suitable* $h_0 > 0$.

Equation 9.83 is in accordance with Abou El-Seoud[1] and Voronin and Cecoho[39]. The stability in connection with convergence Equations 9.27 and 9.81 and with the inverse assumption Equation 9.28 provides us the following convergence result:

Theorem 7.2: *The collocation equations* (9.78) *are uniquely solvable for* $0 < h \leqslant h_0$ *and the approximate solutions converge as*

$$\|\tilde{v} - v\|_{L_2} \leqslant ch^{s+2(\alpha'-\alpha)}\|v\|_{H^s} \text{ if } 0 \leqslant s \leqslant m+1,\ \frac{1}{2} + 2\alpha < s,\ \frac{1}{2} + 2\alpha < m \tag{9.84}$$

The proof follows from the identity

$$I_h A P_h(\tilde{v} - v) = I_h A(v - P_h v)$$

with Inequality 9.83 and the above mentioned properties:

$$\|\tilde{v} - v\|_{L_2} \leqslant ch^{2\alpha'}\|I_h A(v - P_h v)\|_{L_2} + \|v - P_h v\|_{L_2}$$

$$ch^{2\alpha'}\{\|(I_h A - A)(v - P_h v)\|_{L_2} + \|A(v - P_h v)\|_{L_2}\} + \|v - P_h v\|_{L_2}$$

$$ch^{2\alpha'+\sigma}\|A(v - P_h v)\|_{H^\sigma} + ch^{2\alpha'}\|A(v - P_h v)\|_{L_2} + ch^s\|v\|_H \quad \sigma > \frac{1}{2}$$

$$\leqslant ch^{2\alpha'+\sigma}\|v - P_h v\|_{H^{\sigma+2\alpha}} + ch^{2\alpha'}\|v - P_h v\|_{H^{2\alpha}} + ch^s\|v\|_{H^s}$$

$$\leqslant ch^{2\alpha'-2\alpha+s}\|v\|_{H^s}$$

9.8 CONCLUDING REMARKS

Comparing the foregoing results with finite element methods in the domain Ω, it is apparent that for the asymptotic convergence theory of the boundary element methods many questions are still open as maximum norm error estimates, local estimates, mesh refinements and singularities at corners and edges, to name just a few. Moreover we did not worry about the solution methods for the finite dimensional approximate linear equations having a fully distributed stiffnesses matrix. When the number of grid points on Γ is approaching several hundreds — this is easily attained for three-dimensional problems — then the necessity for methods to solve the equations faster than with Gauss's procedure similar to the multigrid methods in connection with finite elements will become very important. These few mentioned problems show that still much can be done in order to improve the boundary element methods.

Since finite element methods can be adapted to variable coefficients, whereas boundary element methods cannot, the *combination* of both methods seems to promise a new generation of extremely efficient numerical methods.

REFERENCES

1 Abou El-Seoud, M. S., 'Numerische Behandlung von schwach singulären Integralgleichungen erster Art', Dissertation, Technische Hochschule Darmstadt, Germany (1979)

2 Abou El-Seoud, M. S., 'Kollokationsmethode für schwach singuläre Integralgleichungen erster Art', *ZAMM*, **59**, T45–T47 (1979)

3 Adams, R. A., *Sobolev Spaces*, Academic Press, New York (1975)

4 Aleksidze, M. A., *The Solution of Boundary Value Problems with the Method of the Expansion with Respect to Non-orthonormal Functions*, Nauka, Moscow (1978) (in Russian)

5 Anselone, P. M., *Collectively Compact Operator Approximation Theory*, Prentice Hall, London (1971)

6 Atkinson, K. E., *A Survey of Numerical Methods for the Solution of Fredholm Integral Equations of the Second Kind*, *SIAM*, Philadelphia (1976)

7 Aubin, J. P., *Approximation of Elliptic Boundary-Value Problems*, Wiley-Interscience, New York (1972)

8 Babuska, I. and Aziz, A. K., 'Survey lectures on the mathematical foundations of the finite element method', in *The Mathematical Foundation of the Finite Element Method with Applications to Partial Differential Equations*, A. K. Aziz (Ed.), Academic Press, New York pp. 3–359 (1972)

9 Baker, C., *The Numerical Treatment of Integral Equations*, Clarendon Press, Oxford (1977)

10 Bramble, J. and Schatz, A., 'Rayleigh-Ritz-Galerkin methods for Dirichlet's problem using subspaces without boundary conditions', *Comm. Pure Appl. Math.*, **23**, pp. 653–675 (1970)

11 Ciarlet, P. G., *The Finite Element Method for Elliptic Problems*, North Holland, Amsterdam (1978)

12 Giroire, J. and Nedelec, J. C., 'Numerical solution of an exterior Neumann problem using a double layer potential', *Math. Comp.*, **32**, pp. 973–990 (1978)

13 Haack, W. and Wendland, W. L., *Lectures on Partial and Pfaffian Differential Equations*, Pergamon Press, Oxford, New York, Toronto, Sydney, Braunschweig (1972)

14 Hämmerlin, G. and Schumaker, L. L., 'Procedures for kernel approximation and solution of Fredholm integral equations of the second kind', CNA Report 128, Center Numerical Analysis, Univ. Texas, Austin (1977)

15 Hildebrandt, St. and Wienholtz, E., 'Constructive proofs of representation theorems in separable Hilbert space', *Comm. Pure Appl. Math.*, **17**, pp. 369–373 (1964)

16 Hsiao, G. C., Kopp, P. and Wendland, W. L., 'A Galerkin collocation method for some integral equations of the first kind', *Computing*, **25**, pp. 89–130 (1980)

17 Hsiao, G. C., Kopp, P. and Wendland, W. L., 'The synthesis of the collocation and the Galerkin method applied to some integral equations of the first kind', in *New Developments in Boundary Element Methods*, C. A. Brebbia (Ed.), CML Publications, Southampton, pp. 122–136 (1980)

18 Hsiao, G. C. and MacCamy, R. C., 'Solution of boundary value problems by integral equations of the first kind', *SIAM Review*, **15**, pp. 687–705 (1973)

19 Hsiao, G. C. and Wendland, W. L., 'A finite element method for some integral equations of the first kind', *J. Math. Anal. Appl.*, **58**, pp. 449–481 (1977)

20 Hsiao, G. C. and Wendland, W. L., 'The Aubin-Nitsche lemma for integral equations', *J. Integral Equations*, to appear

21 Kleinman, R. E. and Wendland, W. L., 'On Neumann.s method for the exterior Neumann problem for the Helmholtz equation', *J. Math. Anal. Appl.*, **57**, pp. 170–202 (1977)

22 Koh, J. J. and Nirenberg, L., 'On the algebra of pseudo-differential operators', *Comm. Pure Appl. Math.*, **18**, pp. 269–305 (1965)

23 Kral, J. and Netuka, I., 'Contractivity of C. Neumann's operator', *J. Math. Anal. Appl.*, **61**, pp. 607–619 (1977)

24 Michlin, S. G., *Variationsmethoden der Mathematischen Physik*, Akademie-Verlag, Berlin (1962)

25 Michlin, S. G. and Prössdorf, S., *Singuläre Integraloperatoren*, Akademie-Verlag, Berlin (1980)

26 Natterer, F., 'Regularisierung schlecht gestellter Probleme durch Projektionsverfahren', *Numer. Math.*, **28**, pp. 329–341 (1977)

37 Nedelec, J. C., 'Curved finite element methods for the solution of singular integral equations on surfaces in IR^3', *Comp. Math. Appl. Mech. Engng*, **8**, pp. 61–80 (1976)

28 Nedelec, J. C., 'Formulations variationelles de quelques equations integrales faisant intervenir des parties finies', in *Innovative Numerical Analysis for the Engineering Sciences*, R. Shaw (Ed.), Univ. Press of Virginia, Charlottesville, pp. 517–524 (1980)

29 Nitsche, J. A., 'Zur Konvergenz von Näherungsverfahren bezüglich verschiedener Normen', *Num. Math.*, **15**, pp. 224–228 (1970)

30 Noble, B., 'A Bibliography on "Methods for solving integral equations"', Math. Res. Center Tech. Report 1176 and 1177, Madison, Wis. (1971)

31 Prössdorf, S. and Schmidt, G., 'A finite element collocation method for singular integral equations', to appear

32 Richter, G. R., 'Numerical solution of integral equations of the first kind woth nonsmooth kernels', *SIAM J. Numer. Anal*, **17**, pp. 511–522 (1978)

33 Riesz, F. and Nagy, B. Sz., *Vorlesungen über Funktionalanalysis*, Deutscher Verlag-Wissenschaftder, Berlin (1956)

34 Le Roux, M. N., 'Resolution numerique du probleme du potential dans le plan part une methode variationelle d'elements finis', These, L'Universite de Rennes, Serie A, No. D'ordre 347, No. De Serie 38 (1974)

35 Schäfer, E., 'Fehlerabschätzungen für Eigenwertnäherungen nach der Ersatzkernmethode bei Integralgleichungen', *Numer. Math.*, **32**, pp. 281–290 (1979)

36 Seeley, R., 'Topics in pseudo-differential operators', in *Pseudo-Differential Operators*, C.I.M.E., L. Nirenberg (Ed.), Cremonese, Roma (1969)

37 Stephan, E. and Wendland, W. L., 'Remarks to Galerkin and least squares methods with finite elements for general elliptic problems', *Lecture Notes Math.*, **564**, pp. 461–471, Springer Berlin (1976); *Manuscripta Geodaetica*, **1**, pp. 93–123 (1976)

38 Triebel, H., *Fourier Analysis and Function Spaces*, Teubner, Leipzig (1977)

39 Voronin, V. V. and Cecoho, V. A., 'An interpolation method for solving an integral equation of the first kind with a logarithmic singularity', *Dokl. Akad. Nauk SSR*, **216**; *Soviet Math. Dokl.*, **15**, pp. 949–952 (1974)

40 Wendland, W. L., 'On Galerkin collocation methods for integral equations of elliptic boundary value problems', in *Numerical Treatment of Integral Equations*, J. Albrecht and L. Collatz (Ed.), Int. Ser. Num. Math., Vol. 53, Birkhäuser, Basel (1980)

41 Wendland, W. L. and Arnold, D. 'On the asymptotic convergence of collocation methods'. In preparation.

Index